FOODS

What is food? This volume, written by a doctor, contains all the ways in which the question might be answered. First published by Kegan Paul in 1883, and reissued because there is nothing quite like it today, this unique study was undertaken at a time when foreign produce and new foods were becoming more widely available, and rising living standards were making familiar foods more accessible to all, prompting an interest in their origins and their nutritive qualities as revealed by scientific research into their chemical composition, preparation and physiological effects. Looking at all generally and some less known foods in this way, Smith's work anticipates the current interest in healthy food. Specifying what the body needs for optimum nutrition and health, and how this can best be supplied, he makes it possible for the reader to think clearly about good, healthy food and how to choose, prepare and eat it. The character and composition of food and the processes of cooking and preservation are described in terms of their biological and chemical processes, and their effect upon the body when eaten. Foods are divided into nitrogenous and non-nitrogenous animal foods, vegetable foods and liquids including dairy products, tea, cocoa, alcohol and water, all treated in detail. Though highly informative, this is not a dry scientific text, because Smith also includes the natural and cultural history of foods whenever possible, noting that so highly was cinnamon esteemed, and so close the monopoly which the Dutch had on it in Ceylon, that the punishment of death was inflicted on those who injured the plant or illegally exported the bark or oil. In addition to examining the way foreign foods are used in their cultures of origin, Smith looks at the way European food was prepared in the past, including fifteen recipes from a rare volume called *Cury* which contains recipes by the master cook of King Richard the Second of England. If you have ever wondered exactly what you are eating and drinking, what happens to you when you do eat and drink, and why exactly you should be eating and drinking some things and not others, you will find this book instructive. For culinary and medical historians, the volume is a detailed and unique description of scientific thinking about food and health in the late nineteenth century, also giving insight into a time when genetically modified produce and fast food were unknown.

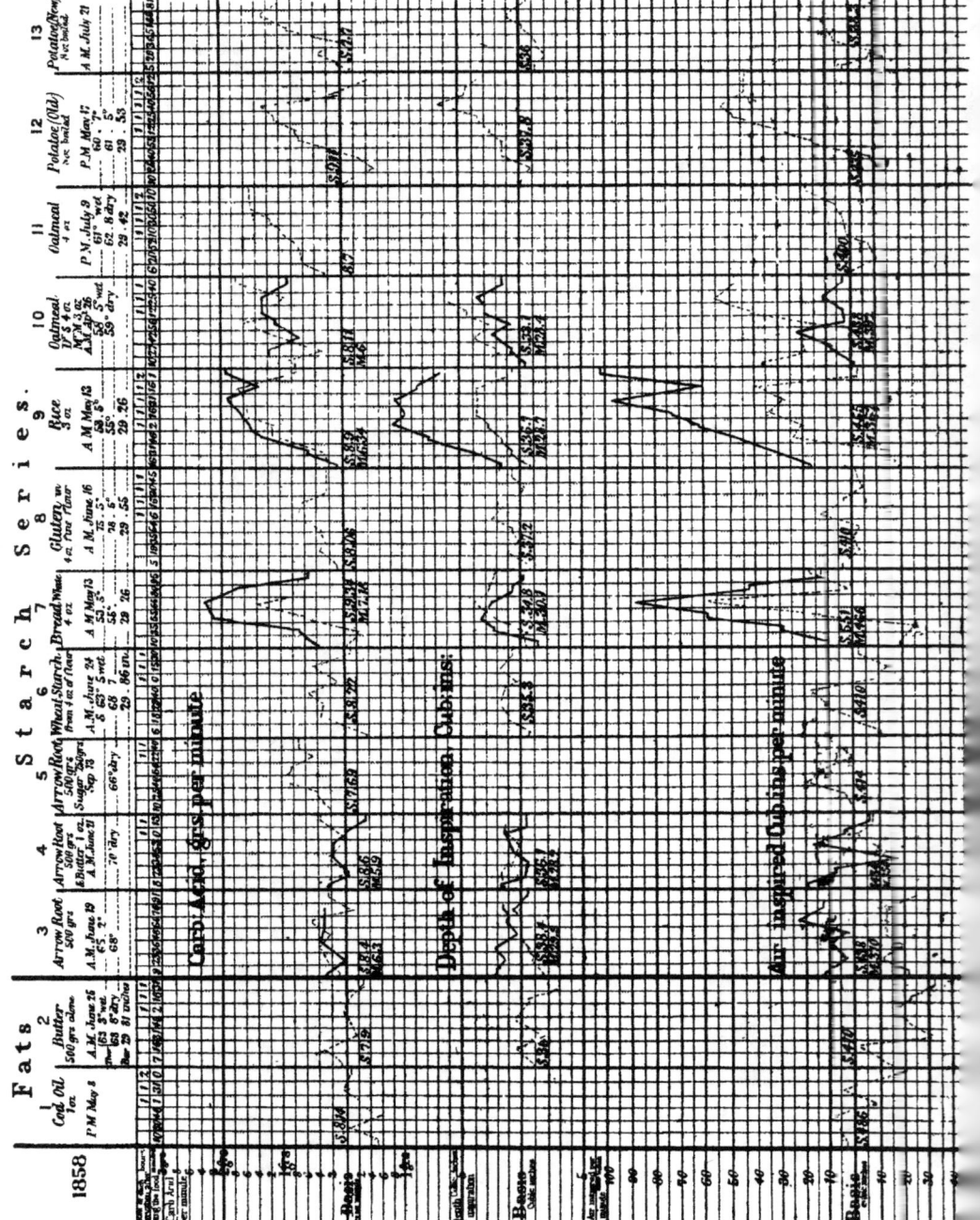

FOODS

BY

EDWARD SMITH

Taylor & Francis Group

LONDON AND NEW YORK

First published 2005 by
Kegan Paul International

2 Park Square, Milton Park, Abingdon, Oxon OX14 4RN
711 Third Avenue, New York, NY 10017, USA

Routledge is an imprint of the Taylor & Francis Group, an informa business

First issued in paperback 2016

Copyright © 2005 Taylor & Francis

All rights reserved. No part of this book may be reprinted or reproduced or utilised in any form or by any electronic, mechanical, or other means, now known or hereafter invented, including photocopying and recording, or in any information storage or retrieval system, without permission in writing from the publishers.

Notice:
Product or corporate names may be trademarks or registered trademarks, and are used only for identification and explanation without intent to infringe.

British Library Cataloguing in Publication Data
A catalogue record for this book is available from the British Library

ISBN 13: 978-1-138-97447-0 (pbk)
ISBN 13: 978-0-7103-1138-2 (hbk)

Publisher's Note
The publisher has gone to great lengths to ensure the quality of this reprint but points out that some imperfections in the original copies may be apparent. The publisher has made every effort to contact original copyright holders and would welcome correspondence from those they have been unable to trace.

PREFACE.

LARGELY-INCREASED commercial intercourse with distant countries, associated with a marked improvement in the purchasing-power of the masses of the people, and the rapid increase of wealth generally, have attracted public attention to the subject of Foods and Dietaries in an unusual degree, so that not only is there a greater importation of foreign productions than formerly, but new foods, or preparations of foods, are produced almost daily, some of which are specially fitted for certain classes of persons, as children, whilst others are of general use. Hence our food supplies, whether natural or prepared, offer increased variety of flavour, if not of nutritive qualities, and foods which were formerly restricted to the few are now commonly found on the tables of the many.

Scientific research in every civilised nation has also diligently busied itself in the elucidation of the subject, and our knowledge has been increased in reference to the chemical composition, preparation, and physiological effects of food.

With so many causes of change since the issue of my work on 'Practical Dietary,' it seemed desirable to produce another which should embrace all generally and some less known foods, and contain the latest scientific knowledge respecting them, whilst at the same time the subject should be treated in a popular manner.

It was originally intended to have included both foods and diets in one work, but the subject has now become so large that it was found necessary to limit the present volume to Foods alone, and to reserve the interesting subject of diets and dietaries for a future occasion.

The following pages contain a large number of tables, which will be useful for reference to both scientific and general readers, and also a number of very interesting recipes which have been extracted from a somewhat rare volume called 'Cury,' containing a copy of ancient manuscript recipes of the master cook of Richard the Second. The latter have been printed in the language of the fourteenth century, and, whilst interesting on that ground, will be understood with careful reading.

The prices at which various kinds of animal food were

sold, at a period somewhat anterior to these recipes, has not been referred to in the body of the work, but it was formerly regulated by law, and in 1315 was as follows:—

'Edwarde by the Grace of God Kynge of England &c. To Shiriffes, Majors Bailiffes of Fraunchises Greeting: For as much as we have heard and understanded the greevous complayntes of Archbishops, Bishops, Prelates and Barons, touching great dearth of victuals in our Realme: We ordeyne from henceforth that no Oxe stalled or corne fedde be sold for more than xxiiiis. No other grasse fed Oxe for more than xvis.; a fat stalled Cowe at xiis.; another Cowe lesse woorth at xs.; a fat Mutton corne fedd or whose wool is well growen at xx*d*.; another fat Mutton shorne at xiiii*d*.; a fat Hogge of two yeres olde at iii*s*. iiii*d*.; a fat Goose at ii*d*., ob., in the Citie at iii*d*.; a fat Capon at ii*d*., in the Citie iii*d*. ob.; a fat Hen at i*d*., in the Citie at i*d*. ob.; two Chickens at i*d*., in the Citie at i*d*. ob.; foure Pigions i*d*., in the Citie three Pigions i*d*. Item xxiiii Egges a peny, in the Citie xx Egges a peny. We ordeyne to all our Shiriffes.'

I have also thought it desirable to somewhat extend the ordinary view of foods, and to include water and air, since they are subjects which command great attention, both in their food and sanitary aspects.

The classification of foods has been made on the simplest basis, since none other seemed equally good, or equally well adapted to the object of this work.

I dare not hope that a volume of so much detail will be entirely free from typographical errors, but great pains have been taken to insure accuracy.

EDWARD SMITH.

LONDON: 140, HARLEY STREET, W.
April 1873

CONTENTS.

PAGE

INTRODUCTORY:—THE NATURE AND QUALITIES OF, AND THE NECES-
SITY FOR, FOODS 1

PART I.

SOLID FOODS.

SECTION I.—ANIMAL FOODS.

a. *Nitrogenous.*

CHAPTER
I. Description and Cooking of Flesh 15
II. Preserved Meat :— By drying ; Cold ; Immersion in Antiseptic Gases and Liquids ; Coating with Fat or Gelatin ; Heat ; Salt 22
III. Bone 38
IV. Characters and Composition of lean and fat Meat . . . 41
V. Beef and Veal 46
VI. Mutton, Lamb, Goat's, and Camel's Flesh 52
VII. Pork, Sucking Pig, Bacon, and Wild Pig 58
VIII. Various Wild Animals and the Horse and Ass . . . 69
IX. Offal 75
X. Sausages, Black Pudding, and Blood 82
XI. Extracts of Meat and fluid Meat 86
XII. Albumen, Gelatin, and Casein 91
XIII. Eggs 95
XIV. Poultry and Game 100
XV. Fish 105
XVI. Shell-Fish and Turtle 114
XVII. Cheese and Cream Cheese 120

B. *Non-Nitrogenous.*

XVIII. Butter, Ghee, Lard, Dripping, and Oils **127**

CONTENTS.

Section II.—Vegetable Foods.
a. Nitrogenous.

CHAPTER		PAGE
XIX.	Analogy of Animal and Vegetable Foods, and General Considerations on Seeds	143
XX.	Peas, Beans, Lentils, &c.	152
XXI.	Maize or Indian Corn, Millet, &c.	156
XXII.	Rice	162
XXIII.	Oats	166
XXIV.	Wheaten Flour and Bread	171
XXV.	Wheaten Bread, Biscuits, and Puddings	181
XXVI.	Barley, Maslin, and Rye; Scotch and Pearl Barley; Gluten	193
XXVII.	Succulent Vegetables	197
XXVIII.	Fruits: Succulent and Albuminous	213
XXIX.	Condiments	229

β. Non-Nitrogenous.

XXX.	Starches, Sago, Arrowroot, Tapioca, Cassava, Manioc, Semolina	239
XXXI.	Vegetable Fats and Oils	246
XXXII.	Sugar, Treacle, Honey, and Manna	249

PART II.
LIQUIDS.

XXXIII.	Water	269
XXXIV.	Milk, Cream, Butter-milk, and Whey	312
XXXV.	Tea, Coffee, Chicory, Cocoa, and Chocolate	330
XXXVI.	Alcohols	371

PART III.
GASEOUS FOODS.

XXXVII.	Atmospheric Air	434
XXXVIII.	Ventilation	459

INDEX	477

LIST OF RECIPES OF THE FOURTEENTH CENTURY,

QUOTED FROM 'CURY.'

	PAGE
Cok a Grees	68
For to kepe Venison fro Restyng	70
For to do away Restyng of Venison	71
Gele of Flessh	81
Lenten Fish Soup	112
Gele of Fyssh	113
Benes yfryed	154
Re Smolle	162
Blank Mang	163
Furmenty	177
Maccoros	191
Salat	207
Apple Tart	218

LIST OF DIAGRAMS, WOODCUTS, AND TABLES.

NO.		PAGE
1.	Thermometric force of Foods. *Frankland*	6
2.	Diagram, showing the rate of pulsation and respiration throughout two nights and one day in three phthisical persons. *E. Smith*	9
3.	Diagram, showing the same for twenty-four hours in a healthy child. *E. Smith*	9
4.	Diagram, showing rate of evolution of carbonic acid in the expired air, with and without food. *E. Smith* . . .	10
5.	Diagram, exhibiting the effect of sudden and marked changes of temperature over the vital functions. *E. Smith* . .	11
6. and 7.	Effect of numerous kinds of exertion over the respiration. *E. Smith*	12, 13
8, 9, 10.	Drawings of Muscular fibre	17
11.	Soda in Blood and flesh-juice. *Liebig*	18
12.	Chemical composition of Bone	39
13.	,, ,, ,, Cartilage	40
14.	Diagram, showing joints in a carcass	42
15.	Proportion of fat and lean in fat and store animals. *Lawes & Gilbert*	43
16.	Chemical elements in Fat	45
17.	Composition of various kinds of Flesh. *Mareschal* . .	47
18.	Chemical elements in Roast and Boiled Beef . . .	49
19.	Loss in Cooking beef	49
20.	Chemical composition of Veal	51
21.	Composition of Boiled Mutton	54
22.	Food stored up in Pigs, Sheep, and Oxen	59
23, 24.	Composition of Bacon	67
25.	Proportion of carcass and offal in animals	75
26.	Composition of Liver	79
27.	,, ,, Tripe	80
28.	,, ,, fresh Blood	84
29.	Salts in Blood	85
30, 31.	Composition of animal and vegetable Albumen . .	91
32.	,, ,, Hen's egg	99

xiv LIST OF DIAGRAMS, WOODCUTS, AND TABLES.

NO.		PAGE
33. Composition of Flesh of Poultry		103
34, 35, 36. ,, ,, Various Fish. *Payen*		107
37. ,, ,, Clean and Foul Salmon. *Christison.*		108
38. ,, ,, Cheese		121
39, 40. ,, ,, ,,		124
41. ,, ,, ,, *Payen*		125
42. Quantity of Water in Butter supplied to Workhouses. *Wanklyn*		134
43, 44. Elements in Butter		135, 136
45. Drawing of Fat Cells		138
46, 47. ,, ,, Starch Cells		147
48. The same expanded by heat		148
49, 50. Drawing of Siliceous Cuticle of Wheat and Meadow Grass		150
51. ,, ,, ,, ,, ,, Deutzia		151
52. Elements in Lignine or Cellulose, and Starch		151
53. Proportion of Starch in Foods		152
54. Composition of Peas and Lentils		152
55. ,, ,, Maize		157
56. ,, ,, Millet		162
57, 58. ,, ,, Rice		165
59, 60. ,, ,, Oatmeal		170
61. Layers of the Skin of Wheat		179
62. Analysis of Hart's whole-meal Flour. *Calvert*		180
63. Composition of Seconds Flour		181
64. ,, ,, Wheaten bread		191
65. ,, ,, Barley meal		194
66. ,, ,, Pearl Barley		195
67. ,, ,, Rye meal		196
68, 69. ,, ,, Potato		199
70. Sp. Gr. of and Starch in Potato. *Pohl*		199
71. Ash in various kinds of Potato. *Herapath*		200
72. Drawing of Section of Sound Potato		200
72. ,, ,, ,, ,, Diseased Potato		201
73. Composition of Chinese Yam. *Frémy*		202
74. ,, ,, Apios and Potato. *Payen*		203
75. Elements in Turnips and Carrots		203
76, 77. ,, ,, ,, ,, ,, and Parsnips		204
78. Composition of Beet-root. *Payen*		205
79. ,, ,, Bread-fruit		206
80. ,, ,, the juices of Plantain		206
81. ,, ,, of Cabbage. *Anderson*		207
82. ,, ,, of dried Mushroom. *Schlossberger & Döpping*		211
83 ,, ,, the juice of the Grape		215
84. Sugar in Rhine Grape juice		216
85. Composition of Mulberries, Bilberries and Blackberries. *Fresenius*		217

LIST OF DIAGRAMS, WOODCUTS, AND TABLES. xv

NO.		PAGE
86.	Composition of Cherries. *Fresenius*	218
87.	„ „ Apples	218
88.	„ „ Strawberries and Raspberries. *Fresenius* .	221
89.	„ „ Plums, Apricots and Peaches . . .	222
90.	„ „ Gooseberries and Currants	222
91.	Drawing of Nutmeg and Mace	234
92.	„ Ginger plant	237
93.	Elements in Sago, Arrowroot and Tapioca	240
94.	Drawing of a Sago-producing plant	241
95.	Diagram, showing the effects of the Starch series of Foods and of Fats on the Respiration. *E. Smith* . . *Frontispiece, faces Title*	
96.	Composition of Sugar Cane	251
97.	Drawing of the Sugar Cane	252
98.	Proportion of Sugar in various Foods	259
99.	Elements in Milk Sugar and Cane Sugar	259
100.	Diagram, showing the effects of the Sugar series of Foods on the Respiration. *E. Smith*	260
101.	Sugar in Syrups	262
102.	Elements in Caramel	263
103.	Water in various substances	270
104.	Drawings of Crystals found in Water	272
105.	Acidity, Ammonia and Albumenised Ammonia in Water .	273
106.	Elements in Thames Water	275
107, 108.	Mineral matters in Water	276
109, 110.	„ „ „ „	277
111.	„ „ „ „	278
112.	Hardness of Water. Soap test	282
113.	Organic matter in Water	285
114.	„ „ „ „	286
115.	Nitrogen and Nitric acid in **Ammonias**	293
116.	Chlorine in Water	298
117, 118.	„ „ „	299
119.	Various substances in Water	300
120.	„ „ „ „	301
121.	Drawing of Charcoal Filter in a cistern	309
122.	Mineral Waters	310, 311
123.	Composition of Cow's Milk	313
124.	Diagram, showing the effects of the Milk series of Foods on the Respiration. *E. Smith*	314
125.	Composition of several kinds of Milk	315
126.	Salts in Milk	316
127.	Composition of Milk of Women in the Siege of Paris . .	317
128.	Salts of Milk	317
129.	Composition of Butter Milk	328

LIST OF DIAGRAMS, WOODCUTS, AND TABLES.

NO.		PAGE
130.	Substitutes for China Tea	341
131.	Weight of Tea to bulk	345
132.	Elements in Theine	346
133.	Diagram, showing the effect of the Tea series of Foods on the Respiration. *E. Smith*	347
134.	Composition of Coffee	363
135.	,, ,, the Cacao bean	370
136.	Diagram, showing the effect of the Alcohol series of Foods on the Respiration. *E. Smith*	377
137.	Alcohol in Wines	390
138.	Œnanthic Ethers in Wines of the Gironde	392
139.	Proof Spirit in Hungarian Wines	397
140.	,, ,, ,, Greek Wines	398
141.	Drawing of the Palmyra Palm	405
142.	Saccharine matter in Beer	412
143.	Oxygen in Atmospheric Air	446
144.	,, ,, ,,	447
145, 146.	Carbonic Acid in Atmospheric Air	450
147.	Estimation of Carbonic Acid in Air	451
148.	Organic matter in Atmospheric Air	454
149.	,, ,, ,, ,, ,,	455
150, 151, 152.	,, ,, ,, ,,	456
153.	Vapour in a Cubic foot of Air at various temperatures	458
154.	Variation of Carbonic Acid with Ventilation. *Dr. Angus Smith*	475
155, 156.	Carbonic Acid and Oxygen in Sick wards. do.	475

FOODS.

INTRODUCTORY.

THE NATURE AND QUALITIES OF AND THE NECESSITY FOR FOODS.

BEFORE PROCEEDING to consider the numerous Foods which will come under review in the course of this work, it seems desirable to offer a few remarks of a general character on their nature and qualities and the necessity for them.

As a general definition, it may be stated that a food is a substance which, when introduced into the body, supplies material which renews some structure or maintains some vital process; and it is distinguished from a medicine in that the latter modifies some vital action, but does not supply the material which sustains such action.

This is certainly correct so far as relates to the substances which supply nearly all our nourishment, and which the Germans class under the term *Nahrungsmittel*, but there are certain so-called foods known as *Genussmittel*, which seem to form a connecting link, in that they increase vital actions in a degree far beyond the amount of nutritive material which they supply.

They thus resemble certain medicines in their action, but as they supply a proportion of nutritive material they should be ranked as foods.

It is essential to the idea of a food that it should support or increase vital actions; whilst medicines usually lessen, but may increase, some of them.

It is not necessary that a food should yield every kind of material which the body requires, for then one might suffice for the wants of man; but that it fulfils one or more of such requirements, so that by a combination of foods the whole wants of the body may be supplied. Neither is it essential that every food should be decomposed or broken up, and its elements caused to enter into new combinations when forming or maintaining the structures of the body, since there are some which in their nature are identical with parts of the body, and, being introduced, may be incorporated with little or no change.

But there are foods which are more valuable to the body than others, in that they supply a greater number of the substances which it requires, and such are known as compound foods, whilst others which supply but one element, or which are incorporated without change, may be termed simple foods. Other foods are more valuable because they are more readily changed into the substance of the body, or act more readily and quickly in sustaining vital actions, and these may be called easily digested or easily assimilated foods. Others are preferred because they supply a greater quantity of useful nutriment at a less proportionate cost, and are known as economical foods; and foods varying in flavour are classed as more or less agreeable foods.

Some foods are classed according to the source whence they are delivered, as animal and vegetable foods;

and others according to the density of their substance, as fluid and solid foods.

There are foods which nourish one part of the body only, and others which sustain one chief vital action, and are called flesh-forming or heat-forming foods, whilst others combine both qualities.

Besides these larger divisions, there are qualities in foods which permit of further classification, such as those which render them particularly fit for different ages, climates, and seasons, and others which possess a special character, as sweetness, acidity, or bitterness.

There are also effects produced by foods apart from or in addition to those of nutrition, which are not common to all; so that some foods more than others influence the action of the heart, lungs, skin, brain, bowels, or other vital organ, whilst others have antagonistic qualities, so that one may destroy certain effects of another.

Foods are derived from all the great divisions of nature and natural products, as earth, water and air, solids, liquids and gases; and from substances which are living and organic, or inanimate and inorganic. The popular notion of food as a solid substance derived from animals and vegetables, whilst comprehensive, is too exclusive, since the water which we drink, the air which we breathe, and certain minerals found in the substance of the earth, are of no less importance as foods.

It is, however, of great interest to note how frequently all these are combined in one food, and how closely united are substances which seem to be widely separated. Thus water and minerals are found in both flesh and vegetables, whilst one or both of the component parts of the air, viz., oxygen and nitrogen, are distributed through every kind of food. Hence, not only may we add food to food to supply the wants of the body, but we may within certain limits substitute one for another as our appetites or wants demand. The necessity for this in the economy of nature is evident, for although a good Providence has given to man an almost infinite number of foods, all are not found everywhere, neither can any man obtain all foods found around him.

Further, there seems to be an indissoluble bond existing between all the sources of food. There are the same classes of elements in flesh as in flour, and the same in animals as in vegetables. The vegetable draws water and minerals from the soil, whilst it absorbs and incorporates the air in its own growth, and is then eaten to sustain the life of animals, so that animals gain the substances which the vegetable first acquired. But in completing the circle the vegetable receives from the animal the air which was thrown out in respiration, and lives and grows upon it, and at length the animal itself, in whole or in part, and the refuse which it daily throws off, become the food of the vegetable. Even the very bones of an animal are by the aid of nature or man made to increase the growth of vegetables, and really to enter into their structure; and being again eaten, animals may be said to eat their own bones and live on their own flesh. Hence there is not only an unbroken circle in the production of food from different sources, but even the same food may be shown to be produced from itself. Surely this is an illustration of the fable of the young Phœnix arising from the ashes of its parent!

Food is required by the body for two chief purposes, viz. to generate heat and to produce and maintain the structures under the influence of life and exertion.

The importance of the latter is the more apparent, since wasting of the body is familiarly associated with decay of life; but the former is so much the more urgent, that whereas the body may waste for a lengthened period, and yet live, it rapidly dies when the source of heat is removed or even greatly lessened.

The production of heat in the body, so wonderful in the process and amount, results only from the chemical combination of the elements of food, whether on the minute scale of the atoms of the several tissues, or on the larger one connected with respiration, and is thence called the combustion of food. As familiar illustrations of the production of heat from chemical change, we may mention that when cold sulphuric acid and cold water are added together the mixture becomes so hot that the hand cannot bear it, and the heating of haystacks, and also of barley in the process of malting, is well known. This action in the body is not restricted to changes in one element alone, but proceeds with all; yet it is chiefly due to a combination of three gases, viz. oxygen, hydrogen, and carbon, and requires for its support fat, starch, or sugar, or other digestible food composed of those substances precisely as coal and wood supply fuel for fire without the body.

This effect is made extremely striking, by Professor Frankland, in the following table, which shows the amount of heat generated from so small a quantity as ten grains of certain foods during their complete combustion within the body, and the force which scientific calculations have shown to be equivalent to that amount of heat. The original quantity used by Prof. Frankland has been reduced by Dr. Letheby to ten grains, for the convenience of English readers. (*Royal Inst.* April 1868.)

INTRODUCTORY.

No. 1.

Food	In combustion raises lbs. of water 1 degree Fahrenheit	Which is equal to lifting lbs. 1 foot high
10 grains of dry flesh	11·23	8,677
,, ,, albumen	10·94	8,453
,, ,, lump sugar	8·61	6,647
,, ,, arrow root	10·06	7,766
,, ,, butter	18·68	14,421
,, ,, beef fat	23·32	18,023

Thus we prove that an ounce of fresh lean meat, if entirely burnt in the body, would produce heat sufficient to raise about 160 lbs. of water 1° F., or a gallon of water about 16° F. In like manner, one ounce of fresh butter would produce five times that amount of heat; but it must be added, that as the combustion which is effected within the body is not always complete, the actual effect is less than that now indicated.

It may thus be shown that the division of foods into the two great classes of flesh-formers and heat-generators is not to be taken too incisively, for whilst a food is renewing flesh it also produces heat, and whilst the heat-generating food is acting it may also produce a part of flesh in the form of fat; but although they are so closely associated in their vital work, the leading characteristic of each kind is so marked as to warrant the classification which Liebig has formulated.

It is understood that the structures of the body are in a state of continual change, so that atoms which are present at one hour may be gone the next, and, when gone, the structures will be so far wasted, unless the process of waste be accompanied by renewal. But the renewing substance must be of the same nature as that wasted, so that bone shall be renewed by bone and flesh by flesh; and hence, whilst the body is always changing, it is always the same. This is the duty assigned to food—to supply to each part of the body the very same kind of material that it lost by waste.

As foods must have the same composition as the body, or supply such other materials as by vital action may be transformed into the substances of the body, it is desirable to gain a general idea of what these substances are.

The following is a summary statement of the principal materials of which the body is composed:—

Flesh in its fresh state contains water, fat, fibrin, albumen and gelatin, besides compounds of lime, phosphorus, soda, potash, magnesia, silica and iron, and certain extractives.

Blood has a composition similar in elements to that of flesh.

Bone is composed of cartilage, gelatin, fat, and salts of lime, magnesia, soda and potash, combined with phosphoric and other acids.

Cartilage consists of chondrin, which is like gelatin in composition, with salts of soda, potash, lime, phosphorus, magnesia, sulphur and iron.

The brain is composed of water, albumen, fat, phosphoric acid, osmazome and salts.

The liver consists of water, fat, and albumen, with phosphoric and other acids in conjunction with soda, lime, potash and iron.

The lungs are formed of a substance resembling gelatin, albumen, a substance analogous to casein, fibrin, various fatty and organic acids, cholesterin, with salts of soda, and iron and water.

Bile consists of water, fat, resin, sugar, fatty and organic acids, cholesterin and salts of potash, soda and iron.

Hence it is requisite that the body should be provided with salts of potash, soda, lime, magnesia, sulphur, iron and manganese, as well as sulphuric, hydrochloric, phosphoric, and fluoric acids and water; also nearly all the fat which it consumes daily and probably all the nitrogenous substances which it requires, and which are closely allied in composition, as albumen, fibrin, gelatin and chondrin. It can produce sugar rapidly and largely, and fat slowly and sparely, from other substances; also lactic, acetic, and various organic acids, and peculiar extractive matters.

So great an array of mysterious substances might well prevent us from feeding ourselves or others if the selection of food depended solely upon our knowledge and judgment; but it is not so, for independently of the aid derived from our appetites, there is the great advantage of having foods which contain a proportion of nearly all these elements; and combinations of foods have been effected by experience which protect even the most ignorant from evil consequences.

Thus flesh, or the muscular tissue of animals, contains precisely the elements which are required in our flesh-formers, and, only limited by quantity, our heat-generators also; and life may be maintained for very lengthened periods upon that food and water when eaten in large quantities. Seeing, moreover, that the source of flesh in animals which are used as food is vegetables, it follows that vegetables should have the same elements as flesh, and it is a fact of great interest that in vegetables we have foods closely analogous to the flesh of animals. Thus, in addition to water and salts, common to both, there is vegetable jelly, vegetable albumen, vegetable fibrin, and vegetable casein, all having a composition almost identical with animal albumen, gelatin, chondrin and casein.

Hence our appetites and the bountiful provision made for us extend our choice to both the vegetable and animal kingdoms, and it is possible to find vegetable

Page 8

Nº 2.

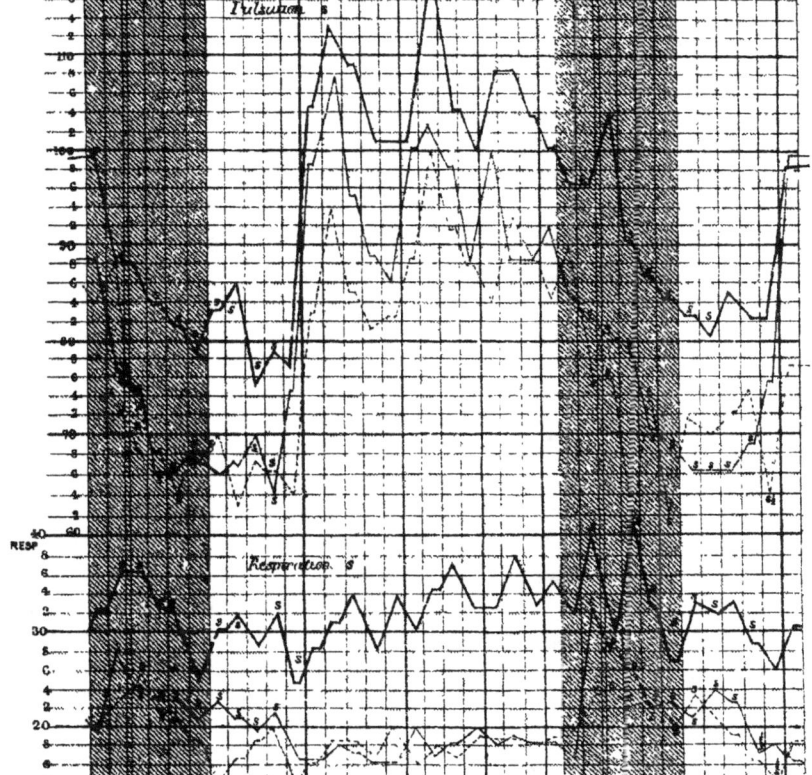

Hourly Pulsation & Respiration in Phthisis
JUNE 12TH

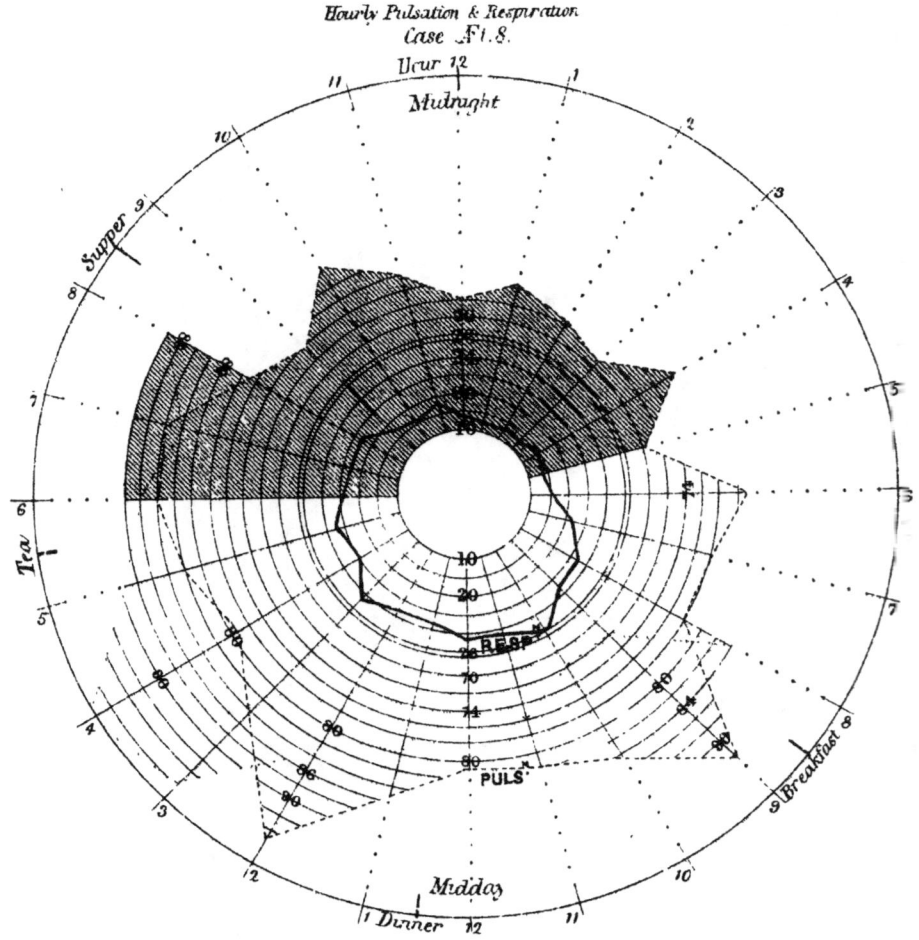

foods on which man could live as long as upon animal food alone. Bread is in vegetable foods what flesh is in animal foods, and each within itself contains nearly all the elements required for nutrition.

When, however, we bring knowledge of a special kind to the aid of our appetites, we are able to discover both the deficiencies in any given food and the kind of food which would meet them. Thus a knowledge of the requirements of the system and of the available uses of food, leads to the proper combinations of food, or to the construction of dietaries.

We have thus placed face to face the requirements of the body and the qualities of the foods to be used to supply them, but it is of very common observation that the effect of the supply is but temporary and needs renewal at definite periods. Hence we show that the needs of the body are tolerably uniform, whilst the effect of the supply is temporary, or that both the need and the supply are intermittent. This may be readily represented by showing the line of change in the degree of vital action on the body during the twenty-four hours, as produced by my own investigations and delineated in the following diagrams.

Diagram No. 2 shows the rate of pulsation and respiration per minute, throughout two nights and one day, in three adult phthisical persons. The pulsation is represented in the upper and the respiration in the lower series of lines, and the letter S indicates that the patient was asleep at the time of observation. (*Brit. & For. Med.-Chir. Rev.* 1856, *and Med.-Chir. Trans.* 1856.)

Diagram No. 3 represents the same facts as observed in a healthy child eight years of age, and delineated in the circle of the twenty-four hours. It includes two lines, which represent the observations on two days. (*Medico-Chirurgical Trans.* 1856.)

In both of these diagrams the shaded part shows the hours of darkness at the period of the year when the observations were made.

Diagram No. 4 shows the quantity of carbonic acid evolved by myself in respiration at each hour from 6.30 A.M. until midnight, on two occasions with food, and on one day of entire abstinence from food, the whole of the carbonic acid being collected. (*Phil. Trans.* 1859.)

Thus during the repose of the night the amount of vital action, as shown by the respiration and pulsation, is low and tolerably uniform, whilst under the influence of food it is high and varies during the day extremely, but the general course is such that a large increase takes place after a meal, and a considerable decrease before the following meal. This increase, followed by decrease, being due to the action of food, proves that the influence is temporary, and that after a sufficient interval another supply of food is required. At the same time it must be allowed that the body is not entirely a passive agent subject to the controlling action of food, for no supply could prevent the vital actions subsiding at night, or make them equal both by night and day.

There is a power inherent in the body which accepts or rejects food as to amount, as well as to quality, and which might at length act through the appetite and refuse the kind supplied. Moreover, the wants of the body vary from many other well-known influences, and cause an increase or decrease in the vital actions which proceeds *pari passu* with the consumption of the transformed or stored-up food in a degree proportionate to the cause, but such effects are often more rapid and transitory than that of food.

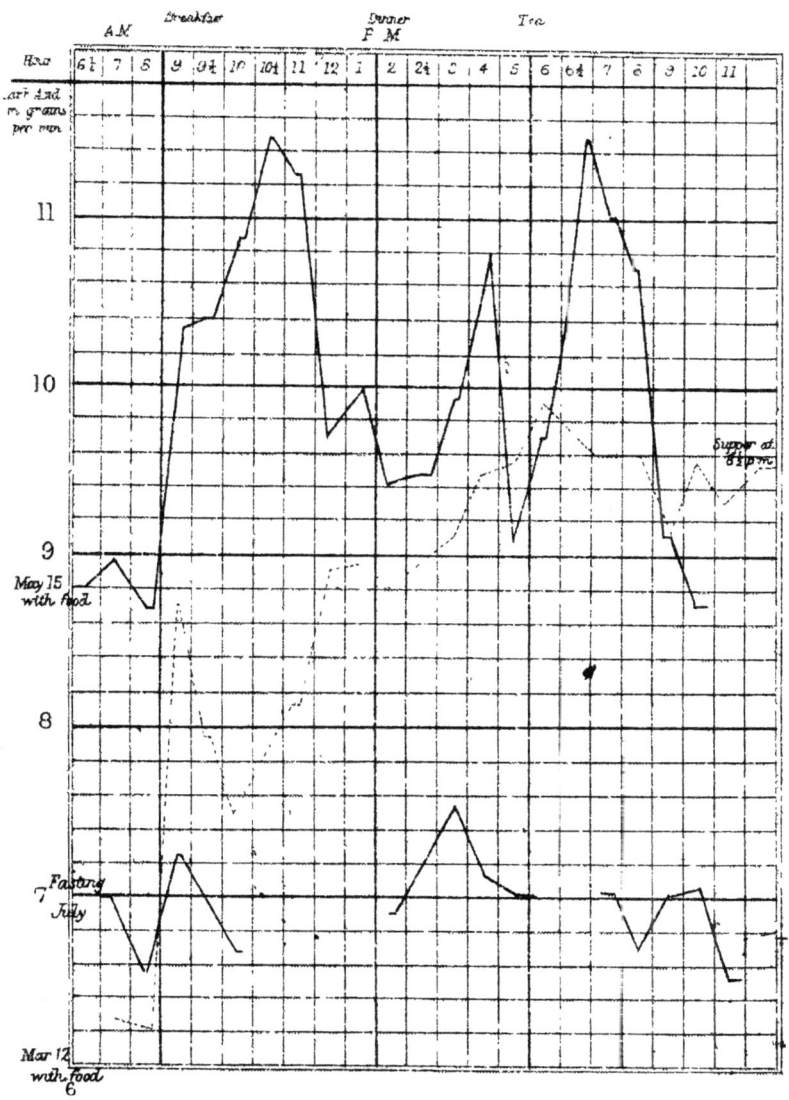

Diagram shewing the quantity of Carbonic Acid evolved during the day with food on two occasions and without food on one occasion 1858

N° 5.

Carbonic Acid and Temperature

Relation of Temperature & Carbonic Acid on April 13th to 27th 1858.

	Temp	49°	47·5°	54°	58·5°	58°	60·5°	62·5°	62·6°	63·6°	59°	58°
Carb. Acid grains per min.	Self	9·43	9·46	8·14	8·3	8·4	8·08	8·04	8·76	8·4	8·11	9·3
	Mr Moul		7·7	6·9	7·2	7·	5·4	6·62	5·38	5·12	6·	6·52
	Mr Moul		482	483	444	400	298	394	306	348	392	394

FIG 3

Relation of Carbonic Acid to Temperature 1858

	April	14	15	16	20	21	22	23	24	26	27
	Temp	47·5°	54°	58·5°	58°	60·5°	62·5°	62·6°	63·6°	59°	58°
Carb Acid (in grains) per min. to each degree of Temp	Self	·2	·15	·142	·144	·133	·128	·14	·132	·137	·16
	M^r M	·162	·127	·123	·12	·04	·105	·07	·13	·1	
Carb Acid (in grains) per min. lessened with each degree of Temp above 47·5°	Self	·2	·1	·1	·1	·093	·046	·065	·128	averages ·104
	M^r M	·12	·045	·066	·177	·072	·153	·098	·148	·11

1st Series.

2nd Series

J. Basire, sc.

The variations in the requirement for food are induced by age, climate, season, and degree of exertion, and will be more fully discussed in the work on Dietaries; but it may now be desirable to give a glance at some of them.

In reference to age, there can be no doubt that all vital processes, including the action of foods, are greater and more rapid in early, and less and slower in later, than in mature life, and in both the former a more frequent administration of food is necessary. In early life, moreover, there is the important function of growth, which demands a large and more frequent supply of food, not only for daily wants, but to promote a due increase in the bulk of the structures of the body.

I have elsewhere[1] shown that the season of the year has also a decided influence over the vital actions, so that they are the greatest in the spring and the least at the end of summer, as is illustrated by the following diagram.

Diagram No. 5 exhibits the effect of sudden and marked changes of temperature over the vital function in two adult persons. The effect was such that the quantity of carbonic acid evolved by the respiration was inverse as the change in the temperature; the vital changes lessening with increase of temperature, and *vice versâ*. (*Phil. Trans.* 1859.)

The action of climate is similar to that of season, and shows that the vital actions are greater in cold than in hot climates, and in the uplands than in close valleys.

The influence of exertion over vital changes is immediate and proportionate, whilst the subsidence with rest is less rapid than the increase. The following

[1] 'Health and Disease: Periodical Changes in the Human Body.'—Henry S. King & Co., Cornhill.

table of experiments upon myself shows the proportionate effect of exertion of varying degrees on the basis of the increased volume of air inspired:—

No. 6.

The lying posture	being 1
The sitting posture	is 1·18
Reading aloud or singing	,, 1·26
The standing posture	,, 1·33
Railway travelling in the 1st class	,, 1·40
,, ,, ,, 2nd class	,, 1·5
,, ,, upon the engine, at 20 to 30 miles per hour	,, 1·52
,, ,, ,, ,, 50 to 60 ,, ,,	,, 1·55
,, ,, in the 3rd class	,, 1·58
,, ,, upon the engine, average of all speeds	,, 1·58
,, ,, ,, ,, at 40 to 50 miles per hour	,, 1·61
,, ,, ,, ,, 30 to 40 ,, ,,	,, 1·64
Walking in the sea	,, 1·65
,, on land at 1 mile per hour	,, 1·9
Riding on horseback at the walking pace	,, 2·2
Walking at 2 miles per hour	,, 2·75
Riding on horseback at the cantering pace	,, 3·16
Walking at 3 miles per hour	,, 3·22
Riding moderately	,, 3·3
Descending steps at 640 yards perpendicular per hour	,, 3·45
Walking at 3 miles per hour and carrying 34 lbs.	,, 3·5
,, ,, ,, ,, 62 ,,	,, 3·8
Riding on horseback at the trotting pace	,, 4·05
Swimming at good speed	,, 4·33
Ascending steps at 640 yards perpendicular per hour	,, 4·4
Walking at 3 miles per hour and carrying 118 lbs.	,, 4·75
,, 4 miles per hour	,, 5·0
The tread-wheel, ascending 45 steps per minute	,, 5·5
Running at 6 miles per hour	,, 7·0

The next table from the same series of experiments shows the same effect on the basis of the amount of carbonic acid evolved by respiration per minute:—

No. 7.

In profound sleep, lying posture	4·5 grains
In light sleep ,, ,,	4·99 ,,
Scarcely awake, 1½ A.M.	5·7 ,,
,, ,, 2½ ,,	5·94 ,,
,, ,, 6¼ ,,	6·1 ,,
Walking at 2 miles per hour	18·1 ,,
,, 3 ,, ,,	25·83 ,,
Tread-wheel, ascending 28·15 feet per minute	43·36 ,,

Thus it is possible that the amount of vital change proceeding in the body may be ten times greater in one state than in another, and it follows that a proportionate quantity of food will be required to sustain it.

PART I.

SOLID FOODS.

Section I.—Animal Foods.

a. Nitrogenous.

CHAPTER I.

DESCRIPTION AND COOKING OF FLESH.

The qualities by which foods may be classified have been already shown to be numerous and open to the selection of the enquirer; but as it is not desirable in a work of this kind to attempt too much refinement of mere outline, we will omit two subjects, and refer the classification on the grounds of economy and chemical action to the work on Dietaries, whilst here we make use of the familiar but comprehensive one of solids, liquids, and gases. Solids will be divided into animal and vegetable, and each subdivided into nitrogenous and non-nitrogenous. We shall proceed first to treat of solid animal foods.

This important series of food might be subdivided into two other classes, viz., those which constitute the substance of an animal and are obtained when it is dead, and those which are the natural product of the animal and are obtained whilst it lives. The former includes the flesh, and as it varies in different classes of creatures, it is popularly subdivided into flesh, fish, and fowl, whilst the latter are milk, eggs, and other products. Those who profess to be vegetarians eat the latter only. It is also divided into lean and fat, both of which abound in animals generally, and this leads to a yet more technical division, viz. into nitrogenous and non-nitrogenous foods, since all lean flesh contains nitrogen, whilst all fats when pure are destitute of it.

Hence, however differing in appearance, all kinds of flesh have certain nutritive qualities in common; but the proportion in which the qualities exist varies, and each large division of the class has its own nutritive value.

The anatomical composition of flesh is very similar in every kind of creature, whether it be the muscle of the ox or of the fly; that is to say, there are certain tubes which are filled with minute parts or elements, and the adhesion of the tubes together makes up the substance of the flesh. This may be represented grossly by imagining the finger of a glove, to be called the sarcolemma, and so small as not to be apparent to the naked eye, but filled with nuclei and the juices peculiar to each animal. Hundreds of such fingers attached together would represent a bundle of muscular fibres. The tubes are of fine tissue, but are tolerably permanent; whilst the contents are in direct communication with the circulating blood and pursue an incessant course of chemical change and physical renewal. The quality of meat consists in the character of the pulp or enclosed substance, whilst the toughness depends chiefly upon the tubes and the structures which bind them and other parts together, and both vary with the age and breeding of the animal. The aim in modern breeding is to produce the greatest amount of muscle and

DESCRIPTION AND COOKING OF FLESH.

fat at the earliest period of life, but it is well known that whilst delicacy of flavour may thus be obtained, fulness and richness can be produced by age only. It is well known also that the connecting tissue, or the substance which binds the parts together, is relatively more abundant in ill-fed, ill-bred, and old animals than in the opposite conditions, and renders the meat tough. Hence it will be readily inferred that young and quickly fed animals have more water and fat in their flesh, whilst older and well-fed animals have flesh of a firmer touch and fuller flavour, and are richer in nitrogen. The former may be the more delicate, the latter will be the more nutritious.

There are, however, two divisions of flesh which must be referred to, for although not differing much in chemical composition, they vary in their value as nutrients. The fibres of flesh generally are crossed by lines invisible to the naked eye, so that all voluntary

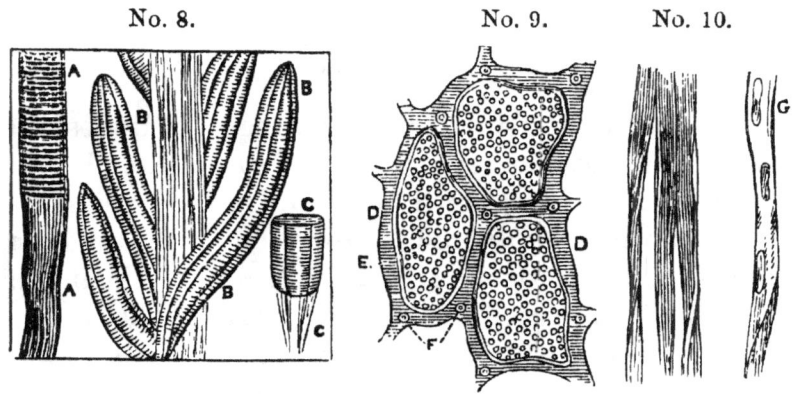

No. 8. Voluntary muscle (striated).—No. 9. Transverse section of three fibres of the teal.—No. 10. Involuntary muscle (non-striated).

A. Of the cat.
B, C. Of the house-fly.
D. The sheath.
E. Round refracting particles.
F. Capillary vessels.
G. Showing nuclei when treated with acetic acid.

muscles are striated (fig. 8); but the other muscular organs which do not move by volition have muscular fibres which are not striated (No. 10), and are termed involuntary muscles. The heart is peculiar, being an involuntary muscle, but with striated fibres. The involuntary muscles are softer in their texture than ordinary flesh, but are not easily masticated; so that, notwithstanding their identity in nutritive elements, they are not so nutritious as ordinary flesh, and never obtain so high a price in the market.

The juice of flesh has an acid reaction, and is more abundant in striated than in plain muscular fibre. It contains albumen, casein, creatine, creatinine, sarcine, lactic acid, inosic acid, and several volatile acids, including formic, acetic and butyric acids, a red pigment similar to the colouring matter of the blood, and inorganic salts, chiefly alkaline chlorides and phosphates. It is much richer in potash than soda, as shown by the following comparison with 100 parts of soda in flesh-juice and blood (Liebig):—

No. 11.

	Ox	Horse
In the blood	5·9	9·5
In flesh-juice	279·0	285·0

It is said by the hunters in southern latitudes that flesh is the most tender and juicy if eaten directly after the animal has been killed, and whilst it is yet soft—nay, even when the animal is alive; but if the stiffening of the flesh, called *rigor mortis*, have begun, it remains less tender until the rigidity has passed away and the changes of early decomposition have set in. Meat is always eaten in hot climates in its first or second state, unless means be taken to preserve it, since decomposition sets in too rapidly to allow it to be safely kept fresh for more than a few hours. On the other hand, flesh is never eaten in colder climates before the second state, and in order to lessen the hardness or toughness of it, it is usual to allow it to enter the third state, when it becomes soft and tender, and has gained a flavour from decomposition which is often approved. The nutritious elements, speaking generally, are the same in each state until the effects of decomposition appear, but the nutritive qualities, being dependent upon the power to masticate and digest the flesh, may be really less in the second than in either of the other states.

The desired effect may be produced at any stage in a rough manner by cutting the flesh into slices and beating it across the cut ends until the fibres are broken and the connecting tissues forced asunder; but to effect this the slices must be thin.

The effect of cooking flesh is chiefly physical, and is chemical in a very limited sense only. When meat is either roasted or boiled, it decreases in bulk and weight, and the cooked food is generally less soft than fresh meat in the first state. The diminution in bulk and weight is owing to the extraction of the juices of so much of the mass of flesh as may have been acted upon by the heat, and these are chiefly water containing salts, and the peculiar flavour of meat, with a proportion of fat in a fluid state, gelatin, and perhaps some albumen. The flesh thus treated becomes contracted in bulk, from loss of the juices and by coagulation of the albumen, whilst the mass is composed of solid fibrin, with a proportion of albumen, and the juices and fat which have not been extracted. The tubes having lost much of their contents, shrink and separate from each other, and so far the meat may be made more tender; but this varies in degree, and is often more than counterbalanced by the hardening of the albuminous contents of the tubes.

The object of cooking is to render the flesh more submissive to mastication and digestion, but it may be entirely frustrated if the substance of the flesh be hardened in any appreciable degree. It is also employed to make the food hot when it is eaten with a view to improve its flavour, and to stimulate the sense of taste. It is only an incident in cooking, however inseparable from the act, that the flesh should diminish in weight by the loss of its fluid parts, but if all that is valuable from the extracted matter be collected, there will be no real loss of nutriment. There is, however, in this respect some difference according to the mode of cooking. If the meat be boiled, the introduction of fluid into the substance of the meat, whether between the structures or within the fibres, aids the extractive process, but at the same time retains and preserves that which is extracted. If it be roasted whilst surrounded on all sides by the air, the heat is not applied so uniformly and gently, and therefore the outside becomes overcooked before the inside is sufficiently cooked, and this occurs to a far greater extent than with boiling. Hence not only is the fluid part of the juices extracted and lost, but the loss is greater than when the meat is boiled. It is, however, to be understood that the matters extracted are only such as may be dispersed by heat; and whilst, therefore, the evaporated water may carry off some of the flavours of the meat, it does not remove the salts which are present in the juices. Hence meat which is properly roasted has lost weight more than that which is boiled; but if no account be taken of the matters extracted, it contains a larger proportion of nutritive elements than the larger mass of boiled meat, and in a given weight is more nutritious. When, however, the extracted matter is collected and used, there is a greater proportion of nutriment in the boiled meat with the broth than in the roast meat with the liquefied fat, and it is clearly desirable that both the broth and the boiled meat should be eaten together.

Stewed meat occupies a position between that of

boiled and roast, for it may have been submitted to a greater heat and for a longer period than boiled meat, and thereby a larger proportion of soluble matter may have been extracted, whilst it differs from roasted meat in that the outside is not hardened and all the extracted material is retained. Boiled meat may be cooked so that the solid part shall still retain nearly all the nutritive elements of flesh, whilst the solid part of the stewed meat may be even less nutritious than the material which has been extracted from it.

The degree in which extraction of the juices takes place in cooking meat depends upon the heat employed, so that the proper application of heat is a fundamental question in cookery. It has been intimated that the extraction of the juices is chiefly from the cut ends of the soft fibres, and that the fibres become harder by the coagulation of the albumen during the process of cooking. When, therefore, the fibres have become hardened, they have lost some of their contents, but this condition prevents or retards the further passage of juices from parts beyond the hardened ends. The sooner, therefore, the hardening process can be effected, the sooner will the loss of juices be diminished or prevented. Dipping the meat to be boiled into boiling water effects this object, for albumen coagulates at a temperature much below that of the boiling point of water; and placing the meat to be roasted very near the fire at first has the same effect. Thus less juices escape (all other parts of the process being equal), and the mass of flesh retains its nutritive elements. This is clearly desirable when the flesh only is to be consumed; but if it be desired to make good broth or beef tea, the opposite course must be adopted, and by keeping the temperature below 160° the tubes may be emptied to a far greater degree than with a higher temperature. Hence the explanation of the saying, that you cannot have good broth and good meat from the same piece of flesh.

But the preliminary point having been settled, the proper mode of cooking is clearly not to coagulate the albumen unduly, but to make the whole mass of meat soft and tender. A slow fire, or water at a temperature of 160°, will suffice to expand the fibres, and in some degree to rupture them, whilst it separates these and other structures and renders the whole mass more fitted for mastication and digestion. To keep meat in boiling water, or to expose the joint to continued heat before the fire, is to make it hard and to extract a greater proportion of the juices.

Flesh thus treated is less susceptible of decomposition than fresh meat, by reason of its harder crust and the diminution of its juices, and may thus be preserved in a state fit for the use of man for some time.

CHAPTER II.

PRESERVED MEAT.

THE art of preserving meat for future use, with a view to increase the supply and lessen the cost of this necessary food, is of very great importance to this country, and all the available resources of science are now engaged in it.

It is of course the most desirable to import meat in its fresh state, both that its nutritive qualities may be retained, and that its flavour and appearance may compare favourably with those of the meat produced at home. It is not, however, possible to obtain a sufficient supply from countries so near to us that it may be brought in a sound state without having undergone some process of preservation, and hence the problem to be solved is what is the best method of preserving it so that it may retain its qualities of fresh meat until its arrival here.

We will now proceed to state briefly only the *typical* methods which have been recently employed in preserving meat on a large scale, viz., by drying, by cold, by immersion in liquids and gases, by coating with fat, by heat, by salting, and by pressure.

A.—By Drying.

A long-known method of preserving meat is to cut it into thin slices, and dry it in the sun, as is practised in South America and in other climates hotter than our own; but unless the atmospheric conditions be favourable, and the drying effected rapidly, the process of decomposition is not prevented, and when it has been prepared it is fit for use for only a very limited period.

Dr. Hassall has applied a method by which meat may be preserved for a lengthened period in a dried state. He thoroughly dries lean meat by the application of a gentle heat, and then grinds it into a coarse powder, which may be made into soup by the addition of hot water, or mixed with flour and made into biscuits. The flavour of it in soup is somewhat rough, and the soup is not equal in quality to that made from fresh meat, but the nutritive qualities of the lean meat are there, except the aromas, which may have escaped under the influence of heat. Hence, if it be entirely digestible, it is the most nutritious of all the preserved substances used in the preparation of soup; and when white of egg, with condiments and vegetable juices, are added, the soup is fit for use, both by the invalid and the healthy.

The method is not adapted to the preservation of meat for separate use, since the cost of preparation would prevent its sale in the market; but it is employed in the preparation of meat biscuits for the use of soldiers or travellers, and for portable soup.

The drying of meat by artificial means may be the most quickly effected over burning coals. Herrings and some other fish are also preserved by the drying process only, as by the heat of the sun or the action of warm air, with or without the fumes of burning wood; but this food will not resist decomposition beyond a very limited period.

Dried Hamburg beef is well known in this country, although not so extensively used as in Germany and other continental states. It is a very nutritious and agreeable food, and will remain good for many months. It is smoked as well as dried, and is used in the preparation of sausages.

M. Tellier of Paris, in his work, entitled 'Conservation de la Viande et autres Substances alimentaires par le Froid ou la Dessiccation,' gives the following method.

He first rarefies the contained air by an air-pump to a tension of two or three centimètres of mercury, and then fills the vessel with carbonic acid gas from a gasholder, so that the atmosphere will consist of only about three per cent. of air. This he removes by the air-pump to the same tension, and the remaining air is almost entirely carbonic acid gas. Once more, however, he fills the vessel with carbonic acid, and again removes it as before. Afterwards he absorbs the carbonic acid by the use of a concentrated solution of potash, by which a very near approach to a vacuum occurs. After leaving the meat thus treated for three days, it is removed, and may be kept sound without any further trouble, but it will have lost eighteen to twenty per cent. in weight.

B.—By Cold.

The application of cold, as is well known, has great influence in retarding decomposition. Flesh packed in ice, from which the water is properly drained, will retain its freshness for several weeks; and in cold climates, where the temperature is constantly below freezing point, will remain fresh for months whilst exposed in the open air. But the effect of cold is to lessen, if not to change, somewhat the flavour of the substance, so that whilst it may be really good food it does not fully equal recent fresh meat. This defect is, however, limited; and if there were sources of meat, which at the same time supplied ice abundantly, it would be possible to import meat at a price and of a quality which might be very acceptable in the markets of this country. Hence the real difficulty is to provide a sufficient quantity of ice at the ports of South America and Australia, in which to preserve the meat for long voyages, without imposing a charge which would unduly raise the price of the meat when imported. There could be no difficulty in obtaining ships that might carry a sufficient quantity of ice from which the water should be perfectly drained, and thus solve the problem in a more satisfactory manner; but so long as our supplies of meat are from hot climates the expense will be a serious impediment to such a commercial enterprise. It is, however, in my opinion, the most fitting mode of solving the problem, and it should be effected either by inducing the inhabitants of countries where ice is abundant in the cold season to grow animals for our markets, or by storing large quantities of ice in an economical manner at the ports of other meat-producing countries. Moreover, it is now possible that ice may be made anywhere at so small a cost as to be available for this purpose; for Messrs. Nasmyth, of Manchester, have constructed machines, on the patent of M. Mignot, by which 50 lbs. of ice may be made per hour at the cost of condensing and then rarefying air, and a company has been formed with the intention of producing as many tons daily.

We need not despair of seeing the time when the whole carcass of an animal will be imported in a state fit to be cut up in our shops for immediate sale, and when the exporters from Australia will supply themselves with ice for this purpose from the southern hemisphere, or when the pastures of the North American continent will become our chief, as they may be our nearest, sources of supply. Canada offers unbounded facilities for this purpose, by reason of its great ice fields, its pastures and agricultural population, as well as its nearness to this country; and should the present high price of meat continue, it will induce commercial men to organise a system, both of feeding the animals and of exporting the meat in ice, which may be very profitable to the Canadians.

M. Tellier of Paris, in his work before mentioned, has described a process which demands great consideration, inasmuch as he removes much of the moisture from the atmosphere.

He proposes to place joints of meat in a chamber through which a current of air, charged with ether or other volatile substance, may be passed with a view to reduce the temperature, so that the vapour in the air shall be frozen, but the temperature not reduced below $-1°$ Centigrade, that is about 30° F. He would not freeze the juice of the meat, but by a low temperature keep the chemical and vital actions at their lowest point. He has given directions for the construction of ships for the voyage, and for the fitting-up of stores, and indeed all the details, from the slaughtering of the animal to the arrival of the food in England. The following is extracted from the work referred to:—

'Ce que j'emploie, c'est un courant d'air froid amené directement un peu au-dessous de 0°, ou des courants liquides à −8° ou −10°, qui, saisissant l'atmosphère, congèlent l'humidité qu'elle renferme, la dessèchent et abaissent rapidement sa température, fournissant ainsi les résultats cherchés.

Dans cette condition, non-seulement l'atmosphère est constamment purifiée des miasmes organiques qu'elle renferme, mais une légère et lente dessiccation se produit, dessiccation qui vient aussi aider à la conservation (environ 10 pour 100 en poids pour six semaines).

Tout le mécanisme de l'opération consiste donc, on le voit, à constituer de simples magasins froids. Ces magasins peuvent être la cale d'un navire, l'intérieur d'un wagon, un local quelconque. *Ce qui importe seulement, c'est que la température y reste* fixe entre 0° et −1°, c'est-à-dire, au point où l'eau en suspension dans l'atmosphère est solidifiée, tandis que celle renfermée dans les tissus se maintient liquide, préservée qu'elle est de la congélation par les substances en solution dans elle.'

C.—By Immersion in Antiseptic Gases and Liquids.

There are numerous gases, such as sulphurous acid and nitrous acid, which have great power in retarding decomposition, and have been used for the preservation of meat; but when the method was practically applied on a large scale, it did not yield satisfactory results. There are two difficulties to be overcome,—first, the removal of the air surrounding the meat; and second, the prevention of a disagreeable flavour, which may be due either to the preserving agent or to the process of putrefaction.

Mr. Jones patented a method by which he first enclosed a quantity of meat in a tin, and after filling the vacuities with water, replaced it by the preserving gas, and then sealed the tin to prevent the re-admission of the atmospheric air. The theory on which the method was founded was good, and the art employed seemed to be trustworthy; but whilst some of the meat thus treated remained fresh for weeks, other joints were partially decomposed and unfit for food. Moreover, when the operation had been successful, the change in the flavour and the appearance of the meat was not agreeable; and although, if the meat were tinned, these considerations might not prevent its sale, it would deter purchasers from giving a price at all approaching that of fresh meat. It is, however, probable that a modification of the method, whether in the gas employed or in the mode of procedure, might obviate both these defects.

Messrs. Medlock and Bailey have patented a process of preserving meat in bisulphite of lime or zinc, which while it arrests putrefaction, is said not to injure the flavour of the meat. It may be adopted in hot climates, where animals are killed only after sunset, and the joints preserved at once by immersion and subsequent drying, or it may be applied in this country, at home or in the market, to a piece of meat with a sponge, and thus retard putrefaction.

D.—By Coating with Fat or Gelatin.

Professor Redwood applied a plan by which the joint would be entirely enclosed in fat, and being thus kept from contact with the air, might not be decomposed.

This appeared likely to be a useful procedure, and it attracted the attention of Professor Gamgee and other scientific men, with a view to utilise it. It has not, however, succeeded in establishing a profitable trade, and probably from one or all of the following reasons. Firstly, when the joint was placed in the heated liquid fat the outside became cooked to a certain degree, and after the fat had been removed it did not present the appearance of a fresh uncooked joint. Secondly, the exclusion of the air was not always perfect, or the putrefactive process had already commenced in some part of the joint previous to the operation, so that on arrival in this country the meat was not fit for food. Thirdly, although the cost of the refined fat where the meat was produced might not be great, its value on being removed from the joint in this country was so lessened that the loss from this cause alone on a leg of mutton was one shilling, and by so much the cost of the meat was increased.

There should be no insuperable difficulty in the removal of the two first-named defects, and so soon as the fat used can be made sufficiently profitable in this country, the third defect should also disappear. We may have to wait for the discovery of a more perfect method, but a step in advance seems to have been taken by Mr. Craig. This gentleman has prepared a fat, almost without flavour, which will remain hard and unchanged for many months, and may be used in cooking after having performed the duty of preserving the joint of meat, whilst its cost in this country is only sixpence to sevenpence per pound.

This method in a modified form is adopted in the preparation of pemmican and potted meats. In the former, the dried meat is minced and mixed with about half its weight of fat, whilst fresh meat is used in the latter, and in both the fat is not only incorporated with the meat, but covers the surface. But although condiments are usually added, and are said to retard the putrefactive process, the latter food will not remain good for a lengthened period, and the method is of little value in reference to the importation of meat.

The application of oil instead of fat as a preserving agent for meat has not hitherto been practicable, on account of its cost and flavour, and the tendency which it has to become rancid, but it has long been employed for more costly substances, as sardines, and when enclosed in hermetically sealed tins, has been found very effectual.

E.—By Heat.

The application of heat has hitherto been the most successful method of preserving meat, and has called into existence a vast machinery for preparation and exportation from distant meat-producing countries. It has two objects,—first, to prepare the meat; and second, to exclude the atmospheric air, which, if allowed to remain, would advance the process of putrefaction.

It is perhaps not necessary that the meat should be boiled in order to acquire a condition in which it is the best fitted for preservation, since a temperature of 200° would coagulate the albumen and reduce the volume of the juices. It is, however, necessary to remove the surrounding atmospheric air from contact with it, and therefore the meat is enclosed in a tin case, from which the air is to be excluded as perfectly as possible, by rarefaction. The degree of heat which is necessary to effect the latter object is more than enough to cook the meat, and hence, whilst both objects are effected by the same operation, the meat becomes overcooked.

The process of preserving meat in this manner is a very simple one. A tin having been prepared, a piece of raw flesh and fat is selected and placed in the tin with or without a small quantity of fluid. The cover is then soldered on, and the tin is closed, except at a small hole through which air and steam may escape when heat is applied.

The following is Mr. Jones' method, as described by himself:—

'We put the meat into tins, either with or without bone, in joints or otherwise. The tins are filled quite full, and are soldered up entirely, with the exception of a small tube, about the size of a quill, which is soldered into the top of the tin. The tins are then put into a bath capable of holding 96 6-lb. tins. Along the centre of the bath runs a tube carrying 12 taps, into each of which may be inserted a tube from 8 tins, there being 8 stuffing boxes to each tap, 4 on each side. The tube communicates with a vacuum chamber. The bath contains a solution of chloride of calcium, which boils at a temperature of from 270° to 280°. In commencing operations, the bath is gradually heated until it gets to about 212°. Communication is then opened with the vacuum chamber, and the result is, that as water boils at about 100° *in vacuo*, the water is carried off in the shape of steam, into the vacuum chamber, where it is condensed. The tap is then turned, so as to shut off communication with the vacuum chamber, and the meat is cooked at a high temperature, until complete preservation is effected. We then go on cooking the meat, occasionally turning the taps, just to do what engineers call priming, in order to draw off any fluid that may be in the tins. After it has been thus cooking for about two hours at 250°, it is in a preserved state.'

This process is not identical with boiling meat, for, as the tin becomes filled with the vaporised liquor, the meat is stewed, and neither boiled nor roasted.

It is not possible, by the application of such a moderate degree of heat as would not destroy the meat, to expel all the air from the tin, and all that can be effected is by expansion of the air to expel as much of it as possible. It has been proved by experiment that the temperature of boiling water does not do this in such a degree as would preserve the meat, and hence a degree of heat which is too great for the cooking of the food must be increased by 20° or more to accomplish the other and yet more necessary object, and by Mr. Jones' process the meat is cooked at a temperature of 250°. After a certain duration of exposure to the action of heat, the hole in the cover is closed by solder, and the tin is hermetically sealed.

The meat thus preserved is overcooked, and a large part of the juices being extracted by the process, the fibres are greatly loosened, and the mass readily breaks up.

On opening such a tin the meat is found enclosed in, and indeed permeated, by a gelatinous substance, the weight of which is about a quarter of the original contents of the tin, and inasmuch as this gelatinous gravy may have been flavoured by condiments and burnt meat, it may be very agreeable, but the meat itself is fibrous, loose, not easily masticated, and not very agreeable to the palate.

The problem of preserving meat by heat for lengthened periods has therefore been solved, but not in a perfectly satisfactory manner. If it be desired to use only the meat and to exclude the gelatinous gravy, the diminution of weight largely increases the original cost of each ration, and when eaten it is neither so agreeable nor so nutritious as meat cooked in the ordinary manner. When, however, all the contents of the tin are used together, as in preparing Irish stew or soup, the extracted matter becomes available for and valuable as food, and the only defect is the difference in sensible qualities, which is perceived when it is compared with fresh meat.

The desideratum in the preserving process is to expel the air sufficiently without over-cooking the meat, but it cannot be effected, except in the manner pointed out when sulphurous acid gas is employed, for the fluid must permanently remain in the tin or the air would be readmitted, and if the quantity of fluid be large the proportion of solid meat would be too small to make the operation profitable in a commercial point of view. It may appear easy to fill the can with the solid meat well pressed down, and then so far exclude the air that but little fluid would be necessary to complete that part of the process; but bubbles of air would be enclosed in the folds of the meat, and would remain notwithstanding the tendency of the liquid to enter every vacuity, and wherever they existed the process of putrefaction could not be prevented.

Yet it is worthy of consideration whether something more cannot be done in this direction, whereby the application of so high a degree of temperature may be rendered unnecessary. Thus, if the meat were first partially cooked at a temperature of, say, 180°, and cut into blocks or slices, and placed carefully in layers until the tin case was nearly filled, and then boiling liquor were added to entirely fill the case preparatory to exposing the tin and contents to a temperature of 200° to 212° for a limited period, I think it probable that the desired end might be obtained without over-cooking the meat in its present degree.

Cooked meat might be preserved by the gaseous and fatty methods already described, and since the bulk would thus be lessened, and the chance of decomposition somewhat lessened also, it may be desirable to further consider the matter. If this could be effected, it would probably be better to roast or bake the meat before it is enclosed in fat, since the latter would more readily attach itself to it than to boiled meat.

The greater proportion of the meat imported from Australia is mutton.

F.—Salted Meat.

The oldest and best known preserving agent is salt, with or without saltpetre. Its chief action appears to be due to its power of attracting moisture, and by thus extracting fluid to harden the tissues. Solution of the salt in water is, moreover, accompanied by the absorption of heat, so that it tends to lower the temperature of the meat with which it is in contact and so far aids in preserving it.

The modes in which it is applied are numerous and deserve a little consideration.

The well-known methods are, simply rubbing the surface of the meat with salt, or immersing the meat in a strong solution of salt with the addition of saltpetre, and in order that they may be effectual it is necessary that the meat to be salted should be of a newly killed animal. When the preservation is to be effected by rubbing the salt into the flesh, the operation should be renewed from day to day, or at longer intervals, and every fold of the meat, particularly near the bones, should be well rubbed. Meat thus preserved by salt alone loses its colour, but when saltpetre is added the flesh becomes of a reddish colour throughout, provided the action be sufficiently prolonged.

PRESERVED MEAT.

A good brine is made of 4 lbs. of salt and ¼ lb. of saltpetre in 6 pints of water.

It is usual to employ salt and saltpetre when preparing the dried strips of flesh, called *charqui* in the South American States and *tasajo* in Nicaragua, for the climate does not allow the meat to be so thoroughly dried in a few hours as to prevent decomposition. The meat selected for this purpose is very lean, and in preparing the food for the table it is essential to use fat of flesh or lard, and to flavour with vegetables, herbs and good gravy.

Salt is also used very generally when preparing dried fish.

Mr. Morgan devised an ingenious process by which the preserving material, composed of water, saltpetre, and salt, with or without flavouring matter, was distributed throughout the animal, and the tissues permeated and charged. His method was exemplified by him at a meeting of the Society of Arts, on April 13, 1854, when I presided, and is described as follows in the *Journal* of that Society of the preceding March:—

'A bullock having been killed in the usual way, the chest was immediately opened, and a metal pipe with a stopcock inserted in connection with the arterial system. The pipe was connected, by means of elastic tubing, with a tub filled with brine, placed at an elevation of about twenty feet above the floor. The stopcock being turned, the brine forced itself through the arteries of the animal and passing through the capillaries flowed back through the veins carrying with it all the blood, making its exit by means of an incision provided for that purpose. About six gallons of brine passed thus through the body, washing out all the blood from the vessels. Having thus cleared all the vessels, the metal pipe was connected with another tub similarly placed, containing the preservative materials to be injected, and at the same time their exit after traversing the body was prevented. On communication being made, the liquid became forced into the vessels, and by means of the pressure it penetrated into every part of the animal.' Much of it proceeded through the minute vessels, called the capillaries, into the veins, but a considerable proportion exuded through the sides of the blood-vessels, and escaped into the cavities of the body and the surrounding tissues. The whole body was thus an incorporation of flesh and brine, and the operation was complete in a few minutes. After a short time, the carcass could be cut up into joints and packed for exportation.

The preservative material which he recommended was 1 gallon of brine, $\frac{1}{4}$ to $\frac{1}{2}$ lb. of sugar, $\frac{1}{2}$ oz. of monophosphoric acid, a little spice and sauce to each cwt. of flesh.

The process was ingenious and sound in theory, and the preservation of the meat was complete when no untoward event occurred, but it failed in certain instances. When performed in this country it was used chiefly to preserve meat for the Navy, and not for home consumption; but there is no difficulty in employing the method in foreign meat-growing countries for the use of any people wishing to eat salt meat.

Salted meat has, however, several defects which will always prevent its general use whenever fresh meat can be obtained.

1. The salt extracts a considerable quantity of the juices, and by so much lessens the nutritive value and natural flavour of the meat, and as these extracted juices are obtained only when mixed with salt they cannot be used as food. The flesh is harder than

cooked fresh meat, in proportion to the strength of the saline solution and the duration of the application of it. This is particularly the case with the meat which is both salted and dried, so that after having been prepared for some months it cannot be rendered soft by any amount of soaking in water and skill of cooking. The spasmodic attempts which were made to introduce *charqui* into common use entirely failed for this reason. This is, however, comparatively slight when the meat is highly salted and intended for early use, and, with subsequent judicious immersion in water and cooking, nearly all the hardness may be removed.

2. The flavour differs very greatly from that of cooked meat, and although when used occasionally it is agreeable, it is not preferred to fresh meat as a regular article of diet.

3. The introduction into the system of so much salt is prejudicial to health, whether by lessening the relish for food or inducing a craving for fluids, indigestion, or skin disease.

4. The capability to nourish the system is lessened by the various effects now mentioned, and a given weight of salted meat is not equal in nutritive value to that of fresh boiled meat with the meat liquid added, or to fresh roasted meat.

Hence it is not desirable to extend the operations of preserving meat by this process, provided a sufficient supply of meat can be obtained, whether preserved or otherwise, in its fresh state, and commercial and scientific men should be encouraged to improve the method by which meat may be preserved unsalted.

It is still the practice to salt or pickle beef and other kinds of meat for the year's supply in Anglesea and in the Highlands of Scotland, and the animals being killed in the cold weather, the meat takes the salt readily, but it is a less prevalent practice than formerly. It is called *mairt* in Scotland to signify Martinmas time—the period when it is prepared.

G.—BY PRESSURE.

The River Plate, and also the Texan Pressure Meat Preserving Companies, have adopted Henley's process of meat preserving, by which the meat is cut into thin slices and subjected to pressure, which causes much of the juice and fat to be removed from the meat, and the meat and juices are treated and preserved separately. It remains to be seen whether meat with the juices pressed out of it can be sold, or will be valuable as food. It will not be meat in the ordinary sense of the word.

CHAPTER III.

BONE.

THE value of bones as food is not a recent discovery, since the knuckle bones of veal and the marrow or marrow bones have long been in request, but it is only of late years that a proper estimate has been made of the nutritive material, which by a careful process may be extracted from them.

Bones consist principally of two substances, viz., gelatin, which may be obtained by immersing them in weak muriatic acid, and mineral matter which may be separated by burning the gelatin, and so great is the proportion of each, that the form of the bone is still retained when either is taken away.

The following is the chemical composition of dry ox-bones, in 100 parts:—

No. 12.

Gelatin	33·3
Phosphate of lime	57·35
Carbonate of lime	3·85
Phosphate of magnesia	2·05
Soda and chloride of sodium	3·45

Hence one-third of the weight of dry bone consists of nitrogenous matter, which when extracted could be used as food. This is a much larger proportion than is found in fresh bread or meat.

But besides these elements, there are others in fresh bones which are of great value in nutrition. Such are oil, nitrogenous juices, and flavouring matters which vary with the kind of bone.

Bones consist of three parts which require notice as foods. The solid shaft, as of the long marrow bones, the cellular structure of the flat bones, and the cartilaginous ends of the bones at the joints.

The solid shaft cannot be used as food by the process of boiling, since it does not disintegrate by that agency, and in order to extract the gelatin, it is desirable to grind the bone before boiling it. The marrow is, however, very valuable both as a fat and for its agreeable flavour, and may be roughly reckoned as equal in nutriment to half of its weight of butter. The cancellated bones may be first roughly broken and then disintegrated by digestion in a closed vessel with hot water for twelve to twenty-four hours. The cells contain fluid which consists of water, fat, and nitrogenous and flavouring matters, which are valuable and agreeable additions to foods, so that this class of bones is the most valuable for food.

The ends of the bones are composed of cartilage, and as in early life they contain but little bony matter, they are easily detached by boiling, and may almost in their entirety be used as food, but in later life they are firmly attached to the rest of the bone, and are filled with bony matter.

Bone cartilage of the ox and calf has the following ultimate composition per cent.—(*Frémy*).

No. 13.

	C.	H.	N.	O.
Ox	49·81	7·14	17·32	25·67
Calf	49·9	7·3	17·2	25·6

It is evident, therefore, that the chemical composition and the nutritive value of bones will vary with the class and age of bones, and particularly with the care which has been taken to extract all the food material that can be obtained from them.

When the shin and leg bones are sawn into small pieces and boiled in an open vessel for 7 hours, they lose 10 per cent. of their weight, and the loss extends to 19 per cent. after 9 hours' boiling. The cancellated bones, as the vertebræ, ribs and flat bones generally, lose 16 and 24 per cent. after having been boiled 7 and 9 hours. The loss of weight indicates soluble matter and, in a general sense, food.

Mr. F. Manning undertook for me a series of chemical enquiries into this subject which have sufficed to show how much greater is the true nutritive value of bones than is ordinarily allowed, and consequently the use which should be made of those structures by all who would not waste, and by the poor who cannot afford to waste food. If we first take a mixture of ordinary flat bones, as the spine, ribs, and shoulder blades, we find that after they have been properly digested in boiling water for about eighteen hours they yield 748 grains of carbon, and 20·1 grains of nitrogen for each pound of bone. The shin bones give a yet higher nutritive value, viz., 817·6 grains of carbon and 28·5 grains of nitrogen. If we compare this with the composition of one pound of beef, we find that the latter is equal to about three pounds of shin bones in carbon, and to six pounds in nitrogen, so that the nutritive value of bones may be reckoned at one-third that of beef in carbon, and one-sixth in nitrogen.

CHAPTER IV.

CHARACTERS AND COMPOSITION OF LEAN AND FAT MEAT.

EACH kind of meat has its own characteristic flavour, so that the tastes of different persons, or of the same person at different times, may be gratified by selection. This depends chiefly upon the juices contained in the fibres of the flesh, and on minute quantities of flavouring matters incorporated with the fat, as well as upon the oily or fatty matters mixed with the juices of the flesh. A fine quality of meat has abundant and full flavoured juices, with a considerable proportion of fatty matter, and appears red and pulpy, but inferior meat is paler and more fibrous in appearance, with but little of the proper flavour peculiar to the animal.

Each animal is also cut up into joints as shown in the following diagram of the side of an ox, and it is well known that different joints or parts of the same animal have different flavours, and not only such parts as are distinct in function as the liver and the flesh, but even those whose function is identical. Thus the flavour of a leg of mutton differs from that of a

No. 14.—Joints of Beef.

shoulder, although both joints are composed of flesh, or muscle, having the same duty to perform. Hence arises the preference for one joint over another, and the agreeableness of variety of joints; and according to the general preference will be the market value of the food.

The flesh of all animals is, moreover, divided into two principal parts, viz., fat and lean in their separate state, besides the oily or fatty matter which is so mixed up with the juices and tissues as not to be evident to the naked eye. This becomes a further ground for preference and selection, as each individual likes much or little fat, for one customer disliking fat, prefers the meat or the joint with little fat, whilst another, liking fat, rejects that kind of meat or that joint which is destitute of it.

The absolute and relative proportions of fat and lean, vary both with the animal and the condition in which it is killed, and it may be convenient to state here the results which have been arrived at by agricultural chemists so far as relates to the carcass or the part of a slaughtered animal which is sold as meat.

No. 15.

	Water per cent.	Lean or Nitrogenous per cent.	Fat per cent.	Salts per cent.
Oxen, store	60·8	18	16	5·2
,, half fat	54	17·8	22·6	5·6
,, fat	45·6	15	34·8	4·6
Calves, fat	62·3	16·6	16 6	4·5
Sheep, store	57·3	14·5	23·8	4·4
,, half fat	49·7	14·9	31·3	4·1
,, fat	39·7	11·5	45 4	3·5
,, very fat	33	9·1	55·1	2·8
Lambs, fat	48·6	10·9	36·9	3·6
Pigs, store	55·3	14	28·1	2·6
,, fat	38·6	10·5	49·5	1·4

Thus the proportionate quantity of fat in an ox may

be doubled according to the condition of the animal when slaughtered, and when quite ready for the market it is probably not less than one-third of the whole weight. This proportion in fat sheep is increased to nearly one-half, and is clearly much greater in sheep and pigs than in oxen, whilst it is the least in calves.

The true nutritive value of the fat and lean respectively is much the same in all animals used as food, so that the same weight of lean meat from one animal should (other things being equal), nourish the body as well as the same from another. When, however, it is used in compulsory dietaries, the influence of appetite or the relish for the food plays an important part in the phenomena of nutrition, and with less relish of even the same weight of food there will probably be less digestion and assimilation of it, and thereby less nutrition. Moreover, when the joint selected is hard to masticate, of coarse grain and poor flavour, the same result will follow for the same reason.

Hence ease of mastication and digestion, and approved flavour, have an influence over the selection of food, and affect its market value.

The question of economy is a complex one, being mixed up with that of selection, nutritive value, and commercial value, whilst the commercial value is based upon the law of demand and supply, which may in only a limited and uncertain degree be based upon nutritive value. Thus whilst the so-called best are probably somewhat more nutritious and agreeable than inferior joints, the limited number of such in an animal may be disproportionate to the number of persons demanding them and a fictitious price may be affixed to them. In like manner, whilst the inferior joints should be sold at a cheaper rate, the supply in different localities may be above or below the demand, and their price will

COMPOSITION OF LEAN AND FAT MEAT.

be above or below that which is fairly due to their nutritive value.

When comparing the nutritive values of different kinds of meat, it is essential to distinguish between fat and lean, and the nutritive elements of both in a given joint or whole animal will be proportionate to the combination of fat and lean. Fat is heat-generating alone, whilst flesh is both flesh-forming and heat-generating; and the difference between the two in this respect is the absence of nitrogen in fat and its presence in lean flesh.

As this question will frequently arise in the following pages, it may be desirable to anticipate a little and to state here the elements of fat and lean flesh respectively.

Fat, when entirely deprived of water, consists of three elements only, viz.: carbon, oxygen, and hydrogen in the following proportions, in 100 parts:—

No. 16.

C. 77.   O. 11.   H. 12.

When the fat is decomposed in the body these elements unite, so that the carbon takes a part of the oxygen and becomes carbonic acid, whilst the hydrogen takes another part of the oxygen and becomes water, any deficiency in the quantity of oxygen for this purpose being supplied by the inspired air.

Lean flesh entirely deprived of fat consists of four elements, viz.: nitrogen, carbon, oxygen, and hydrogen, and the proportions of each are almost identical with those of albumen, to which reference will be made in a future page.

Besides the combination of the three latter elements already mentioned in reference to fat, the nitrogen unites with the hydrogen in the formation of urea and other compounds which are thrown out of the system, and ultimately transformed into ammonia.

As heat is generated by every chemical combination, it is evident that both fat and flesh are heat-generators, but as nitrogen enters into the composition of flesh in the body, the lean flesh and not the fat can supply that element, and therefore the former and not the latter is the flesh-former. It must not, however, be supposed that this line of division is so strongly drawn that no fat is found in lean flesh, for a proportion of fat moving in the circulation must enter into and pass through the tissues of muscles as of other parts of the body.

CHAPTER V.

BEEF AND VEAL.

BEEF.

BEEF is popularly regarded in all parts of the world as the most nutritious kind of flesh, and although this opinion was formed without the aid of science, it is so far true that in the carcass of the ox there is a larger proportion of flesh or flesh-forming materials than in that of the sheep or hog. It is of closer texture than many other kinds of meat, so that if the measure be bulk, there is more nutritive material in a given quantity of beef. It is also the fullest of red blood juices; so that Byron, seeing Moore eating an underdone beef-steak, asked if he were not afraid of committing murder after such a meal.

This is clearly shown by the proportion of lean flesh which exists in the muscular fibre of various animals used as food. The following is the analysis of Maréchal, in 100 parts:—

No. 17.

	Ox	Fowl	Pig	Sheep	Calf
Muscular fibre, free from fat	25·0	24·9	24·3	23·4	22·7
Fat	2·5	1·4	6·0	3·0	2·9
Water	72·5	73·7	69·7	73·6	74·4

Moreover, the flavour of beef is fuller and richer than that of other meats, so that its use not only gives greater enjoyment, but a sense of satisfaction is obtained from a less volume of that kind of flesh. The first is, however, the best reason; and it is based upon two facts, viz., that the proportion of lean to fat in moderately fed beasts is less and of both to bone is greater than in either of the other animals referred to.

The composition of beef is as follows:—carbon 34·3, and with the free hydrogen reckoned as carbon 45·02, nitrogen 2·4, besides oxygen and hydrogen in proportions to form water.

When considering the average amount of lean and fat on the whole carcass, Lawes and Gilbert found the proportion in 100 parts of the carcass to be 34·4, and of the offal 21, in pigs, sheep and oxen together, but it varied in each according to the statement on page 43.

This is no doubt the most practical mode of determining the true value of the food, for when the carcass is cut up, the proportion of fat will be preserved in all the joints, whilst only the loose fat will be detached, and of the latter whatever is fit for food will be used as suet and fat in the preparation of other foods.

When we refer to the several joints the estimation is less satisfactory, for it is possible to select some which may consist of lean meat almost exclusively, and others where the fat will represent one-fourth to one-sixth of the whole weight. This can be properly determined only by ascertaining the weight of fat and lean in a given ration, or on the average of many rations, and is essential in all scientific calculations in the construction of dietaries.

It also fails when we consider the different degrees in which animals are fed, since a prize ox and an ill-fed ox differ, as already shown, in the proportion of fat and lean flesh, and a special determination of the fat and the lean would be necessary to give the required information; but there is a condition of an animal in which it is usually offered for sale to the butcher which is tolerably uniform, and from such a fair practical average estimation may be made.

The proportion of bone in the carcass of a moderately fed ox is from 10 to 15 per cent., and is the least in the round and thick flank, whilst it is the greatest in the head, shins, and legs (where it amounts to about 50 per cent.), and the aitch bone. Of the best pieces it is in the most moderate proportion in the loin, and next in the chine, or the thick ribs.

Fat is found the most abundantly on the inside of the loin, which being hard, is usually detached and sold separately as suet, also on the thin ribs and the brisket; and in over-fed animals it is laid up in considerable quantity on the outside of the whole carcass, except on the legs.

The lower-priced joints are those which are coarse in texture, as the neck and brisket, or which have much fat, as the thin ribs, or are bony, as the head, legs, shins, and aitch bone. The parts which are preferred by the meat-preserving companies are those thick in flesh, as the thick flank, round, loin, and thick ribs.

The legs and shins are richer in gelatin than any other joint of the body, whilst the largest proportion of oily fat, or of fat having the least degree of consistence, is found in the flesh of the face. Hence, both of these parts are especially fitted for the preparation of soup.

The loss of weight in cooking beef is less than of mutton, by reason of the greater solidity of the flesh and the smaller proportion of fat.

At the Leipsic Soup Kitchen there were $9\frac{1}{4}$ oz. of cooked meat left after boiling 1 lb. of fresh beef and bone, but the joint is not stated. The solid matter in the broth was about 27 per cent. of the weight of the raw meat and bone. In my own experiments the solid matter derived from 1 lb. of meat without bone, boiled in the usual way, was, per cent., 28·4 on lean beef, 57·6 on fat beef, and 34·3 on mutton.

The following may be accepted as the composition of one pound of roast and boiled beef respectively:—

No. 18.

	Carbon grains	Nitrogen grains
Roast beef, say	3,600	262
Boiled beef ,,	3,240	215

If, as is probable, the liquor in which the beef was boiled has been preserved, and will be eaten by those who consume the beef, there will be a further addition for each pound of meat of 490 grains of carbon and $18\frac{1}{2}$ grains of nitrogen, besides salts and other products, so that the total carbon and nitrogen obtained from one pound of beef will be practically the same whether the meat be roasted or boiled.

If we select a piece of beef which is devoid of separated fat, the following will be the loss per lb., and the composition of the boiled flesh and the broth respectively:—

No. 19.

1 lb. of fresh beef lost in boiling . . . 30 per cent.

	Carbon grains	Nitrogen grains
1 lb. of boiled lean beef	3,240	215
Broth from do.	360	47

In my experiments (*Phil. Trans.* 1859), the effect of 6 oz. of beef-steak reduced to $4\frac{1}{2}$ oz. when cooked, was to cause a maximum increase of carbonic acid evolved in respiration of ·7 grain per minute.

Ten grains of raw lean beef, when burnt in the body,

produce heat sufficient to raise 3·66 lbs. of water one degree Fah., which is equal to raising 2,829 lbs. one foot high.

Dr. Beaumont proved that it required 2¾ to 3 hours for the digestion of beef.

The amount of carbon and nitrogen in 1 lb. of raw beef, selected on the average of the carcass, is 2,401 grains and 168 grains, but if the free hydrogen be reckoned as carbon, the quantity of carbon will be increased to 3,221 grains.

Veal.

Calves are killed for the market at various ages in different countries, and even in different parts of the same country. Thus, on the Continent of Europe they are six to nine months old, whilst in the country parts of England it is usual to kill them whilst still fed with milk, and often when not exceeding one month old. Since 1855 it has been unlawful in Boston, United States of America, to kill calves under one month old.

The mode of killing has also an influence over the nutritive quality of the meat. In one, much of the blood is left in the flesh, whilst in that which consists of blanching the flesh by repeated bleedings, the blood with its valuable salts has been almost entirely extracted.

It is thus evident that the character of the flesh varies very greatly in delicacy, nutritive value, and digestibility.

Perhaps the most delicate food derived from the flesh of a mammal is that of a very young calf, well fed on new milk, and cooked by roasting; but its nutritive qualities are much below those of beef. Veal is popularly known to be difficult of digestion: a fact which I years ago showed to be due to the difficulty of masti-

BEEF AND VEAL. 51

cating it, not because the fibre is harder, but that it eludes the teeth; but there can be little doubt that it is more easy of mastication when well roasted or broiled than when boiled, and when very young and well fed.

It is not, however, a food which should be regarded otherwise than as a luxury, and the use of it should be much more limited than fashion now dictates.

For the above-mentioned reason, it is very difficult to define the chemical composition of veal, and the following is only an approximation thereto.

In 100 parts there are:—

No. 20.

Water 63. Nitrogenous 16·5. Fat 15·8. Salts 4·7.

Ten grains of raw lean veal, when burnt in the body, produce sufficient heat to raise 3·01 lbs. of water 1° F., which is equal to lifting 2,324 lbs. one foot high.

The time required for the digestion of veal varies, but it is not much less than that of pork, and may extend to 5 hours or upwards.

The bones of the calf at the early period of life contain little earthy matter, but yield a large proportion of gelatin and chondrin, whilst the flavour of the juices is very delicate, and almost entirely destitute of fat.

Hence the reason for selecting calves' feet for the preparation of jelly, and the value of this kind of food to invalids. If, however, with so great a scarcity of animals, it should become as unfashionable as it is undesirable to kill calves for food, this product could be readily obtained from other sources in a state of sufficient purity.

The sweetbread of the calf is the most expensive part of any ruminating animal ordinarily eaten by man—far more expensive than its nutritive qualities and even its flavour warrant; but in accordance with the fashion of the day, it is in great request for dinners, and commands an extravagant price. At the same time it should be added, that it is probably the most delicate in flavour of any meat with which we are acquainted, and is perhaps equally so whether boiled or fried.

CHAPTER VI.

MUTTON, LAMB, GOAT'S, AND CAMEL'S FLESH.

MUTTON.

MUTTON is popularly regarded as a lighter food than beef, and it has doubtless a more delicate flavour, less red-blood juices, a looser texture, and a larger proportion of fat. Although an agreeable and valuable food for all classes, it is not so well fitted as beef to sustain great exertion, but is rather a food for those of sedentary and quiet habits, including women and the sick. It is said that Kean suited the kind of meat which he ate to the part which he was about to play, and selected mutton for lovers, beef for murderers, and pork for tyrants.

The joint in a sheep of the best breed and in fair condition, which contains the least proportion of fat, is the leg, and next to that the shoulder, whilst the loin, neck, and breast have the largest proportion. It is most unusual to cut off any fat from the leg, and not usual to cut any from the shoulder before they are sold, but in all instances the hard fat from the inside of the loin is removed, and very frequently a part of that on the loin and neck is not eaten.

The least proportion of bone to meat is found in the leg, and having regard to that fact as well as to the absence of excess of fat, this must be regarded as the most useful joint in the sheep, and justifies the higher price demanded for it. The neck has the greatest proportion of bone, with a large proportion of fat, and is not so economical as the price of it might indicate.

The most solid meat is the leg, and the least the shoulder and breast. The latter has the further disadvantage of having more fat than lean, whilst the lean is not easily masticated, and has but little flavour, but as the price at which it is sold is less than that of any other joint of mutton, it is not without its advantage to the poorer classes.

The liquor in which mutton is boiled has less nutritive value, except in the fat, than the broth of beef, but having a delicate flavour is preferred by many persons. It is, however, too rich in fat, unless a large portion of that substance be first removed.

The loss on cooking mutton is generally believed to be greater than that on beef. At the Soup Kitchen in Leipsic only $9\frac{1}{3}$ ounces of cooked meat remained after boiling one pound of fresh meat and bone, whilst the broth contained nearly one-third of the weight of the fresh meat. In my own experiments the loss on the knuckle of mutton was 30 per cent. on the meat, and 14 per cent. on the bone; or $4\frac{3}{4}$ ounces and $2\frac{1}{4}$ ounces in the pound of raw meat and bone. The solid matter in the broth was 34·3 per cent. of the raw weight.

This varies much with the breed of the sheep. Thus, mountain sheep have the least fat, and the mutton loses the least in cooking, whilst the Leicestershire breed produces a very large proportion of fat, and causes the greatest loss in cooking.

The mixed breed of the Lincolnshire and Leicester sheep is much approved in the midland counties, as increasing the proportion of lean flesh; whilst the South Down is preferred in London and neighbourhood, as offering the least proportion of bone and fat, and the finest flavour.

The loss in cooking varies also with the food of the sheep, and is the least when they are fed on cake or dry food.

The following may be accepted as the average composition of one pound of boiled mutton with the usual proportion of fat:—

No. 21.

Carbon grains	Nitrogen grains
3,175	192

If a piece of the leg of mutton be taken without any separated fat or bone, the composition will not differ much from that of boiled lean beef, but will probably have a somewhat larger proportion of nitrogen.

The composition of an average sample of raw mutton is as follows per cent. :—

Carbon 41·45, and with free hydrogen reckoned as carbon 55·18, nitrogen 2·0, besides oxygen and hydrogen in proportions to form water.

The quantity of carbon and nitrogen in one pound of raw mutton is 2,900 and 127 grains, but if the free hydrogen be reckoned as carbon it will increase the quantity of carbon to 4,108 grains per pound.

The time required for the digestion of mutton is three to three and a quarter hours.

Braxy Mutton.

The frequent use of braxy mutton in the northern parts of Scotland renders it desirable that reference should be made to it, since it is the only instance in which the habits of our fellow countrymen tolerate the consumption of the flesh of an animal which has died a natural death.

The eating of such flesh was forbidden to the Jews. Thus in Leviticus, c. xxii. v. 8 : 'That which dieth of itself, he shall not eat.' In Leviticus, c. vii. v. 24 : 'And the fat of the beast that dieth of itself may be used in any other use, but ye shall in no wise eat of it.' But it is not stated that such food would be unwholesome, and it is added in Deuteronomy, c. xiv. v. 21, that one not being a Jew might eat it : 'Ye shall not eat of anything that dieth of itself; thou shalt give it unto the stranger that is within thy gates, that he may eat it, or thou mayest sell it unto an alien.' It was probably associated with the prohibition of the eating of blood, for it is added in Leviticus, c. xxii. v. 8, 'he shall not eat to defile himself therewith,' and in other passages the like prohibition is directed against eating an animal which has been torn by wild beasts.

There is no doubt a prevalent objection to eating the flesh of animals which have died naturally, and in consequence much flesh is lost as food, but such is not necessarily unwholesome, or other than good food. So far as it is well grounded it is based upon the probability of the animal having died of disease, which might have injuriously affected the flesh, and rendered it either less valuable as food, or likely to induce disease if eaten as food. Whilst there is a *primâ facie* objection to such food on this ground, it is clear that each case should be considered separately, and the flesh tested on its own appearance and merits. When flesh is unnaturally light or dark in colour, unusually soft and watery, and has evident marks of disease, it should be rejected; but mere emaciation of an animal from want of feeding does not make the flesh unwholesome.

The flesh of animals which have died naturally is very commonly eaten by the low caste tribes of India, as, for example, the *Bowrees, Swalghur,* and *Kucker* castes of Barrackpore, and the *Chhura* caste of Googaira, and by some African nations.

The flocks of sheep which constitute an important part of the property of the Scotch farmers are liable to 'braxy,' which is said by some to be a disease of the brain known as the staggers in horses, and by others an inflammation of the lungs. Whatever may be the cause of death, the duration of the disease is very short, so that there is a presumption that the general structures of the body have not been tainted by it, and that the disease has been limited to the vital organs immediately affected. The duration of the disease is believed to be about twenty-four hours; but during my enquiries into the dietary of the Highlanders,[1] I had reason to believe that neither the nature nor the duration of the disease is very carefully enquired into, and that a sheep found dead, without marked evidence of long-continued disease, is eaten.

This food enters into the contract between the farmer and his shepherd, so that, in addition to the advantages of house, oatmeal, cowfeed, and money, he becomes entitled to a given number of braxy sheep yearly. It is also a part of the ordinary food of the farmer in those localities, and also of his men, whether fed singly or on the bothy system.

I made the most careful enquiries, but could not learn that any disease or disorder of the human system had been known to follow the use of this food, and it is almost universally believed to be good in flavour and wholesome in quality.

From necessity the shepherd salts the meat, so that it may be kept fit for use, and the diseased parts are cut away before the salting process is effected.

---

[1] Sixth Report of the Medical Officer of the Privy Council.

Assuming that the disease was of short duration, and of the kind named, it may be inferred that the chemical and nutritive qualities will be the same as those of a healthy sheep at the same stage of feeding, but it does not follow that the animal when dying was in a state fit for the market. This, however, would tend to increase the proportion of nitrogen, and to lessen the quantity of carbonaceous elements as the lean flesh predominated.

Lamb.

Lamb varies in its nutritive, chemical and digestible qualities in proportion to its age, as well as in those peculiar to its breed and feeding.

The meat is deficient in strength, as compared with mutton, however it may excel it in delicacy of flavour, and it possesses a larger proportion of water, and a less proportion of nitrogenous matter. Hence it is really a luxury; and having regard to the deficient supply of meat, its use should be much more limited. The fry of this animal, including the sweetbread, is also somewhat of a luxury. The time required for its digestion is less than that of the flesh of a mature sheep, viz., two and a half hours.

Goat's Flesh.

Goat's flesh is very commonly eaten in Switzerland and other mountainous regions of the world. In its general characters it resembles mutton, but is harder and tougher and has a stronger flavour, so that it is not preferred to it. It is, however, much more nutritious than mutton, so far as nitrogenous elements are concerned, but is inferior in carbonaceous, for the animal does not fatten to so great a degree as the sheep. It must be regarded as a stronger food than mutton.

The flesh of the kid is more esteemed than that of the goat, both in Europe and India, as it was in Arabia 3,600 years ago, when Jacob made a savoury dish for his father Isaac, and it has a flavour not very unlike that of venison.

Camel's Flesh.

The flesh of the camel is not unfrequently eaten in the countries where the animal is produced, but it was forbidden to the Jews—'The camel, because he cheweth the cud, but divideth not the hoof; he is unclean to you,' Lev. c. xii. v. 4. It is usually lean and hard, so that, whilst it offers a large proportion of nitrogenous matter and is good food, it is not so agreeable as a well-fed ox. In estimating its nutritive qualities, we may take those of lean beef as already described.

CHAPTER VII.

PORK, SUCKING PIG, BACON, AND WILD PIG.

Pork.

Pork is a food which was forbidden to the Jews ('and the swine although he divide the hoof and be cloven-footed, yet he cheweth not the cud; he is unclean to you,' Lev. c. xi. v. 7), and was not eaten by many African and Asiatic nations, but the force of this prohibition and objection is sensibly diminishing even amongst the Jews, so that the use of this food is more general over nearly the whole world. Indeed, in certain parts of this country, as Hertfordshire, Cambridgeshire, and Bedfordshire it is eaten by no inconsiderable proportion of the inhabitants in preference to other kinds of meat: a fact due partly to taste, but chiefly perhaps to the universal habit among the peasantry of feeding pigs, which has descended from Saxon times. Moreover, there is a convenience in the use of it, which does not exist with regard to beef and mutton, for in such localities the pork is always pickled and kept ready for use without the trouble of going to the butcher, or when money could not be spared for the purchase of meat.

The preference for pork is not, however, restricted to the old, but is found in all new countries, and may be due chiefly to the ease with which pigs are bred and reared, and the meat preserved, whilst there is great difficulty in obtaining a sufficient number of persons, in a thinly populated country or a small village, to eat a sheep or an ox whilst the meat is fresh. This I found in my journey through the western and south-western States of America, and particularly in Texas, where hog and hominy was the only food of that class which could be ordinarily obtained.

Pork differs from beef and mutton, not in flavour only, but in the larger proportion of fat to lean flesh. This is due both to the nature of the animal in its tendency to store up fat, and to the habit of so feeding and treating it that this tendency may be fully developed. This is strikingly shown by contrasting the proportion of food which the pig stores up within its body as compared with the sheep or the ox during the process of fattening for the market, as appears in the following table:—

No. 22.

In the food	Per Cent. Stored up		
	Pig	Sheep	Oxen
Of 100 Nitrogenous	13·5	4·2	4·1
„ 100 Carbonaceous	18·5	9·4	7·2
„ 100 Mineral	7·3	3·1	1·9

The pig thus stores up in its body three times more of its food than the ox, and by so much is it more cheaply and quickly grown and fattened, but in its wild state or in the inferior breeds of semi-domesticated pigs (such as still form the stock of a backwood settler, and until recent years the property of the cottagers in this country), which have opportunity to run about and in the main find their own food, this tendency is greatly diminished, and the proportion of lean to fat is much greater.

The smallest proportion of fat is found in the leg and the largest in the belly and face, but so great is the amount of it which is laid up on the outside of the animal that it could not be eaten in its fresh state, but must be cut off and salted, so as to become bacon, or cut up with the lean meat and pickled before it is used.

It is, therefore, exceedingly difficult to ascertain the fair average composition of the pork which is actually eaten, and it can be only approximately arrived at by taking the whole carcass and estimating it as food to be consumed by a number of people. On this basis, the average composition of 1 lb. of fresh pork will be 4,200 grains of carbon and 79 grains of nitrogen, but if the free hydrogen be reckoned as carbon, the total quantity of carbon will be 5,809 grains. It is, however, clear that this must represent too much carbon and too little nitrogen; or, in other words, too much fat and too little lean for such pigs as are eaten altogether as pork.

If, however, we take the leg, and exclude all separate fat, the composition will be nearly that of lean beef already described, or of albumen which is more closely allied to it than the difference in appearance indicates.

Pork having so very large a proportion of fat, cannot be regarded as equal to beef or mutton in nourishing the system of those who make much muscular exertion. Moreover, there is a peculiarity about pork by which it is believed to be less digestible than other kinds of flesh, and it appears to me that this is due to the greater hardness of the muscular fibre, by which the mastication of it is rendered so difficult that much of it is swallowed in pieces too large for immediate solution in the juices of the stomach. This attends the eating of pork by all persons, but particularly by those who habitually masticate quickly, or who have defective powers of mastication, or who are careless in performing the act of mastication—classes embracing the old and the young, and no inconsiderable proportion of those of intermediate ages.

There is nothing known as to the chemical composition of this food, which accounts for this undesirable quality, and the universal belief in its existence.

The experience of the Leipsic Soup Kitchen on the solubility of pork in water affords a striking illustration of these facts, and shows how great is the difference between that flesh and beef and mutton, for whilst the solid matters in the broth of beef and mutton were 27 and 33 per cent., those from pork were not 19 per cent. The proportion of the cooked meat on 1 lb. of raw meat and bone was over 12 ozs., or nearly $\frac{1}{3}$ more than with beef or mutton.

It is stated by Sir A. Brady, that the loss on cooking American pork is 50 per cent., whilst on Dutch and Irish pork it is from 25 to 30 per cent.; the difference being due to the nature of the food of the animal.

Dr. Beaumont found, that a piece of roasted pork required five and a quarter hours for its digestion; but the time varies materially with the proportion of fat and lean, and the age, breeding, and condition of the pig. Young pickled pork will probably be digested in about three hours.

In addition to the convenience of pickling this food, as already mentioned, there can be no doubt that the use of the salted part of it, whether as ham or bacon, is convenient and agreeable in a high degree, and if it were unattainable its loss would probably be more widely and deeply felt than that of any other single food. It is a luxury to the rich man, whilst to the poor man and his children it is an agreeable necessary which cannot be universally supplanted by fresh meat until the pecuniary means at their command are much greater than at present.

There is, however, a greater danger in the use of pork than of any other kind of meat, since, so far as is known, it is more frequently diseased, and the nature of the disease is such as to be very injurious to man. Thus measly pork—a disease consisting of cystocerci as large as hemp-seed—is known to have produced fatal results to many of those who have incautiously eaten it, and although the characteristics of the disease may be recognised by those who understand it, they are neither known nor observed by the great majority of the poorer classes. Further, the terrible pest of the small worm, called the *Trichina spiralis*, is much more frequent in this than in other kinds of flesh in its uncooked state, and the power which the creature has to penetrate the tissues of the body of those who eat it has been vividly described by German and American writers.

Many instances of this terrible disease, isolated or in numbers, have now been recorded and particularly in Germany. Of 103 healthy people who ate diseased pork, which had been made into sausage meat, at

Helstadt, in Prussia, 20 died within a month. In Massachusetts a family was thus poisoned, with symptoms of pain and swelling in the eyes, stomach, and bowels, and also in the limbs, which became rigid and could not be moved without giving excruciating suffering. There was also vomiting, diarrhœa, and profuse perspiration. In all the fatal cases the worm was found to have penetrated the whole muscular system, and upwards of 50,000 were computed to exist on a square inch.

This diseased state may not be evident to the naked eye, so that as a precaution all pork should be well cooked. The instances of the disease occurred chiefly after eating uncooked sausages or uncooked ham —a habit not confined to Germany.

Yet formidable as this objection may appear, it is of very little weight, when we consider it on the great numbers of pork consumers, since the known instances of injury are infinitesimally few.

There is a very large importation of pickled pork into this country, which is usually divided into three classes, or qualities, known as mess, prime, and cargo. The 'mess' quality consists altogether of sides, the 'prime' of 3 shoulders (without feet) and other joints, and the 'cargo' of 30 lbs. of head and four shoulders and other joints in each barrel.

The importance of this article of commerce may be best estimated by the trade which is carried on with America, and the magnitude of the establishments devoted to the killing and curing of the flesh.

Sucking Pig.

Sucking pig is one of the choicest foods, and has a delicacy and richness of flavour which is perhaps unsurpassed. Charles Lamb wrote of it as follows:

Of all the delicacies in the whole *mundus edibilis*, I will maintain it to be the most delicate—*princeps obsoniorum*.' But, although he may be quite correct, this food is not one of the established delicacies of the dinner-table of the rich, and it is, perhaps, now more commonly eaten by the middle and labouring classes on festive occasions. The flesh of the very young pig has still some of the characteristics of the older animal, and although so easily separated when cooked, is not easily masticated, and is often swallowed in pieces too large for digestion. The fat is far more delicate than that of an adult pig. This is essentially a luxury, since its nutritive value is small in proportion to the price paid for it, and it should be as essentially the rich man's food. The time required for its digestion is the same as for lamb, viz., two and a half hours.

Bacon.

This term is applied to the sides of the pig which have been prepared by the removal of some of the lean flesh and ribs, and preserved by means of salt and saltpetre. It is usual to rub the saline mixture into every part of the pork and to repeat the process in a modified degree regularly for about three weeks, during which time the flesh is kept in a cool place, and afterwards it is removed and dried. Hence the first effect is to abstract a certain portion of the juices of the flesh, which with the moistened salt forms a pickle which may be applied by the hand from time to time; but during the whole of this part of the process the flesh is not only moist but wet, and has rather increased than decreased in weight. The second action is to dry the flitch of bacon, with or without wood smoke, by which it is reduced in weight.

The first process is solely for the purpose of preventing decomposition of the flesh, and it is requisite that it be begun very soon after the death of the animal, and be carried into every part of the flitch. If too long continued the lean flesh will shrink unduly and become hard, and if it be insufficient, the subsequent drying will allow the process of decomposition to begin. The second process is effected to render the flesh more portable and convenient to the consumer, and to give a different flavour to the food. As fresh meat is not attainable in a large part of the country more frequently than once a week, it is most convenient to the poor to have food in the house which may be used at any time, and which also may have been obtained by previous savings in anticipation of a period when poverty might be more urgent and ready money scarcely attainable. Moreover, dried bacon divides itself during the process of cooking into two parts, of which the labourer and his wife may have the solid and the children the liquid part, and thus both be in a degree pleased, if not satisfied. So far, it may be said, that bacon is the poor man's food, having a value to the masses which is appreciated in proportion to their poverty, and it is a duty to offer every facility for its production in the homes of the poor.

It is also the rich man's food, for the flavour, which is naturally or artificially acquired by drying, is highly prized, and although it may not be taken as a necessary by the rich, it is in universal request as a luxury. When the drying process has been continued too long the lean part of the bacon becomes hard and does not soften by cooking, whilst the fat becomes rancid, and is no longer a luxury to the rich or easily digestible by the poor.

Commerce has of late years sought the aid of art in effecting these processes with the greatest rapidity, so that the food may be ready for the market with the least delay. It is now usual to place the prepared flitch in a fluid pickle, and after a certain time to dry it by artificial heat, so that the whole process occupies three to six weeks as against three months under the old mode of procedure. There is, however, some risk lest the preserving process should not be complete, and particularly in the thicker parts, as the shoulder and ham, or where there are folds in the flesh, and also that the bacon should after drying retain an undue amount of moisture. The practical effect has been to render the flavour of the bacon milder and more delicate, and to cause it to retain a larger amount of moisture, which so far lessens the proportion of nutritious matter in a given weight. Moreover, as the artificial mode of drying cannot be carried on in every cottage, the pork is removed to kilns properly constructed, and a monopoly has thus been created, by which the price has of late years been unduly increased.

A very large quantity of bacon is imported into this country from America and Ireland, and a small portion from Strasburg and Hamburg, the latter having acquired a peculiar flavour by being smoked. That from America used to be of inferior quality, owing to the defective system of feeding, and particularly to the habit of allowing the pigs to feed on acorns, by which the fat was less solid and more oily, and shrank on being boiled.

The finest bacon in our market is the Wiltshire and Cumberland, but some of the Irish is nearly equal to it.

The nutritive value of bacon is within certain limits a varying quantity as the two processes are more or less prolonged, but that of wet bacon is the more stable of the two. Newly dried bacon will have an excess and old bacon a defect of water, as compared with the usual standard. Bacon from which much lean meat has been removed, or which was obtained from a well fattened pig, will leave a less proportion of nitrogenous matter.

The following is the approximate composition of green (or wet) bacon and dried bacon in 100 parts:—

No. 23.

Green bacon . .	Water 24	Nitrogenous 7·1	Fat 66·8	Salts 2·1
Dried bacon . .	,, 15	,, 8·8	,, 73·3	., 2·9

There are the following quantities of carbon and nitrogen in a pound of an average specimen of dried bacon:—

No. 24.

Carbon 4,340 grains   Nitrogen 97 grains

But if the free hydrogen be reckoned as carbon the total quantity of the latter will be 6,006 grains in the pound.

The digestibility of bacon varies with its dryness and cooking, but it may be assumed that the time required will be less than that of fresh pork. From its large proportion of fat, and from the ease with which it may be masticated, it may be inferred that the process will be usually accomplished in less than three hours, and at any rate that it is not less digestible than fresh beef or mutton.

THE WILD PIG.

The pig in its wild state is still found in France and other parts of continental Europe, in India, Ceylon, Zanzibar, and nearly all the great eastern and northern countries of the world.

Its flesh, as compared with that of the domestic pig, is harder, tougher, and stronger flavoured, and would not be preferred to it. The proportion of gelatinous matter is greater, as that of fatty matter is less than in the flesh of the domestic pig, so that scouse made from the gelatinous parts of the animal as the ears and face, is stiffer and harder than that from the domestic pig.

The cause of these differences is that of the habits, breed, and food of the two classes.

The boar's head has been for ages in repute in the halls of the great for its flavour when roasted and stuffed with nutritious and appetising compounds. It was anciently the first dish on Christmas-day, and was carried to the principal table with great solemnity. In the 'Collection of Christmas Carols,' published in 1521 by Wynkin de Worde, is 'A Carol bryngyn in the Bore's Head.' This ceremony is still performed at Queen's College, Oxford, and the following ditty sung:—

> The boar's head in hand bear I,
> Bedecked with bays and rosemary;
> And I pray you my masters be merry,
> *Quot estis in convivio.*
>
> The boar's head, as I understand,
> Is the bravest dish in all the land;
> When thus bedecked with a gay garland,
> Let us *servire cantico.*
>
> Our steward hath provided this
> In honour of the King of Bliss,
> Which on this day to be served is
> *In Reginensi Atrio.*

Refrain after each verse :—

> *Caput apri defero*
> *Reddens laudes Domino.*

One of the most sumptuous dishes which have been handed down to us, is that of roasted wild pig, which was in use in the Norman and Early English periods of our history. The following recipe is of the fourteenth century :—

'Cok a Grees (Cock and Wild Pig).

'Take and make the self fars; but do thereto pyn and suḡ. Take an hóle raosted cok, pulle hȳ (in pieces) and hylde (cast) hym al togg'd, saue (save) the legg̃. Take a pigg and hilde (skin) hȳ fro the mydd' doñward, fylle hī ful of the fars and sowe hȳ fast toged'. Do hȳ in a panne and seethe hȳ wel, and when thei bene isode, do hē on a spyt and rost it wele. Colô it wᵗ zolkes of ayren (eggs) and safron. Lay theron foyles (leaves) of gold and of silu', and sūe hit forth.'

The so-called boar's head of the present day is usually that of an animal obtained with less prowess and delay than that of the wild boar of the middle ages.

CHAPTER VIII.

VARIOUS WILD ANIMALS AND THE HORSE AND ASS.

As a general expression, it may be stated that the points of contrast between the flesh of wild animals and that of domesticated and artificially fed animals are the greater hardness and solidity of the flesh, the greater proportion of solid fibre to juices, the less proportion of water and fat in the juices, and the greater proportion of lean to fat. Hence it follows that under the same circumstances the mastication of the flesh of wild animals is less easy, the flavour more concentrated and less luscious, and the proportion of nitrogenous or flesh-forming compounds greater. They are therefore strong foods, requiring good powers of mastication and digestion, and if well digested, are highly nutritious. Their intensity of flavour is also a recommendation to those who are satiated with ordinary food and to the convalescent who seek something to stimulate defective appetite or to satisfy a sudden desire.

But in all these particulars there is a difference with

the habits of the animal. Deer kept in a park where they take but little exercise and are fed in the winter by the aid of man, approach in character the domesticated animal, whilst the sheep which lives on mountains and travels far and wide to procure food will approach the wild animals in the character of its flesh if it be allowed to live until it is five years old or upwards. Thus, much of the flesh of park deer when eaten soon after the death of the animal differs little from Highland mutton, whilst that of the true mountain sheep has a flavour resembling that of park deer, and in dryness after cooking and in the absence of fat, differs extremely from the meat of a well-fed South Down sheep.

In treating venison it has been found necessary to allow it to hang for a considerable period in order to separate the bundles of fleshy fibres; and when cooked, to add gravies, condiments, and sweets, so as to give a quality to the flavour, which it naturally lacks, and it may be added that no part of the animal is cooked by being boiled.

Two recipes of the year 1381, one of which is to prevent venison from becoming tainted, and the other to remove the taint, cannot be without interest:—

'FOR TO KEPE VENISON FRO RESTYNG.

'Tak Venisoñ wan yt ys newe and cuver it hastely wyth Fern that no wynd may come thereto, and wan thou has ycuver yt led yt hom and do yt in a soler that sonne ne wynd may come ther'to and dimembr' it and do yt in a clene water and lef yt ther' half a day and after do yt up on herdeles for to drie, and wan yt ys drye tak salt and do after thy venisoñ axit and do yt boyle in water that yt be so salt als water of the see and moche more and after lat the water be cold that it be thynne and thanne do thy Venisoñ in the water and lat yt be therein thre daies and thre nyzt, and after tak yt owt of the water and salt it wyth drie salt ryzt wel in a barel and wan thy barel ys ful cuver it hastely that sunne ne wynd come thereto.

'FOR TO DO AWAY RESTYNG OF VENISON.

'Tak the venison that ys rest and do yt in cold water, and after mak an hole in the herthe and lat yt be thereyn thre dayes and thre nyzt, and after tak yt up and spct it wel wyth gret salt of peite (saltpetre) there were the restyng is, and after lat yt hange in reyn water al nyzt or mor'.'

The composition of venison differs from that of beef and mutton in that there is a large proportion of nitrogenous and a less proportion of carbonaceous matter and less water.

Dr. Beaumont found that broiled venison steak was digested in one and a half hour, but I cannot think that that represents the true period of chymification. It is probably nearly the same as mutton.

BISON AND OTHER WILD ANIMALS.

The Bison may be regarded as the representative of the ox in its wild state, and might equally well be used for food. The flesh is, however, tough, hard, and dark in colour, and strong in flavour, so that it would not be preferred to that of the domesticated ox, but its fulness of flavour when in good condition makes certain parts of the animal very acceptable if not luxurious to the hunter. The parts usually selected are the hump and tongue, and it is probable that if they could be imported into this country at the price of ordinary meat, they would be reckoned amongst our luxuries.

The relative proportions of fat and lean as regards the bison and the domestic ox vary with the condition of the animals, and therefore, as respects the wild animal, with the season of the year; but speaking generally, the wild animal yields a larger proportion of nitrogenous matters, and particularly of gelatin, and is therefore the stronger food.

The Eland was introduced into this country by a former Earl of Derby, with a view to test its capability to compete with the ox as a meat-producing animal, but the experiment failed on the ground of economy.

The Antelope is used as food, but the flesh of some species, as the *Antelope Saltiana* has a smoky flavour like that of a rat.

The Kangaroo, Zebra, Springbok, and many other large herbaceous and graminivorous animals are eaten as food in Africa and Australia when obtained by hunting. The preserved flesh of the Kangaroo (*Macropus*) is now imported from Australia, as is also the very excellent soup prepared from the tail. It has the character of the flesh of wild animals, but probably a larger proportion of fat.

The Dog, Cat, Rat, Fox, Wolf, Leopard, Jackal and other carnivorous animals are eaten by the low caste inhabitants of India and other countries.

The Horse and Ass.

The flesh of the horse and the ass has been long known to be good and nutritious food, and in quality it occupies a place rather among wild than domesticated animals. The objection to its use does not rest upon either chemical or physical grounds, for the carcass has a larger proportion of nitrogenous material than the well-fed ox, and the mastication of it is not more difficult than that of fresh venison or an ill-fed ox. There is not the richness and fulness of flavour which belongs to good beef, and in its preparation for the table it demands the same additions as those required by venison.

Several reasons have no doubt prevented the general use of horseflesh. 1st. The sentiment with which the horse is regarded as a useful, intelligent, high-spirited and so-called noble animal, by which it is esteemed a degradation to feed it and fatten it for the table. 2nd. The necessity which exists for its use as a living creature—a necessity which increases and cannot be supplied by any other animal. 3rd. Its high price as compared with that of oxen and cows which are reared for feeding only. 4th. The toughness of the flesh, and the use as food for the lower animals of such horses as are worn out by work and then killed. It would require a far greater necessity for the supply of meat than any which has occurred in our country for ages past, to overcome the first, second, and fourth objections, but the value of the third has greatly diminished with the present high price of meat.

Having regard to the greater cost and risk of production of the horse than the ox, it is not probable that there will ever be so great an excess in the supply of horses as to make the rearing of them for the butcher a commercial success at even the present price of meat, and it seems impossible that it could ever successfully compete with the production of oxen.

Hence the introduction of this flesh as a food must be that of the worn-out and disused horses, and a rigid system of supervision and examination of the animal before death and of the flesh after death would be necessary to prevent the occurrence of disease from its use. Such food could not, moreover, attract the rich man, and if used at all, it would be by the poorest classes—by those who in this country strongly object to eat anything which is regarded as of inferior quality or which is commonly rejected by their richer fellow citizens. It is, therefore, really useless to bring the subject before the public attention of this country, and horse flesh must be reserved for such terrible circumstances as those which recently led to its use on a large scale in Paris.

The composition of this kind of flesh may be reckoned as that of lean beef.

The following is the *menu* of the horse-flesh dinner which was eaten at the Langham Hotel, on February 6, 1868:—

Potages.

Consommé de cheval.   Purée de destrier.
Amontillado.

Poissons.

Saumon à la sauce arabe.   Filets de Soles à l'huile hippophagique.
Vin du Rhin.

Hors-d'œuvres.

Terrines de Foie maigre chevalines.   Saucissons de cheval aux pistaches syriaques.
Xérès.

Relevés.

Filet de Pégase rôti aux pommes de terre à la crème.   Dinde aux châtaignes.   Aloyau de cheval farci à la centaure et aux choux de Bruxelles.   Culotte de cheval braisée aux chevaux-de-frise.
Champagne sec.

Entrées.

Petits pâtés à la Moëlle Bucéphale.   Kromeskys à la Gladiateur.   Poulets garnis à l'hippogriffe.   Langues de cheval à la Troyenne.
Château Pérayne.

Rôtis.

Canards sauvages.   Pluviers.
Volney.
Mayonnaises de Homard à l'huile de Rossinante.   Petits pois à la française.   Choux-fleurs au parmesan.

Entremets.

Gelée de pieds de cheval au marasquin.   Zéphirs sautés à l'huile chevaleresque.   Gâteau vétérinaire à la **Ducroix**.   **Feuillantines aux pommes des Hespérides.**   Saint-Péray.

Glaces.
Crème aux truffes. Sorbets contre-préjugés.
Liqueurs.
Dessert.
Vins fins de Bordeaux. Madère. Café.
Buffet.
Collared horse-head. Baron of horse. Boiled withers.

CHAPTER IX.

OFFAL.

THE offal of animals consists of the skin, feet, tail, horns, head, and tongue; of the lungs, liver, spleen, omentum, pancreas, and heart, which constitute the pluck or the fry, and the intestines and other internal organs.

It was in daily use among our Saxon and early English ancestors, for pigs' feet and head, calves' feet, sheep's trotters, the maw or paunch, and intestines of the wild and domestic pig and other animals and the blood, enter into the recipes of the fourteenth century.

The proportion per cent. of the weight of the offal to that of the carcass varies with each kind of animal, as shown in the following table:—

No. 25.

	Carcass	Offal
Store oxen	59·3	38·9
Fat oxen	59·8	38·5
Fat heifers	55·6	41·3
Fat calves	63·1	33·5
Store sheep	53·4	45·6
Half-fat sheep	59	40·5
Very fat sheep	64·1	35·8
Store pigs	79·3	18·8
Fat pigs	83·4	16·1

Thus it appears that generally one-third of the weight of animals is offal, and is not sold as meat, and in certain animals the offal approaches to the same weight as that of the carcass. The greatest proportion is found in sheep, and the least in pigs, and the extreme may be one-half and one-sixth of the whole weight of the animal.

It must not, however, be inferred that the offal is valueless as food, or that it should be otherwise than carefully consumed, since it has much nutritive value, although not so much as would induce the purchaser to give the price of meat for it. Its flavour is also peculiar and not equal to that of meat, so that for several reasons it is rather the poor than the rich man's food.

If we take the whole offal of various classes of animals which are used as food, we find that the proportion of nitrogenous compounds in it is greater than that of the carcass, viz. 17·2 per cent. against 13·5 per cent., whilst that of fat is less, viz., 21 per cent. against 34·4 per cent. The percentage of salts is about equal in each of the two divisions of the animal, that is to say, 3·7 per cent. in the carcass and 3 per cent. in the offal.

It must not, however, be assumed that the whole of this is available food for man, and it must be added that the nitrogenous part is in a less nutritious form than in flesh, since it consists largely of gelatin and chondrin.

The skin, so far as it is useful as food, is consumed in the form of gelatin, and is probably the greatest source of that article.

The tongue of all animals used as food is in request, and is regarded as a delicacy. That of the ox is frequently salted or pickled until it assumes a red colour on being boiled. It is sometimes dried, and must afterwards be soaked well in water before it is boiled; or it is taken direct out of the pickle and may be at once washed and boiled. The muscular part of the tongue has a rich and somewhat luscious flavour, whilst the fat which lies under the tongue is very agreeable. Both fat and lean may be eaten when either hot or cold with equal gratification.

A large trade is established between Russia, South America, and England in the import of dried and smoked tongues, which are commonly no doubt those of the ox or cow, and not unfrequently of the horse. From their preparation more than from their other qualities they are agreeable foods when well soaked and boiled and eaten cold. The tongues of reindeer and probably of other animals are also imported and eaten as a relish.

The head is a somewhat favourite dish. That of the sheep is boiled or grilled, and particularly in Scotland and amongst the poorest classes. Calves' head is more especially the food of the richer classes, and is a very delicate and agreeable dish when boiled and properly served. In both cases it may be sold without the skin, or the skin may be left on and the hair scraped off.

Ox head is used both for the preparation of soup and as ordinary meat. It consists of about 30 per cent. of meat which contains much oil, and some solid fat, and produces a rich and nutritious soup. It is a convenient dish for the poor man's wife, since it enables her to make good and cheap soup for the children, whilst the adults eat the solid meat.

The proportion of meat to bone is much greater in the pig's head than in the heads of other animals, since the pig lays up much fat about the jaws. Instead of being 30 per cent. as in the ox, it is 60 to 70 per cent.

There is less oil but more solid fat than in the head of the ox, so that it is not so well fitted for the preparation of soup, yet it is a very good addition to beef or ox heads for that purpose, and supplies a delicate and agreeable flavour.

The upper part being composed chiefly of bone is but little used as food, but the sides and front of the face are eaten either when fresh or pickled. More commonly they are pickled, dried and sold as cháps or chowls, which are boiled and eaten cold, but not unfrequently they are eaten fresh after having been roasted or taken out of pickle and boiled.

When served as a Christmas dish they are stuffed with a great variety of valuable condiments, so that a slice consists more of stuffing than flesh, and thus prepared is a costly and highly prized dish. The bones are removed, the skin is left on and coloured and decorated, and the whole is boiled.

The liver of certain animals is a favourite dish, as that of the pig by the poor, and of lamb, calf, and Strasburg goose by the rich, and although it is not equal to flesh as a food, it furnishes a considerable proportion of nutritive elements.

The liver of the ox and sheep is a less agreeable food, and indeed is rarely eaten in England, except by the very poor; but in Scotland the latter is more frequently consumed in the preparation known as *haggis*. It is often eaten raw by the Arabs of Mesopotamia.

The liver of all animals is apt to be infested by a parasite of the hydatid class, but as it is evident to the naked eye, it may usually be avoided. It is desirable that the food should be cut into slices and examined, and also that it should be well cooked. The usual mode of cooking it is frying, by which the flavour is greatly improved.

The composition varies somewhat with the nature and degree of fattening of the animal, so that its

OFFAL.

albuminous and fatty elements will not always bear the same relation to each other.

The following is the composition, in 100 parts, of the most solid kind, viz., ox liver, in which the nitrogenous elements are usually in excess, as compared with the liver of other animals:—

No. 26.

Water 74   Nitrogenous 18·9   Fat 4·1   Salts 5

An average specimen of the liver of the ox yields $17\frac{1}{2}$ per cent. of carbon and 3 per cent. of nitrogen, or 1,225 grains of carbon and 210 grains of nitrogen in the lb.; but if the free hydrogen be reckoned as carbon, the total quantity of carbon will be 1,338 grains per lb.

The lungs, or as they are vulgarly termed lights, are eaten as a part of the pluck or fry, and as they are composed almost exclusively of membranes and vessels they contain a high proportion of albumen and other nitrogenous matter. They are not, however, very easily masticated or digested, and could scarcely be eaten alone. It is desirable that they should be well cleansed, and any diseased portion and the glands removed.

The omentum consists partly of membrane and vessels, and partly of fat, and is an agreeable addition to the otherwise lean fry. That from an old animal is not so tender or so readily masticated as from a young one, and it is always desirable to masticate it well. A part of it is eaten as tripe.

The pancreas, and the thyroid and sublingual glands, pass by the name of sweetbread, and command a very high price. That of the calf is the most esteemed, but that of the lamb is not unfrequently substituted for it. It contains a considerable proportion of water and some fat, and has a delicious flavour when properly prepared.

The intestines are used as food by man in the preparation of sausages and black puddings, whilst the thicker and fatter parts are eaten as tripe.

Tripe is prepared from the stomach and intestines, with the fatty structures attached thereto, of the ox and cow, and consists of two parts, viz., the walls of those organs and the enclosed fat. It is prepared simply by thoroughly cleansing the organs from every adherent substance, and from the flavours of bile or other disagreeable matters, and then gently boiling them in clean water for some hours. When thus prepared, it is a food of somewhat delicate and agreeable flavour, and of very easy mastication and digestion.

Its chemical composition necessarily varies with the proportion of fat; but, taking an average of the different cuts, it may be represented as follows, in 100 parts:—

No. 27.

Water 68   Nitrogenous 13·2   Fat 16·4   Salts 2·4

Hence, it is a food affording considerable nutriment, but not very satisfying, for it is fully digested in about one hour, and the stomach soon calls for another supply of food. Moreover, the nitrogenous compounds are rather those of gelatin than of albumen, and are perhaps, therefore, somewhat less valuable as nutriments than the quantity of nitrogen might indicate.

The ease and rapidity with which tripe is digested seem to render it a very proper food for the sick, but in practice it is found that the absence of pronounced flavour, and perhaps the unusual nature of the food, prevent its selection by the sick generally. It is rather the food of the poor, and is not in general use by any class of the community. There are 1,705 grains of carbon and 143 of nitrogen in each lb.

The feet of animals consist of two chief chemical constituents of food, viz., oil and gelatin, and hence we have neat's foot oil and calf's foot jelly.

The oil has too strong a flavour to be used as food, and must be removed before the foot is eaten. This is effected by the application of heat after the free use of the knife, and as the foot is cooked by being boiled in water it is necessary that the oil should be skimmed from the broth that the latter may be fit for food.

When properly prepared the broth contains jelly and oil, both of which may be separated and used, whilst the insoluble part of the foot contains a portion of skin and cartilage, which is nutritious and agreeable. The flavour, however, of both cow-heel and tripe is but slight, and does not stimulate the sense of taste, so that vinegar and other condiments are commonly eaten with them.

Sheep's trotters, called *shepis talon*, were made into a dish in the fourteenth century, with eggs, pepper, salt, saffron and raisins.

Pigs' feet were found by Dr. Beaumont to be digested in about one hour, and it is probable that cow-heel would require the same time, except such tendinous parts as are masticated with difficulty and may be only partially digested after the lapse of several hours.

Collared pork is made from the gelatinous parts of the pig, as the ears, feet and face, and was in use in the fourteenth century according to the following recipe:—

'Gele of Flessh.

'Take swyn' feet and snowt' and the cerys, capons' conyin'g, calu' fete, and waische he clene, and do he to seethe in the thriddel of wyne and vyneg' and wat and make forth as biforo' (see recipe on fish jelly, page 113).

CHAPTER X.

SAUSAGES, BLACK PUDDING, AND BLOOD.

THERE are two kinds of sausages, viz., those made from fresh meat and those from preserved meat, and both are placed in pieces of intestine.

The former are frequently prepared at home, and consist chiefly of meat which has been chopped finely, bread, and condiments; and as they are moist and fresh, they cannot be kept long without decomposition. Their value depends upon the kind, quality, and quantity of the meat, and when made properly are agreeable and valuable foods, but they may be made of inferior and even diseased meat, and should there be any disagreeable flavour it is disguised by an excessive use of pepper. Hence, their composition is so doubtful as to be a byword; yet because it is possible to have them pure, and their flavour is agreeable, as well as for their convenience at table, they are in very general use.

The latter kind are prepared on the Continent, where they are intended to be kept for use. Their nutritive value is much greater than that of fresh sausages, since the meat is very dry, and they are composed of meat only, and, as a general expression, it may be stated that they are equal to three times their weight of fresh meat. Their flavour is agreeable and varied by cloves of garlic. They are particularly adapted to the use of travellers, soldiers in camp, and labourers, who cannot cook their meat, but the cost in this country precludes their use by the latter. Great care should be taken lest diseased pork may be thus eaten and cause the trichinous disease.

In the island of Majorca each family makes a supply large enough for a year. They are made of pork exclusively, stuffed into large skins, and flavoured with spices, and are not considered ready for use until they have been kept six months. With bread, they constitute the ordinary food of all classes.

The Prussians, during their recent war with France, prepared sausages, which were in a high degree nutritious and not disagreeable. They consisted of a mixture of bacon, pea-flour, onions, salt, and condiments. The pea-flour was prepared after a patented method, by which it did not become sour, and the food remained good. The daily ration per man was 1 lb., and it was only necessary to boil it in water for a short time before eating it.

Black puddings and blood are much more frequently eaten in large towns than in villages, since the frequency with which animals are slaughtered in the former renders the daily preparation of the latter possible. They are prepared with blood, and chiefly that of pigs, to which groats and various herbs with lumps of fat are added. The whole is enclosed in a piece of the intestine of the pig and boiled, but it is usual before eating it to cook it further by frying it, with or without previously warming it by immersion in hot water. When quite fresh they are savoury and agreeable, but as blood rapidly decomposes, they may become tainted before being cooked, and when yet apparently fresh, and if kept long after having been cooked, they lose their agreeable flavour and contain ammonia.

There is no doubt an objection to, perhaps a prejudice against, eating blood, based in some degree upon the prohibition to the Jews as contained in the Old Testament: 'Ye shall eat the blood of no manner of flesh: for the life of all flesh is the blood thereof: whosoever eateth it shall be cut off.'—Lev. xvii. 14; and also on the common belief that the blood may be diseased without offering evidences whereby the disease might be recognised. As to the former, it may be scarcely necessary to add, that we eat a portion of blood in flesh, and that even when the animal is killed by cutting its throat after the Jewish fashion, it is not possible to extract all the blood from the body, and that even the Jews must eat some of it. Moreover, blood contains nutritive elements of great value, and is inferior only to the flesh which is made from it.

The following is the composition of fresh blood in 1,000 parts:—

No. 28.

Water	779·00
Fibrin	2·20
Fatty matter	1·60
Serolin	0·02
Phosphorised fat	0·49
Cholesterin	0·09
Saponified fat	1·00
Albumen	69·40
Blood corpuscles	141·10
Extractive matters and salts	6·80
Chloride of sodium	3·10
Other soluble salts	2·50
Earthy phosphates	0·33
Iron	0·57

Besides salts of sodium, potassium, calcium, magnesium, and iron, there is sugar; and in certain cases, salts of copper and lead have been discovered.

The salts in blood perform an important *rôle* in nutrition, and it may be well to indicate their nature in the pig, sheep and ox, whose blood is used as food. The following are the quantities per cent. of all the salts:—

No. 29.

	Pig	Sheep	Ox
Phosphoric acid	36·5	14·8	14·04
Alkalies	49·8	55·79	60
Alkaline earths	3·8	4·87	3·64
Mineral acids and oxide of iron	9·9	24·54	22·32

If there be any ground for fear lest diseased germs should exist in the blood, it may be set aside by the consideration that a temperature at and above 212°, if fully applied, will suffice to destroy all known elements of disease, and that blood when fresh and so cooked may be eaten with impunity. Having regard to these facts, and to the high price which is now paid for meat, I think it would be folly to object to the use of blood as a food under proper restrictions, one of which should be that the animal from which it was taken should not be in a state of disease.

The common snail has been occasionally eaten in England and the vineyard snail on the Continent, whilst slugs are commonly consumed in China. So far as respects the use of the snail as food in England, it was limited to the sick, and is probably a harmless and not an ill-flavoured substance, but the quantity of nutriment which it can afford is very small. The snail, in common with other creeping things, was a forbidden food to the Jews: 'These also shall be unclean unto you among the creeping things that creep upon the earth; the weasel, and the mouse, and the tortoise after his kind, and the ferret, and the chameleon, and the lizard, and the snail, and the mole.'—Lev. xi. 29, 30.

It is not desirable to occupy the pages of this book with reference to foods which are of a somewhat disgusting nature, or such as are rarely used by civilised people, but it may be proper to name a very few of them.

The iguana, an enormous lizard, is commonly eaten in Zanzibar; and lizards of smaller size, as the sanda, are frequently consumed by the low caste people of India. Lizards were prohibited to the Jews as food (Lev. xi. 30).

The edible frog has long been used as food, and when properly fed the hind legs are accounted by the French as a luxury.

Snakes of many kinds are eaten in India and in other hot countries where the inhabitants are extremely poor.

A small monkey (*Cercopithecus griseoviridis*) and a frugivorous bat are eaten as delicacies in Zanzibar.

The bat (but of what kind is not stated) was prohibited to the Jews as food (Lev. xi. 19).

The large white ants are eaten by many African nations, in the absence of more agreeable foods.

Locusts are commonly eaten by the Bedouins of Mesopotamia and other Eastern peoples, who string them together and eat them on their journeys with unleavened cake and butter. It is worthy of note, that this food was not prohibited to the Jews. 'Even these of them ye may eat; the locust after his kind, and the bald locust after his kind, and the beetle after his kind, and the grasshopper after his kind.'—Lev. xi. 22. John the Baptist ate locusts and wild honey.

CHAPTER XI.

EXTRACTS OF MEAT AND FLUID MEAT.

THESE are prepared in two forms, viz., in a thick semi-fluid state and as solids.

The very large proportions which the trade has attained since Baron von Liebig lent his name to the process by which the former kind is prepared, demands that the precise value of these foods should be well known, and with the observations which have been made by numerous scientific men, as well as the aid afforded by popular experience, the task is far less difficult than when I first called attention to the subject.

The extracts are prepared by boiling down the flesh of animals, so that thirty-two pounds of flesh are said to be required to prepare one pound of Liebig's extract.

It is desirable to select lean cattle for this purpose, and if we assume that the net flesh weight of an average beast in Australia, the Brazils, and other meat-producing countries in America, is 300 lbs., one animal will yield only about ten pounds of the extract.

During the process, all the fat and as much of the gelatin and albumen as can be extracted, are removed from the solution of flesh, whilst the fibrin, being insoluble, is necessarily left behind. Hence there remain water, salts, osmazome and the extractives of flesh, or in general terms, the flavouring matters and the salts of meat—thus leaving out all that is popularly regarded as nutritious.

This substance varies in value according to the amount of water (usually about 20 per cent.) which is allowed to remain in it. The intense flavour of meat which it possesses is like that of roasted flesh, and is always the same.

The following are tests of its value:—

1. Sixty per cent. should be soluble in strong alcohol.
2. When dried at a low heat it should not lose more than sixteen per cent. of its weight.
3. There should be no fatty matter or albumen.
4. There should be nine to ten per cent. of nitrogen.
5. The ash should be about twenty per cent., and contain from one quarter to half its weight of potash, besides about half that proportion of phosphoric acid.

What is necessary to render this extract as valuable as the meat itself for the purposes of nutrition is to restore the substances which were rejected, and those have been shown to be almost equivalent to the whole meat. There is but little left in the extract to nourish the body, and the elements which it really possesses are salts which may be obtained otherwise at an infinitely smaller cost, and the flavour of meat which disguises the real poverty of the substance.

If it be asked why so much of the flesh is thus unused, we answer that only the soluble parts of the meat could be obtained in this form, whilst the insoluble but most nutritious parts are left behind, and only such of the soluble parts are retained as do not put on the putrefactive process, and hence nearly all nitrogenous matters are excluded. If it be further asked whether the popular belief in the value of this food is altogether based upon a fallacy, we answer, No; for it is a valuable addition to other foods, since it yields an agreeable flavour, which leads to the inference, however incorrect, that meat is present, and when prepared with hot water and properly flavoured, gives a degree of exhilaration which may be useful to the feeble, and is as useful to the healthy as tea and coffee.

If, however, it be relied upon as a principal article of food for the sick it will prove a broken staff, except to those extremely feeble persons who can take very little food, and are favourably influenced by slight causes.

Its proper position in dietetics is somewhat more than that of a meat-flavourer, but all that is required for nutrition should be added to it. Thus, in the preparation of ordinary soup and beef tea, the extract may be added to the stock to increase the flavour, or it may be mixed with white of egg, gelatin, bread and other cooked farinaceous substance. Liebig, in a letter to the 'Times,' stated that 'it is not nutriment in the ordinary sense;' and Professor Almen has shown the small nutritive value of this substance in the 'Transactions of the Medical Society of Upsala, 1868.'

It is not, however, to be entirely removed from the list of foods and classed with condiments, for the salts of flesh which it contains are valuable in the process of nutrition; but it should be classed with such nervous stimulants as tea and coffee, which supply little or no nutriment, yet modify assimilation and nutrition. Used alone for beef-tea it is a delusion.

The solid preparations contain a considerable proportion of gelatin, and do not putrefy because the gelatin has been dried. It is needful to use a much larger quantity of these than of the semi-fluid extracts, in order to obtain the same amount of meat-flavour and salts, but by so much is the gelatin and thereby nutriment increased. Hence its qualities and uses are not identical with those of the semi-fluid extract, and when derived from the same class of animals, both may be advantageously used together. These solid foods are not exclusively prepared from beef, but represent the flesh of other animals, and offer a variety and delicacy of flavour adapted to the wants of the sick.

Fluid Meat.

A preparation of lean meat, by Mr. Stephen Darby, has been introduced under the title of fluid meat, which differs from the various extracts of meat in retaining the fibrin, gelatin, and coagulable albumen. This is effected by dissolving them as they would be in the process of digestion in the stomach, and thus both advancing them a stage in the process of digestion, and enabling them to resist decomposition. One pound of the fluid meat is obtained from four pounds of lean flesh, and assuming that all the nitrogenous compounds in the flesh as well as the salts are present in it, it must be a convenient as well as valuable food. The following is Mr. Darby's description of it :—

'Fluid meat contains all the constituents of lean meat, including fibrin, gelatin, and coagulable albumen. By the process pursued these are all brought into a condition in which they are soluble in water and are not any longer coagulable on heating—in which state they have been designated peptones. This change is effected, as in ordinary digestion, by means of pepsin and hydrochloric acid. Lean meat, finely sliced, is digested with the pepsin in water previously acidulated with hydrochloric acid, at a temperature of 96° to 100° Fahr., until the whole fibrin of the meat has disappeared. The liquor is then filtered, separating small portions of fat, cartilage, or other insoluble matters, and neutralised by means of carbonate of soda, and finally carefully evaporated to the consistence of a soft extract. But this process, whatever care may be taken, leaves the fluid meat with a strong bitter taste. This bitterness attaches always to meat digested with pepsin; and this, in the opinion of medical men, would wholly preclude its acceptance and adoption as an article of food. In order to remove this bitter taste, and to obviate the objection to fluid meats on that ground, I have made many experimental researches, and have at length discovered that the purpose is completely and satisfactorily effected by the addition, in a certain part of the process, of a small proportion of fresh pancreas. The fluid meat so prepared is entirely free from any bitter flavour.'

Mr. Darby regards these changes as exactly analogous to the action of the pepsin and pancreatic juice on food in the body.

CHAPTER XII.

ALBUMEN, GELATIN, AND CASEIN.

Albumen.

THIS substance is familiarly represented by white of egg, in which it appears as a bluish white, transparent, and semi-fluid substance, at the ordinary temperature, but when heated becomes opaque, yellowish white, and hard, and when the drying process is complete it is greatly reduced in bulk and resembles horn. Its chemical composition is as follows:—

No. 30.

C. 53·4   H. 7·0   O. 22·1   N. 15·7   Sulphur 2

It is also found in all compound animal structures and food substances, as well as in the most important vegetable productions in a somewhat modified form.

The composition of vegetable albumen is as follows:—

No. 31.

C. 54·0   H. 7·8   O. 22·4   N. 15·8

Albumen is by far the most important single element of food, since it contains nutritive matter in a compendious and easily digestible form, and being almost without flavour may enter into the composition of foods very diverse in other respects, whilst it is adapted to every variety of taste. Its composition is identical with that of the same substance in the blood and tissues of man and other animals, and may be used by the body or incorporated in it with great ease, although it is necessary that it should first be broken up, so as to form new combinations. The following is the effect on the vital processes as shown by my experiments. Two good-sized eggs caused an average increase in the quantity of carbonic acid evolved in respiration of ·27, ·45, and ·13 grain, and maxima of increase of ·88, 1·12, and ·38 grain per minute on different persons. The quantity of air was increased 17 and 38 cubic inches per minute. (*Phil. Trans.* 1859.)

Ten grains of dry albumen when burnt produce heat sufficient to raise 12·85 lbs. of water 1° Fahr., which is equal to lifting 9,920 lbs. one foot high.

Gelatin.

This substance, although differing from albumen in appearance, is very similar in chemical composition. It is always found in a solid state, as in the tendons and bones, and may be readily seen in the latter by immersing them, for some days, in hydrochloric acid and water for the removal of the mineral matter.

It is usually prepared as food from the tendinous and horny structures of the hoof and heels and from the skin and bones of animals, and when properly purified it is equally valuable from any of these sources. Care and skill are, however, required to perfectly separate the gelatin from other animal substances which are found with it and would injuriously flavour it, and hence the cost of one specimen may be thrice that of another.

When pure it is almost destitute of flavour, and, unlike the albumen in white of egg, requires wine and other substances to render it palatable. When, however, it has been properly prepared it is a very agreeable food, and particularly for the sick, from its flavour and readiness of mastication and digestion.

Its nutritive properties have been unjustly called in question, because an elaborate series of experiments which were made in France, on the feeding of dogs with ground bones showed that it could not alone sustain life, but these experiments proved only that life and health could not be sustained on one food alone, and not that gelatin, when properly mixed with other foods, is not a valuable aid to nutrition.

My experiments, and those of others, have proved that it is a valuable food, since it increases vital action in the same direction, if not in the same degree, as albumen, as shown by the following details (*Phil. Trans.* 1859) :—

120 grains of dry isinglass prepared with 12 ounces of water gave a maximum increase of ·84 grain of carbonic acid per minute in respiration, and an average increase during the period of its action of ·45 grain. The effect was quite equal to that of albumen.

It is time that the fallacy referred to should be exploded, and that it should be no longer necessary to allude to it, for it is the experience of all people that jelly is a valuable food.

The purest kind is isinglass, and is that obtained from the intestines of the sturgeon, and commands a price far above that of ordinary gelatin. Its advantages are a lighter colour, less flavour, and greater thickening power, so that it will produce a jelly clearer and in greater quantity than ordinary gelatin. The improvements which have been effected in the preparation of gelatin, and particularly in France, have, however, provided new facilities for the adulteration of isinglass.

Although dried gelatin contains much nutritive matter in a small compass, the same substance when prepared for the table is greatly diluted by the addition of water, and a considerable quantity must be eaten in order to supply as much nutritive material as would be found in a good-sized hen's egg, and hence persons in health partake of it as a luxury or as an unimportant addition to other foods, rather than as a useful article of diet, and its merits as a food are not fairly tested.

Vegetable jelly is found in the juices of plants, as in that of the vine, and differs entirely in composition from animal jelly, as it contains no nitrogen. A substance somewhat similar is found in various sea weeds, as the common green moss; whilst gum, which is widely distributed throughout the vegetable kingdom, has an analogous composition.

Birds' nests, which are used in China and some other Eastern countries as food, resemble isinglass in their quality and flavour. They contain 90 per cent. of animal and nitrogenous matter, which is secreted by the bird itself in the act of making the nest. It is, however, a dear mode of obtaining gelatin.

Casein.

Casein is readily obtained by throwing down the curd from milk and washing it repeatedly in pure water, and is employed in the preparation of cheese cakes and other agreeable and well-known foods. When dried the composition resembles that of albumen and gelatin.

The effect upon myself of eating the casein thrown down from one pint of good skimmed milk was to increase the carbonic acid in the expired air by 1·34 grain, and the volume of air inspired by 28 cubic inches per minute. (*Phil. Trans.* 1859.)

CHAPTER XIII.

EGGS.

IT would not be possible to exaggerate the value of eggs as an article of food, whether from their universal use, or the convenient form in which the food is preserved, presented, and cooked, and the nutriment which they contain.

There is no doubt much difference in their flavour according to the kind and feeding of the bird, for that of the large egg of the sea gull is much stronger than that of the duck, and both than that of the common fowl or the plover, and that kind is preferred which gives the sweetest and richest flavour. The feeding of the fowl has far greater influence over the flavour of the egg than is usually acknowledged, for the egg of the stray fowl on the Irish bog is far less rich than that of the well fed barn-door fowl or a fowl which is in part fed on kitchen scraps. It is also well known that the breed of the fowl greatly influences the quality as it does the size of the egg, and the intermixture of breeds which is carefully carried on is of service in this respect.

But it is only within narrow limits that there may be said to be variety, for there is no egg of a bird known which is not good for food or which could not be eaten by a hungry man. This is due to their similarity in chemical composition, for there is always a white portion and a yelk, the former consisting of nearly pure albumen with water, and the latter of albumen, oils, sulphur and water.

The oil is separated from the yelk of hens' eggs in Russia and used for medicinal purposes, each yelk yielding about two drachms. The yelk was also used in the middle ages for the painter's art before the discovery of oil colours, as in the Chapter House at Westminster.

The weight of an ordinary fowl's egg is one and a half to two ounces, whilst that of the duck is two to three ounces, of the sea gull and turkey three ounces to four ounces, and of the goose four ounces to six ounces.

When eaten by persons in health, it is customary to use both the yelk and the white, but many sick persons eat the yelk only on account of its flavour and digestibility. When used in puddings or in the preparation of a thousand delightful culinary compounds, it is more common to separate the yelk from the white, and to use one without the other.

When boiled it is usual to eat them before the white or albuminous part has become entirely opaque by the consolidation of the albumen, and there cannot be a doubt that they are more readily digested in that state than when further hardened by heat. It is said that an unboiled egg is more easily digested than one that is cooked, but this may be doubted if the egg be not overcooked, and unless the person eating it had been accustomed to eggs in their uncooked state he would not prefer it. Perhaps the most agreeable form is the flaky state in which the egg may be obtained when placed in cold water and eaten very soon after the water has been boiled, but the effect of heat thus applied varies with the time required to raise the water to boiling point.

When, however, it is cooked by dry heat, as in preparing omelettes, the albumen is more consolidated, and the merit of the dish will consist in its retaining much of the softness and all the flavour of the egg, notwithstanding the solidification of the albumen. There is no food in which egg is required to be dried or in which the drying and hardening process should be carried far beyond that of the boiled egg.

A very favourite compromise is the use of the raw egg with a stimulant both to render the compound more agreeable to the palate and more easily digestible, as for example, wine and egg or brandy and egg, but of the two the former is to be preferred. Another is the admixture of it with milk to render the compound more nutritious, but if the milk be new and good, it is very doubtful whether such a combination is not more fitted to hinder than to promote digestion and nutrition. When it is used it should be in a cooked form, as that of pudding, and thereby be rendered more digestible.

A most delicious food is that of poached egg, as prepared in France, Mexico, and the East generally, where a pottery dish is used instead of an iron pan as in this country, and the heat kept moderate by using charcoal fires. The dish is very thick, so that it must be placed upon the fire for a little time to become well warmed through, after which butter, with pepper and salt, are placed in it, by which the surface is well lubricated and a savoury mixture prepared to receive the egg. The egg is then broken and dropped into it, and in a short time turned so that both sides may be slightly browned, but without breaking the yelk bag. When cooked, it is served on the same dish and in the hottest state possible, and then the flavour is the most delicate and perceptible. This food will often be eaten by the invalid when others are rejected.

Eggs are not improved by keeping, but should be eaten as early as possible; yet they may be kept for days or even weeks without material change by placing them on end in net trays and in dark corners and reversing them daily. When intended to preserve them for months, it is necessary to adopt special means, and the simplest and perhaps the best is to place them in lime or lime water in a dark room, but as this renders the shell more brittle, many are broken on being removed or during the process of cooking.

The rapidity of intercourse between countries has been so greatly increased of late years, that Irish and French eggs are frequently brought and sold in the shops in England within seven or fourteen days after they were laid and whilst they still retain much of the character of fresh eggs. This proceeds throughout the warmer parts of the year, but the French and some English farmers have the habit of accumulating eggs laid at that season in order to send them to market when they are scarcer and dearer, and hence eggs sold in winter are not only dearer but inferior to those of summer.

The consumption of eggs in this country is now enormous. Over six million great hundreds (120) of eggs are received annually from the Continent alone, of the value of more than 2,500,000l. Of the receipts in 1875 nearly 5,000,000 great hundreds came from France, and the rest from Germany, Belgium, Denmark and Spain.

The digestion of egg is not so rapid a process as might have been anticipated, for the time is that of the digestion of mutton, viz., three to four hours.

The offensive character of decomposed eggs is such as to prevent their use under ordinary circumstances, and therefore the instances are few where disease is known to have originated from that cause. The effect of frying is more satisfactory than that of boiling eggs which are in a state of decomposition, but such eggs are generally cooked in puddings by which the flavour is disguised. Eggs in such a state should certainly be well cooked if eaten at all.

It may be doubted whether as much attention is paid in England to the production of eggs as the utility of the food demands, and particularly by the poor to whom their value is a consideration. Efforts should be made to induce all persons conveniently circumstanced to keep hens and ducks, and there is reason to believe that ducks are more profitable than hens, having regard to the number and size of the eggs laid by them. The solid matter and the oil in a duck's egg exceeds that of a hen's egg by so much as one-fourth.

An egg weighing 1¾ ounce consists of 120 grains of carbon and 17¾ grains of nitrogen or 15·25 per cent. of carbon and 2 per cent. of nitrogen.

The following is the composition of a hen's egg, according to Lawes and Gilbert:—

No. 32.

Fresh weight	1·8 ounce
Dry weight	·45 ,,
Fat	·198 ,,
Mineral matter	·025 ,,
Nitrogen	·036 ,,
Carbon	·275 ,,

or, per cent.—

Dry matter	25·0
Mineral matter	1·4
Dry fat	11·0
Nitrogen	2·0
Carbon	15·25
or Carbon and Hydrogen reckoned as Carbon	20·56

CHAPTER XIV.

POULTRY AND GAME.

This class of foods is an exceedingly large one, and although it might not be difficult to name those which are sold in the markets of this country, it would be impossible to include all which are eaten as food by the inhabitants of other, and particularly of the Eastern, countries.

The animals which are regarded as game by our laws are hares, pheasants, partridges, grouse, heath-game, moor-game, black-game, and bustards; but the following are also protected: woodcocks, snipe, quails, landrails, and conies.

The flesh of birds differs very considerably from that of mammals in two respects, viz.: in the relative quantity of fat and the quality of the juices. The fat of birds is laid up in various parts of the interior of the body as well as under the skin, but is very sparingly found in the fibres or the juices of the flesh; and as it has a flavour which is not regarded as agreeable, it enters but little into the food of man.* The juices are deficient in red blood, and have a more delicate flavour than those of adult mammals, but they differ less from those of young mammals.

The muscles of fowl differ little in their anatomical structure from those of the mammalia, and are quite as rich in nitrogenous elements, but are relatively poorer in fat and salts. They are regarded as light foods, and more fitted for invalids than strong men, or as an adjunct to flesh rather than a food to sustain man. This is due, however, rather to the delicacy or absence

* Goose-grease, however, is a good deal used in Germany.

of flavour, than to any real deficiency of nutriment as a food, so that a person could not be induced to limit himself to this kind of flesh food alone, if he could obtain beef or mutton. It should, however, be understood that it is a nutritious food, and one which might be further produced with advantage to our food resources.

There is a general similarity of character by which the flesh of this class of creatures is readily distinguished, but there are, at the same time, very appreciable differences, according to the nature of the bird, and its breed, food, and feeding. The flesh of the domestic fowl differs very greatly, both in fulness and delicacy of flavour, in different specimens, and the flesh of a graminivorous is very readily distinguished from that of a carnivorous bird. The flesh of a fish-eating bird, as the sea-gull, and of a carrion bird, as the crow or buzzard, is disagreeable; and even a domesticated fowl, as the duck, may be rendered disagreeable by being fed on fish.

The flavour of wild birds is fuller and stronger than that of domesticated birds, and the flesh is richer in nitrogenous, as it is generally poorer in carbonaceous, matter. The structure is also closer and firmer, so that in the fresh state it is regarded as hard, and tough, and it is desirable, if not necessary, to allow the process of decomposition to commence, in order to cause a separation and softening of the fibres. Hence, whilst a domestic fowl is eaten when quite fresh, a wild fowl is kept for many days, or for weeks, before it is cooked.

There is, however, much difference in this respect according to the habits of the birds, as well as to their nature; for larks, sparrows, and other small wild birds, feeding on insects or grain, are both rich and delicate food when properly cooked.

It may also be added, that the young of flesh-eating birds, as the rook, may be eaten, when the adult bird would be repulsive. Rook-pie is a dish which but few despise, and has a fulness and lusciousness of flavour which excels any dish of graminivorous birds.

The crane, bustard, curlew, and heron were eaten by our Norman invaders and their descendants; and it is related of the Conqueror that he attempted to strike his favourite, William Fitz-Osborn, for serving up a crane when half roasted. As the crane and heron feed chiefly on frogs, fish, and other animal food, they would not be an acceptable dish at the present day. They were directed to be larded with fat of pig or bacon, and eaten with ginger.

The peacock and swan were in great repute as a dish at high festivals, and may be seen carried in state in drawings of the middle ages. As a matter of historical interest, it may be recorded that on the visit of the Prince and Princess of Wales to Chatsworth, they lunched with the Duke of Rutland in the ball-room of Haddon Hall, on December 20, 1872, and were served with the two ancient dishes already referred to, viz., boar's head and peacock pie. The pea-hen is still occasionally eaten, and has a flavour intermediate between that of the common fowl and the pheasant.

Many of our smaller birds, as quails, rails, teal, woodcock, and snipe, were not used as food in the fourteenth century, and the aborigines of Britain were deterred from eating hens or geese by superstitious fear.

There is no reason to believe that the flesh of any bird is injurious to man; but it is only certain people, as the low caste natives of India, who would eat carrion birds, except as a last resource; and even in reference to these it may be doubted whether poverty has not led to this degradation of taste.

The flesh of the male bird has usually a fuller flavour than that of the female, whether we refer to domesticated or wild animals, but particularly to the latter, as will be remembered by one who has shot and eaten the magnificent wild turkey of the Western States of America.

In reference to domesticated fowls, it may be added that the capon retains some of the strength of flavour of the male bird with much of the delicacy of the female. The merits of this mode of preparing poultry for the table are not sufficiently appreciated in our day, but were well known to our ancestors in the middle ages.

There were numerous birds, and particularly the flesh and fish-eating birds, which were prohibited to the Jews. The list, as given in Lev. xi., is as follows:—The eagle, ossifrage, ospray, vulture, kite, raven, owl, night-hawk, cuckow, hawk, little owl, cormorant, great owl, swan, pelican, gier eagle, stork, heron, and lapwing.

The blood of the common fowl differs very considerably from that of red-blooded animals, since the salts of iron, upon which the red colour depends, are in the fowl only 2·11, whilst they are 24 or 25 per cent. of the salts in beef and mutton. So far it is inferior, but in another quality it is superior; for the phosphates, which are said to play so important a part in regenerating nervous tissue, are three times more abundant in the latter than in the former.

The following is the average chemical composition of the flesh of poultry when fit for the market, in 100 parts:—

No. 33.

| Water 74 | Nitrogenous 21 | Fat 3·8 | Salts 1·2 |

The goose, wild goose, gosling, duck, wild duck,

canvas-back duck, widgeon, teal, quail, water-hen, grouse, black game, prairie hen, partridge, pheasant, woodcock, snipe, cygnet, capercailzie, pea-hen, Guinea fowl, turkey, and larks, are amongst the birds with which we are well acquainted as food. They have their own special flavours, and are not eaten indiscriminately; but they are too well known to render it necessary to particularly describe them apart from the general characters of the whole class.

The flesh of rabbit has a nearer resemblance, in general and nutritive character to that of the hen than of its wild congener, the hare. It is pale in colour, somewhat loose in texture, without a well-defined flavour, and is digested with ease. It is eaten by the convalescent, from the delicacy of its flavour and its digestibility; whilst it is consumed by the healthy, with condiments to provoke the appetite, and rather as a change than as a staple article of food.

The constant use of this food soon induces satiety, and even disgust.

The flesh of the hare and leveret is much darker in colour, harder in texture, and fuller of flavour than that of the rabbit, and is one of the most nutritious kinds of flesh. Its digestibility is less than that of rabbit, and it is a food for the healthy rather than for the sick. It is the practice to hang this meat for a considerable period before it is cooked, not only because the flavour is thus improved, but it is rendered much more easy of mastication and digestion.

The Jews were prohibited from eating the rabbit and the hare, but the sanitary reason, if any, for so doing is not clear:—' And the coney, because he cheweth the cud, but divideth not the hoof; he is unclean unto you. And the hare, because he cheweth the cud, but divideth not the hoof; he is unclean unto you.'—Lev.

xi. 5, 6. It may also be added, that our British ancestors refused to eat hares from a religious objection to them.

Squirrels are not eaten in this country, but they are, nevertheless, an agreeable and highly nutritious food. It is not unusual in the Western parts of America, as, indeed, in all new woodland countries where they abound, to eat them occasionally. The flesh is very dense and gelatinous, and, judging by our own experience, may be described as luscious and satisfying.

The flesh of the racoon is sometimes eaten.

CHAPTER XV.

FISH.

THE varieties of Fish which are used as food throughout the world are almost infinite. No fewer than forty-five are quoted by Dr. H. Simpson as eaten in only one locality in India, viz., Dacca. (Report on Dietary, 1862.)

Those with which we are the best acquainted are, cod, ling, plaice, turbot, sole, sturgeon, haddock, whiting, herring, sprat, mackerel, pilchard, eel, mullet (red and grey), skate, halibut, pike, carp, tench, roach, perch, salmon, trout, bream, anchovy, whitebait, and smelt.

The use of fish is doubtless greatest in those countries where it is the most readily caught, and where poverty so abounds that many can obtain only that kind of flesh. This is particularly the case with many parts of India, the Mediterranean coast of Spain, and the Western coast of Ireland.

Many kinds of fish were eaten in this country in the fourteenth century, as the sturgeon, salmon, pike, haddock, codling, roach, tench, turbot, plaice, eel, the conger eel, mackerel, sole, lamprey, loch, &c.; and, although this has been doubted, recipes for the cooking of them at that period still exist.

It is said that a fish-eating people are ill nourished, and in Eastern countries are particularly liable to become leprous. If this be so, it must be associated with such poverty as prevents the inhabitants from obtaining a proper variety of flesh food and vegetables. Certain districts on the Mediterranean coast of Spain are cited in illustration of these facts, where there is a fish-eating and a poverty-stricken population; and if we turn to another fish-eating community—that of parts of the west coast of Norway—we find the disease, although some of the people are characterised by robustness of health and great physical strength. In the latter case, however, the consumed fish is chiefly the red-blooded salmon, which approximates in character to the flesh of the mammalia, and there is not an universal exclusion of ordinary flesh from their dietary.

It is not desirable that fish should be the sole kind of nitrogenous animal food eaten by any nation; and, even if milk and eggs be added thereto, the vigour of such a people will not be equal to that of flesh-eating nations. At the same time, the value of fish as a part of a dietary is indicated by the larger proportion of phosphorus which it contains, and which renders it especially fitted for the use of those who perform much brain work, or who are the victims of much anxiety and distress.

Fish are generally divided, for the purposes of food, into two classes—viz., into white-blooded and red-blooded, of which cod may represent the former and salmon the latter. It also varies in flavour, not only on this ground, but from the presence of oil in the flesh; so that some kinds of fish, as the sole, contain but little, whilst others, as the eel, contain much. Moreover, in some, as the salmon, the oil is distributed throughout the muscular tissue, whilst in others it is stored up in the liver, as in the cod. As a general expression, it may be stated that the flesh of white fish contains far less oil than that of red fish.

It is therefore evident that, whilst the term fish may be applicable to an infinite number of creatures, it represents very different nutritive values—a difference far greater than exists in the flesh of the mammalia—and it is very difficult to give an adequate expression by simply stating the composition of one or two specimens. The following must, therefore, be accepted as a general expression only: per cent.—

No. 34.

White fish . .	Water 78	Nitrogenous 18·1	Fat 2·9	Salts 1
Salmon . . .	,, 77	,, 16·1	,, 5·5	,, 1·4

The following table, arranged to show points of contrast, indicates the composition of several kinds of fish in their fresh state, according to Payen: per cent.—

No. 35.

	Water	Dry substance	Fat	Nitrogen
Conger eel . . .	79·909	20·091	5·021	2·172
Eel	62·076	37·924	23·861	2·000
Whiting . . .	82·950	17·050	0·383	2·416
Mackerel . . .	68·275	31·725	6·758	3·747
Pike	77·530	22·470	0·602	3·258
Sole	86·144	13·856	0·248	1·911

The same authority gives the following as the ultimate elements of dried substance, free from fat, of four well-known fishes: per cent.—

No. 36.

	C.	H.	N.	O.
Eel	52·999	7·474	14·644	19·296
Mackerel	51·515	6·902	15·836	19·608

	C.	H.	N.	O.
Sole.	48·795	6·581	15·460	20·032
Barbel	45·927	6·800	15·535	22·783

The effect of white fish upon the vital functions is much less than that of the flesh of the mammalia, and is probably less than that of poultry, but is greater than that of eggs. The nutritive value of the flesh of a red-blood fish, as salmon, differs but little from that of the red-blood flesh of other animals. Thus, in my experiments, eight ounces of fresh salmon caused a maximum increase in the quantity of carbonic acid in evolved respiration of ·84 grain per minute in sixty-five minutes, and so far it was equal to flesh meat. (*Phil. Trans.* 1859.)

It is well known that fish are out of condition in the spawning season, and are then less fit, or even unfit for, food. Sir Robert Christison shows the deteriorated state of salmon at that period by the following analysis of equal parts of the dorsal and abdominal flesh: per cent.—

No. 37.

Clean	Oil 18·53	Nitrogenous 19·69	Salts 0·88	Water 60·89
Foul	„ 1·25	17·07	„ 0·88	„ 80·8

Hence, whilst there is a diminished proportion of nitrogenous matter in the flesh of the foul fish, the chief deterioration is in the quantity of oil, which is reduced by 17 per cent., and is replaced by water. Whether, therefore, we consider the reduced nutritive value of the food, or the destruction of an animal which in a few months would have yielded twice the amount of food, the use of foul fish is properly forbidden.

Nothing, perhaps, could be more striking than the contrast of the same fish in its clean and foul state, as shown by the painted casts of Mr. Frank Buckland.

A member of the family of *Salmonidæ*, the *capelin*,

which is as large as a smelt, abounds in shoals in the Arctic seas, and might be largely imported into this country.

Prof. Gulliver, F.R.S., makes the following remarks on edible sharks, better known as 'Canterbury gurnets':—

'Risso, the ichthyologist of Nice, says that the porbeagle is good eating, and thus much used and esteemed by the people of the Mediterranean. Some of the smaller members of the shark family afford an almost constant and very bountiful supply of valuable food to the poor people of the Shetlands and Hebrides and other parts, and at Canterbury may occasionally be seen loads of skinned fresh fish, each about 18 in. long and 2 in. thick. They are commonly brought up Northgate into the city, where they are sold, under the fictitious name of "gurnets," as cheap food, which is said to be agreeable and wholesome. They belong to a small species of shark, known at Hastings as Robin Hursts, elsewhere as rough Hounds, and to naturalists as the small spotted Dog-fish—*Scyllium canicula*. At Canterbury these fish always arrive decapitated, gutted, and skinned, probably to conceal what they really are, and to serve the cabinet-makers, who are said to use the skins to smooth down or polish the surface of their work. This fish, under the name of *morghi*, is commonly used in the west of Cornwall for soup, which is much liked by the natives. The Pricked Dog-fish or Hoe (*Spinax acanthias*), so abundant as to be contemptuously rejected on the Sussex coast, is considered valuable as diet in the Scottish islands, where these fish are dried for this use, and a large and profitable quantity of oil obtained from their livers; and in the west of England the same fish is used and much valued as excellent aliment, both fresh and salted, by the fishermen and others. The smooth-hooped or skate-toothed Shark, a fish about a yard long, is esteemed as delicate food in the Hebrides. And indeed, when we consider the constant abundance of food and oil for man offered by the small sharks, it seems lamentable that they are so much despised and wasted on most parts of our coasts.'

The market value of fish differs extremely with the season, weather, and kind of fish; but, under ordinary circumstances, fresh herring offers the largest amount of nutriment for a given sum of money of any kind of animal food. A fresh herring weighing $4\frac{1}{2}$ ounces, and costing $\frac{1}{2}d.$, contains 240 grains of carbon and 36 grains of nitrogen; and a dried herring, weighing 3 ounces, and costing $\frac{3}{4}d.$, contains 269 grains of carbon and 41 grains of nitrogen. The largest and best flavoured herrings are obtained from Loch Fyne.

Hence, herring is of great value to the poor man, whilst salmon, which centuries ago was the poor man's food, is of all fish the most to be desired for the richer classes; and whilst the former is abundant without the aid of man, the latter demands the protection of the law that it may increase. The cost of the former is scarcely under the control of man, whilst that of the latter may be greatly reduced as the supply increases, and the present monopoly in its capture and sale is broken up. Perhaps no subject, in reference to food, demands and will repay greater care in production, and freedom in sale.

The most delicate of all kind of fish is probably the whitebait, when properly cooked and served; but it is said that the smelt, when newly caught in the Humber, is almost its rival in that quality.

The lamprey, which was so commonly eaten in the middle ages, is now rarely mentioned, but it may be obtained, and is an agreeable food.

Eels are a luscious and favourite dish whenever they can be obtained. Those from Ballyshannon, and other Irish rivers, are accounted the best in this country, inasmuch as the flavour of them is rich and agreeable and their size convenient. Those of the fenny districts of Lincolnshire and Cambridgeshire, and from Holland, which are found in the dykes, are also good, but their size is commonly less than those above mentioned. They are found chiefly in comparatively still waters, where there is a deep deposit of mud, and particularly in the great lagoons of the Adriatic, whence, under the name of *Capitone*, an enormous quantity of large repulsive-looking eels are obtained. More than one million pounds are imported into Naples alone every year, for the fish is the national dish of both rich and poor. A more striking idea of numbers of living things than that derived from the young eels as they ascend the rivers in Ireland can scarcely be conceived. The sides of the river seem to be a solid mass of wriggling lines, stretched out for miles in length.

The use of fish by the Jews was greatly restricted, for the prohibition extended to all fish without scales and fins. 'These ye shall eat of all that are in the waters: all that have fins and scales shall ye eat: and whatsoever hath not fins and scales ye may not eat; it is unclean to you.'—Deut. xiv. 9 and 10. 'Whatsoever hath no fins nor scales in the waters, that shall be an abomination unto you.'—Lev. xi. 12. Hence cod, whiting, eel, and all kinds of shell-fish are forbidden.

It may also be added, that the aborigines of Britain objected to eat fish, whilst they were not unwilling to eat seals and porpoises (doubtless without enquiring into classification), and recipes of the fourteenth century are still extant, which prescribe the proper mode of cooking those creatures.

Fish is prepared for the table in every known mode of cooking, but most frequently by boiling, and much pains are taken by cooks to garnish it and to provide suitable sauces. Very commonly it is cooked in the mass, but sometimes in slices. The preparation of the slices is called crimping and should be performed immediately after the fish has been caught, and before the *rigor mortis* has set in. The flesh is then easily divided, and there is a slimy or creamy fluid lying between the layers of muscles. In this state, the flesh of salmon is said to be curdy, and is believed to be more digestible and to possess a more delicate flavour than at any other; but it passes off in a very short time.

Opinions differ as to the advantage of keeping any kind of fish for twenty-four hours before it is eaten; and whilst some affirm that the flavour is increased by the delay, the majority of people prefer to eat nearly all kinds at the earliest possible moment after capture. Our own opinion coincides with the latter view; and in order to retain fish alive until they are required for the table it is desirable to adopt the plan in use in certain parts of Wales, and on the Continent in reference to fresh-water fish—that of placing them in a box or tank, through which a stream of water constantly flows, or in an enclosed part of the sea, in reference to sea-water fish.

Salmon when eaten almost immediately after capture is still called *Calver Salmon* in Lancashire, and in that state it was eaten by our forefathers. Recipes of the fourteenth century mention *Calwar Salmon*, which was prepared with almonds, rice, milk, and pepper.

The recipe for a Lenten fish-soup of the same period may be of interest:—

'Take the blode of pykes oth' of cong̃ (conger eel) and nyme the pañc'h of pykes, of cong̃ and of grete code lyng, and bolle hē tendre and mynce hē smale and do hē ī that blode. take crust' of white brede and stȳne it thurgh a cloth. thenne take oynoñs iboiled and mynced. take pep and safroñ wyne. Vyneg̃ aysell oth' aleg̃ and do th'to and sūe forth.

Jellies of fish and flesh were in common use in the

middle ages, and probably at the Conquest. The following recipe of the fourteenth century possesses an interest, since it shows that fish was in use earlier than Henry VII.'s reign:—

'GELE OF FYSSH.

'Take Tench, eelys, pykes, turbut and plays, kerue (carve) hē (to) pecys. scalde hē and waische hē clene. drye hē w^t a cloth. do hē ī a pāne. do th'to (thereto) half vyneḡ and half wyne and seethe it wel. and take the Fyssche and pike it clene. cole (strain) the broth thurgh a cloth īto aɔ erthen pāne. do th'to powdŏ of pep (pepper) and safroñ ynowh. lat it seethe and skym it wel whan it is ysode. dof (do of) th^e grees clene, cowche fisshe on chargeŏs and cole the sewe thorow a cloth onoward and sūe it forth.'

The roe of fish whether fresh or dried is esteemed a luxury, and is also nutritious. The dried roe of the salmon and cod is very commonly eaten in this country; whilst in Russia the roe of the sturgeon and allied fishes is preferred, and the preparation and sale of it is now a large branch of trade, under the name of *caviare*. A similar preparation, from the roe of the cod and other fish, is also eaten in Sweden and other parts of northern Europe, under the same name.

Fish are generally pickled in a very simple manner; they are rubbed with salt, and thrown into a cask, with layers of rough salt, until the cask is filled, and in this state they will keep for some weeks. This is the usual mode of treating herrings, but large fish, as cod, are cut open, and salted both on the inside and outside.

A favourite mode of pickling pilchards in Cornwall is much more complicated. Those fishes are cut open, and have their heads, tails, and fins cut away, after which they are washed, and rubbed on the inside with salt, and various kinds of pepper and other spices. They are then placed in layers in a jar, separated by bay leaves, salt, and spices, and so laid, layer on layer, until the jar is filled, after which they are covered with vinegar, tied over, and exposed to a gentle heat for several hours. So prepared, they will keep several weeks, and are a savoury as well as good food.

Salmon is often preserved by being cut open, salted, and dried in smoke, and is a good and agreeable food. Haddocks are slightly salted and dried, and herrings are smoked and dried, as well as salted.

According to Frankland's experiments, ten grains of whiting, when entirely consumed in the body, produce heat sufficient to raise 2·03 lbs. of water 1° F., which is equal to lifting 1,569 lbs. one foot high.

CHAPTER XVI.

SHELL-FISH AND TURTLE.

THIS class may be conveniently subdivided as foods into gelatinous and fibrinous, the former being molluscs, and represented by the oyster, mussel, whelk, and cockle, which are soft, and easily masticated and digested; and the latter crustaceans, represented by the lobster, crayfish, and crab, which abound in true muscular fibre, and are less easy to masticate or digest.

The whole class is very extensive, and embraces members unknown to our markets, but everywhere regarded as a luxurious rather than as a necessary food. Oysters, although one species, the *Ostrea edulis*, is chiefly used here, differ greatly in size and flavour. They were eaten by our forefathers in the middle ages, if not earlier, and are found in many parts of the coast of Great Britain and France, where they are more carefully protected and cultivated than heretofore. The most delicate in flavour are called Natives, and grow more particularly off the coast at Whitstable; but owing to their great scarcity and very high price, a variety has been imported from America, the *Ostrea Virginica*, which is an excellent substitute, and costs scarcely half so much. The importation has already assumed considerable proportions, and it is desirable for both countries that it should be greatly increased.

The historians of Alexander's expedition state that there were oysters a foot long in India, and Sir J. Emerson Tennent found the mother of pearl oyster in Ceylon 11 inches by 5 inches. The largest which are known in our markets do not attain to more than half that size, and although containing a far greater amount of nutritive matter than the small Whitstable oyster, their flavour is not so delicate, and they are much cheaper than the smaller variety.

The tree or parasitic oyster abounds in the mangrove swamps of the West Indies, Africa, and Eastern countries.

Oyster-beds are found in estuaries, or other places on the coast, where the sea is tolerably quiet and the water not very salt, and where there are supports to which the oysters may attach themselves. Their extent may be very great, and whilst it is probable that there are an infinite number of such beds at present unknown to us, we are acquainted with enormous banks on the coast of Georgia, and know that even mouths of the sea have been closed by them. The oyster is hermaphrodite, and produces young by millions, so that the shallow seas might well be filled with them if all the product lived and in their turn produced oysters; but, in fact, only a few in a million succeed in attaching themselves to the parent shells, or to rocks, or to pieces of wood, seaweeds, or clumps of reeds or grass, so as to remain in the bed, whilst an infinite number become detached, and die, or are eaten by fish, reptiles, and birds. The great oyster-beds of Georgia are called racoon-banks, because the racoon frequents them in search of the oyster as food.

When an oyster-bed has been discovered, it has usually been impoverished, and sometimes destroyed, by continued and careless dredging, so that too great a proportion of oysters has been removed to enable the bed to be maintained. Laws regulating this fishery have been in force for nearly 500 years; but more stringent ones have now been established, and, with the influence of public opinion and self-interest, will probably restore our supply of this favourite luxury.

Artificial beds have been prepared from the time of Sergius Orata, who established them at Baiæ, to our own day, when the French have converted the shores of a whole island (the Isle of Ré) into oyster beds, and our own countrymen are adopting the same expedient.

The oyster is not a food of high nutritive value, but is nevertheless useful to the sick, whilst its delicacy of flavour leads to its selection when other foods are rejected.

The more usual mode is to eat it when uncooked, and it is very doubtful whether cooking increases its digestibility. It is, however, possible that the flavour of scalloped may be preferred to that of the raw oysters, or that the vinegar which is usually eaten with the latter may be disliked or may disagree with the stomach, but with such exceptions the usual method of eating them raw is to be preferred. When oysters are to be cooked it is needless to obtain those of the finest quality, for the flavour is in great part changed by the cooking, and a larger and coarser oyster is equally good.

Mussels, cockles, whelks, limpets, and scallops, are cheap and abundant, and are commonly eaten by the poorer classes.

Lobsters and crayfish were known to and eaten by our ancestors at least 400 years ago, and in a recipe of the date of 1581, it is directed to roast the lobster in its shell in an oven or in a pan and eat it with vinegar. They rank higher in price and are certainly more delicate in flavour than crabs, but at the same time they are tougher and more difficult to masticate and digest. It may be doubted whether there are any foods which are so little desirable in a sanitary point of view, or which so frequently cause indigestion, yet they are extremely popular as a change of food and a luxury, and are as agreeable to the eater as useful to the doctor. They consist chiefly of muscular fibre which is not rich in oil, so that they are most frequently eaten with oil and condiments in the form of lobster salad, and should be cut into very thin slices and well masticated.

The crab offers a much larger proportion of non-muscular material than the lobster, and is fuller of flavour and more easily masticated. It is believed to be also somewhat difficult of digestion, and the flavour is apt to repeat itself in the mouth for many hours after the food has been eaten, but it cannot be so difficult of digestion as the tough tail of the lobster or crayfish. The muscular structures in the limbs are also less tough than in lobsters, but they elude the teeth and are not perfectly masticated with ease.

Hence we are of opinion that whilst crab is cheaper, it is much to be preferred to lobster as a food, if not as a luxury.

All the kinds referred to require the aid of condiments, if not of a stimulating liquor, to promote their digestion, and unless they are perfectly fresh it is desirable to correct the ammoniacal salts by an acid. A composition of olive oil, pepper, mustard, and vinegar is a very suitable adjunct to the flesh when eaten.

Turtle (*Chelonia*), which is so costly and favourite a food here, is neither scarce nor good in the tropical countries where it is produced. The number of these creatures lying on the sandy banks when depositing their eggs, or floating in the shallow bay, is almost infinite, so that they might be the sole animal food of the inhabitants of those regions, but neither the people who live among them, nor sailors who remain there temporarily, can continue to eat them.

Mr. Bates, in his very interesting work, 'The Naturalist on the River Amazon,' writes:—

'The abundance of turtles, or rather the facility with which they can be found and caught, varies with the amount of annual subsidence of the waters. When the river sinks less than the average, they are scarce; but when more, they can be caught in plenty, the bays and shallow lagoons in the forests having then only a small depth of water. The flesh is very tender, palatable and wholesome; but it is very cloying; every one ends sooner or later by becoming thoroughly surfeited. I became so sick of turtle in the course of two years, that I could not bear the smell of it, although at the same time nothing else was to be had, and I was suffering actual hunger.'

With such testimony, how may we explain the favour with which it has always been received by civilised nations, and the price which is paid for it? Simply by the mode of preparation for the table. The flesh is never served separately, but is made into soup with a great variety of condiments, *recherché* wines, like Madeira, and other agreeable adjuncts, and with high culinary skill. The soup so prepared is doubtless luscious and rich, if not easily digestible; but if, instead of being rare and costly, it were a constant and cheap dish, as it might be on tropical coasts, the appetite would soon reject it, and disease rather than health would follow its use. It must also be added that, as at present consumed, it is accompanied by costly viands and wines, which lend a gourmand's charm to the entertainment.

The expedients which are adopted by the natives in cooking it, with a view to keep their appetite for it, are worthy of note. Mr. Bates says:—

'The native women cook it in various ways. The entrails are chopped up and made into a delicious soup, called *sarapatel*, which is generally boiled in the concave upper shell of the animal, used as a kettle. The tender flesh of the breast is partially minced with farina, and the breast shell then roasted over the fire, making a very pleasant dish. Steaks cut from the breast and cooked with the fat, form another palatable dish. Large sausages are made of the thick-coated stomach, which is filled with minced meat and boiled. The quarters cooked in a kettle of Tucupi sauce, form another variety of food. When surfeited with turtle in all other shapes, pieces of the lean part, roasted on a spit and moistened over with vinegar, make an agreeable change.'

There are numerous varieties of turtle which are fit for food, but that which is the most agreeable in flavour and is alone imported into this country is the **edible green turtle** (*Chelonia Midas*) which abounds in the West India Islands. The turtle (*Chelonia virgata*) and the Hawksbill turtle (*Chelonia imbricata*) inhabiting the Indian Ocean, the estuaries of the Indian coast as well as the southern shores of the Atlantic, are also well known and valued both for their flesh and shell.

The eggs of the turtle are eagerly sought after in the countries where the turtle is found, and while fresh are doubtless good food, but as they are hatched in the warm sand they soon lose their freshness and are no longer fit for food.

CHAPTER XVII.

CHEESE AND CREAM CHEESE.

CHEESE is obtained exclusively from the milk of animals, and its quality varies with the class, breed and appropriateness of the food of the animal and the process of manufacture. It was well known to the ancients, and is mentioned by Homer, Euripides, Theocritus, and Aristotle, and in the 1st Book of Samuel, where David, nearly 3,000 years ago, was directed to take ten cheeses to the captain of his brethren (1 Sam. xvii. 18).

The most ordinary source of cheese is the milk of the cow, and there are certain varieties of cows which produce much cheese and little butter, as there are others of the contrary qualities. The kind of pasture is so important that certain farms in the cheese-making districts are prized for the cheese which they yield, and obtain greater rentals accordingly. Cheese is also somewhat largely made from goats' and sheep's milk on the continent of Europe.

The mode of preparation varies as to the application of heat, the duration of the process, and the kind of milk which is used; and it is said that to this more than to any other separate cause is to be attributed the excellence of the Stilton cheese and the product of other famous cheese-producing districts.

The essential constituent is casein or the curd; which when pure consists of the following elements in 100 parts, and is the sole source of the nitrogen in which cheese so much abounds:—

No. 38.

C. 53·83 O. 22·52 H. 7·15 N. 15·65

Casein alone is very difficult of mastication and digestion when dry and eaten alone, and is not very tempting to the palate. It is, however, rarely made for use in that form, for when prepared from the most carefully skimmed milk there will be some admixture of butter and the usual proportion of salts in milk, and when fresh and moist is an agreeable food.

Samples of cheese differ chiefly in the quantity of butter which they contain, and in the flavour which is due to the food as well as to the nature of the cow. The best is made from new and unskimmed milk, so that nearly all the butter of the milk is incorporated with the casein in the production of a fat and rich cheese, and in making Stilton cheese a further quantity of cream is or should be added. When skimmed milk is used the cheese contains very little butter, and is called poor, so that according to the richness of the milk in casein and cream is the richness of the cheese. A proper admixture of the milk of different cows is one of the arts of cheese-making.

The process of cheese-making really consists in gently warming the milk and causing the curd to be separated by the addition of an acid. The acid used in this country is that of the rennet, which is the dried stomach of the calf, whilst in Holland it is muriatic acid. The curd is connected with the cream or globules of oil which pervade new milk, and the whole is separated by means of a sieve and placed in a mould, where it is subject to pressure until it has become consolidated into one mass, after which it is turned out of the mould and laid upon a shelf or the floor to dry. The temperature of the cheese room should not be too low, and in order to mature the cheese more perfectly and rapidly, it is customary, as in the Stilton-cheese district, to heat the room artificially and to maintain a tolerably equal temperature of 65° to 70° by night and day for several months. During the process the cheese is turned over every day or two and kept perfectly clean.

The stock of cheese is sold off at the end of the summer, and although that which was made in the early part of the winter may then be ready for use, that made during the summer is improved by being kept still longer. It is matter of opinion as to when cheese is in its most perfect condition; but when it is new it is tough and when old it is rancid, and experience has assigned an age varying from nine to twenty months.

The poorer and cheaper kinds of cheese are largely eaten in substitution of meat by the poor labourers in South Wales, Somerset, Wilts, Dorset, and other counties; but in districts where meat is obtainable by the poor and the quality of the cheese is good, as in Yorkshire, this habit does not prevail.

Luxurious kinds of cheese, as the Gruyère and Parmesan, are made from goats' milk or an admixture of goats' and cows' milk, and a certain degree of fermentation is allowed in the process of manufacture. They are not in any degree superior as food to our Cheshire or Stilton cheese; but their mildness of flavour and a certain character of texture fit them for the preparation of macaroni and other foods.

The natural colour of cheese is a dingy white, and when cheese is highly coloured there has been an addition of annatto or other colouring matter. The green colour of Stilton cheese is due to a vegetable growth; but there are some kinds of cheese which are rendered green by the addition of powdered sage leaves.

The value of cheese as an article of diet has not been entirely established. If we consider it in its chemical composition we find it very rich, richer than any other known food, in nutritive elements, provided we select a good specimen; but this varies, as has already been pointed out, with the conditions of its manufacture. The poorer the cheese the greater is the proportion of the casein or nitrogenous element, whilst the richer the cheese the greater is the proportion of fat or butter which it contains; but in either case the proportion of nitrogenous matter in a given weight far exceeds that of meat.

There is, however, a long-standing belief that cheese is not easily digested, and in fact, that of any given quantity much will pass off by the bowels unused, and so far as this is true its nutritive value is lessened. That it is true has been repeatedly proved by us when determining the amount of nitrogenous matter which appears in the fæces after eating cheese; but it is very probable that this will be less in one who from infancy has eaten it daily than in one but little accustomed to its use.

In my own experiments it was shown that the use of cheese caused an increased excretion of nitrogen by the kidneys as well as by the bowels, which proved that part of it was digested and transformed in the system and part passed off unchanged.

There is a popular belief formulated by Shakspeare,[1] when he makes Achilles say, 'Why my cheese, my digestion,' that cheese although difficult of digestion promotes the digestion of other foods, and this was

[1] *Troilus and Cressida.* act II. scene 3.

shown in my experiments by the increased quantity of carbonic acid which was eliminated by the lungs under its influence.

There is, therefore, good reason to consider that this belief is well founded, and that whilst it may be proper to eat a small portion of cheese, both for the nutriment which it supplies and for the promotion of digestion, it is not proper to eat a large quantity or to make it a principal article of food and a substitute for meat.

The chemical composition of cheese varies so much that it is not easy to state it with great precision, but if we take a poor cheese made from skim milk and a good cheese made from new milk we shall obtain tolerably correct results. Thus the proximate composition, in 100 parts, is as follows :—

No. 39.

Skim milk cheese . . . Water 44 Nitrogenous 44·8 Fat 6·3 Salt 4·9
Very good new milk cheese „ 36   „ 28·4   „ 31·1   „ 4·5

The quantity of carbon and nitrogen in 1 lb. of moderately good cheese is 2,660 grains of the former and 315 grains of the latter, showing how rich is this substance in nitrogen. If the free hydrogen be reckoned as carbon the former quantity of carbon becomes 3,283 grains per pound.

Professor Voelcker found that the poorer kind of skim milk cheese yielded 2,345 grains of carbon and 364·3 grains of nitrogen per pound, the former made up of the following items :—

No. 40.

Butter . . . 611·2 grains of carbon
Casein . . . 1,250·5   „   „
Lactic acid . . . 484   „   „

Payen gives the following as the composition of the cheese of different counties, in 100 parts :—

No. 41.

	Water	Nitrogen	Fat	Salt
Cheshire.	30·39	5·56	25·41	4·78
Marolles	40·07	3·73	28·73	5·93
Roquefort	26·53	5·07	32·31	4·45
Holland.	41·41	4·10	25·06	6·21
Gruyères	32·05	5·40	28·40	4·79
Parmesan	30·31	5·48	21·68	7·09
Brie	53·99	2·39	24·83	5·63
Neufchatel	61·87	2·28	18·74	4·25

In my experiments (No. 124) the effect of eating the casein from one pint of new milk was to cause a maximum increase of 1·34 and ·92 grain of carbonic acid in the expired air, with an increase of 28 cubic inches per minute of the air inspired. It is, however, curious to note that my colleague in the experiment had no increase; and, in explanation, it may be added that he dislikes cheese and says that it does not agree with him. It is not at all improbable, from what has already been stated, that cheese may produce different effects on different persons, and that the effect will bear some relation to the desire for it. (*Phil. Trans.* 1860.)

The time required for the digestion of cheese varies with its age and as the fat more or less abounds; and in a fairly good cheese of medium age it is from three and a half to four hours. New cheese and poor cheese require a longer time for digestion, inasmuch as they are masticated with greater difficulty. Old poor cheese also requires a longer time, for it is so hard as to be almost incapable of solution in the gastric juices, and if a good cheese be old and greatly decayed it plays the part of an irritant in the stomach which may cause a form of indigestion, and be itself hurried through the stomach into the intestines so rapidly as to almost prevent its digestion.

It is probable that the establishment of cheese

factories in America and in this country will tend to produce cheese of more uniform quality. There are now nearly 2,000 such factories in the United States, and three or four have been opened in England during the last five years. It is necessary for their success that there should be good pasture land and plenty of water in the vicinity, and that the farmers should be able to take their milk to the factory while it is yet fresh and new. The manufacture of cheese by small farmers is not always effected in the most cleanly manner, neither with the uncertainty of seasons is it always lucrative.

Ten grains of good cheese when consumed in the body produce heat sufficient to raise 11·2 lbs. of water 1° F. which is equal to lifting 8,649 lbs. one foot high.

Cream Cheese.

Cream cheese is also practically found of two qualities and in two conditions. It usually appears in the London market as new curd containing but little fat, and placed upon straws. If eaten in this state there is but little flavour of cheese, whilst the prevailing flavour is that of the whey which is allowed to abound in it. After a few days the process of decomposition begins, and the cheese assumes a less solid but more creamy appearance, whilst its flavour is richer than before. As this proceeds an odour of a very disagreeable kind arises, and ultimately the cheese assumes a semi-fluid or fluid aspect, and is offensive to both the senses of smell and taste. The proper course is not to eat it until the process has begun.

The cream cheese which is produced in Lincolnshire is far richer in quality than that now referred to, and has a very agreeable flavour. It may also be kept for a longer time without becoming offensive, but as it is ready for use early it is not usual to keep it.

Some foreign kinds, as Neufchatel cheese, which is imported in little pillars, somewhat resembles the Lincolnshire cheese in quality, and when exposed to the air does not run into a liquid but gradually shrinks and dries up. Yet the outside may become soft and yield an offensive odour. This cheese is rarely obtained until it has been kept for some time, and not unfrequently vegetable growths occur in it and on it before it can be eaten.

Cream cheese is more digestible than ordinary cheese, both because it is softer and may be readily masticated, and has a less proportion of casein. It is, moreover, probable that the process is effected in from two to three hours.

β *Non-Nitrogenous.*

CHAPTER XVIII.

BUTTER, GHEE, LARD, DRIPPING, AND OILS.

The non-nitrogenous animal foods are very numerous, but belong to only one class, viz., fats, and as they have much similarity in composition and action, it will not be requisite to refer to them at great length.

It is desirable to premise that fats which are used as food are naturally divided into two sub-classes, differing as to the temperature at which they remain fluid; so that at one temperature we have solid fats and at another oils; but it is needful to add further, that even such fats as are solid at a given temperature contain also oil which, when separated, remains fluid. Hence the division of the two great classes of fats into stearine or margarine and oleine.

In treating this subject it will not be necessary to refer to the great advances which chemistry has made in determining the nature of fatty compounds, and which have brought to light a vast number of substances which were unknown at the time when Chevreul made the researches which are still the foundation of our knowledge of fats, for they have little bearing upon the subject of food. It will suffice, if we proceed to indicate those animal fatty substances which are commonly used as food.

BUTTER.

Butter was unknown to the ancient Greeks, as may be inferred from the absence of any reference to it by Homer or Aristotle. It was also unknown to the aborigines of these islands, and was not in common use in England until after the fourteenth century, whilst to this day it is less frequently eaten by barbarous than civilised nations.

It is obtained from milk, and chiefly from that of the cow, in which it exists as minute globules scattered through the whole substance so long as the fluid is in motion, but when at rest it rises to the surface and constitutes cream. It is separated generally by churning, but in a rougher and readier way by heating the milk and shaking it, as in India, Texas, and other very hot countries, where the labour of churning is not desired and waste of butter and milk not important.

It is the best known of all this class of substances, but it is eaten in very different quantities; from the large cupful before breakfast, as drank by the Bedouins near the Red Sea and the Persian

Gulf, to the scarcely perceptible layer on the bread eaten by the needlewomen of London, and the supply is limited by pecuniary means rather than desire. It is also the form of separated fat which is less frequently disliked by consumptive people and invalids generally, as was shown by me in an enquiry into the state of 1,000 patients at the Hospital for Consumption, Brompton.[1]

The flavour, however, differs according to the animal from which the milk was derived, so that butter from the buffalo in Egypt and India, and from the goat, has a strong taste; but it also varies with the food of the animal, and is much stronger when the cow is fed on turnips than on grass or hay. In like manner the colour varies both with the animal and its food, so that it may be nearly white or very yellow; but a false colour is sometimes given by the addition of annatto.

Some of the milk-preserving factories in the United States of America are also butter factories, and may ere long supply no inconsiderable part of our best butters. On their mode of preparation and preservation of butter, Mr. Willard has supplied information in the columns of the Journal of the Royal Agricultural Society for 1872, from which we shall make a few extracts.

He states that during the churning it is thought desirable to keep the cream from rising above 60° F. When the churns are started the temperature of the cream should be about 56° F.; and it has been found that the best results are obtained when the dashers make from 40 to 42 strokes per minute. At this rate of stroke and no less than one hour being consumed in the process of churning, if the temperature of the cream be kept below 60° F., or no higher than that, the butter will come of good colour and texture,

---

[1] 'Tubercular Consumption: its early and remediable stages.'—Henry S. King & Co., Cornhill.

and will be in the right condition for a first-class fancy product,' at least, so far as it can be made by the operation of churning. It is important, of course, that the cream be in the proper condition when it goes to the churn; but the manner in which the churning is conducted has a much greater influence upon the product than many people imagine.

The agitation of the cream over the whole mass should be as even and uniform as possible, in order that all of it may be turned into butter at about the same time. If the agitation be too rapid, or if it be unevenly distributed through the mass, a part will be transformed into butter whilst some will remain unchanged, and by the time the whole mass is churned the particles of butter first formed will have been beaten up in the agitation so as to injure the texture; or portions of unchanged cream may become mingled with the butter, thereby not only lessening the quantity of butter from a given quantity of cream, but materially injuring its quality. Again, in order to preserve a nice flavour and colour, as well as fine texture, the mass of cream while churning must not be allowed to rise to a high temperature.

Great care is taken, by the use of very simple means, to wash all the buttermilk out of the butter. 'The batch of butter,' or the 'churning,' say of about twenty to twenty-five pounds in weight, is laid upon the butter-worker and water applied from a sprinkler or small watering-pot. It is provided with a rose-nozzle, so as to distribute the water over the mass in numberless small streams. The watering-pot is held in the left hand, and the butter is worked with the right hand at the same time by applying the lever, going rapidly from one side of the mass to the other. The butter being on the inclined slab or bed-piece of the butter-worker, the butter-milk flows off readily, and by a few movements of the lever the butter-milk is expelled. When the water flows from the mass without being discoloured, the process of washing is completed. The water falling in a spray over the whole surface of the butter cools it, and gives the proper degree of hardness for working with the lever, a point of considerable importance, especially in hot weather.

On the subject of preserving butter by salt and saltpetre, he states that at the Orange County factories, the following recipe is used, viz. :—For every 22 lbs. of butter, 16 ounces of salt, one teaspoonful of saltpetre, and a tablespoonful of the best powdered white sugar; and the butter makers of that part of the country claim, that by the use of saltpetre the butter will retain its flavour and keep sound longer in hot weather than when it is not used.

The mode in which butter is packed for transport and keeping materially affects its preservation, and Mr. C. H. White, of Michigan, has invented a tub which deserves notice. The tubs are 14 inches in diameter at the top, and 9 inches at the bottom, and about 16 inches high. They are well made, of oak, with strong hoops, and with heads at both ends. A sack of cotton is made to fit the tub for the reception of the butter. It is placed in the tub as it stands on the small end, the sides of the sack being long enough to extend over the top of the tub.

The butter is packed firmly in this sack until within $1\frac{3}{8}$ inch of the top of the tub, when a circular piece of cloth is laid on the top of the butter, and the sides of the sack are brought over and nicely plaited down over the circular cover. A layer of fine salt is now laid on the top, the head is put in, and the hoops are driven so as to make a perfectly tight fit, that will admit of no leakage. The tub is then turned upon the larger head, and the butter in the sack drops down upon the larger end, leaving a space between it and the sides and top of the tub. Strong brine is then poured into it through a hole in the head, till it fills the intervening space between the tub and the butter, when the hole is closed tightly with a cork. The brine thus floats the butter, so that it is completely surrounded by the liquid, and effectually excluded from the air.

When the butter is to be used, the tub is turned on its small end, the hoops are started, and the large head is taken off, after which the butter may be lifted entirely off the tub by taking hold of the ends of the sack. It may be placed upon a platter or large earthen dish, the cloth removed from the top, and the butter cut into desirable shapes for the table. If any portion remains, it may be returned to the tub, and in this way it can be preserved for future use.

Although butter, when pure, is a fat destitute of nitrogenous elements, it is rare to meet with it in that state, for if any portion of the whey be present there will be a nitrogenous and decomposing compound. Moreover, as the flavour of butter changes with the food on which the animal is fed, and chemical changes rapidly proceed after its production, its elements vary within somewhat narrow limits. If butter be so well made that no fluid remains in it which could be extracted without other artificial means than that of pressure, its composition will be tolerably uniform, and it will not differ much in its ultimate elements from that of the fat of flesh removed from the fat cells. In that state it may be kept in a cool place without much chemical change for several weeks; but, in order to preserve it, it is customary not only to make it as solid as possible, but to mix it in layers with salt, and when well prepared and the surface kept from exposure to the atmosphere by salt and water, it may be preserved for a year, or indeed for a much longer time, without such changes as would render it unfit for food.

If, for the purpose of preservation, it is desired to obtain butter free from water, it may be effected by a proper application of heat, as is accomplished in Turkey and other Eastern countries. When the butter is quite fresh it should be placed in a long necked vessel over a heat not exceeding 140° F., by which it will rise to the surface, as a clear transparent fluid, and should be carefully removed. It is then strained through a cloth and cooled by immersion in cold water, after which it may be mixed with salt and packed closely. The action of salt seems to be chiefly that of attracting any water which may remain, and in Spain and America a little saltpetre is added for the same purpose. It is said that sugar has power to prevent the decomposition of butter, as will be more particularly stated in our observations on preserved milk.

The proportion of water which is commonly met with in butter is from half an ounce to one ounce in the pound, according to the care, and perhaps, the honesty of the manufacturer; but a larger quantity is said to exist in the cheaper kinds. In a report recently made by Mr. F. W. Rowsell, on the supply of provisions to the metropolitan workhouses, Mr. Wanklyn found that the quantity of water exceeded two ounces in the pound, or twelve and a half per cent., in more than one half of the samples submitted to him, as stated in the following table:—

No. 42.

Workhouses	Per-centage of water	Character of butter
St. Saviour's	12·6	Fair
Stepney	12·7	Middling
St. Pancras	12·8	Bad
Poplar	12·9	Very bad
Shoreditch	13·2	Bad
St. Giles	13·2	Tolerably good
Lambeth	13·2	Exceedingly bad
Fulham	13·1	Good
Wandsworth	13·3	Very good
City of London	13·7	Good
Hackney	14·2	Good
St. Olave's	14·3	Fair
St. Luke's, Chelsea	14·5	Fair
Camberwell	14·7	Exceedingly bad
Shoreditch	15·3	Bad
Wandsworth	15·3	Bad
St. George's in the East	15·4	Bad
Paddington	16·5	Bad
Stepney	16·5	Nasty
Hackney	16·6	Tolerable
Marylebone	18·2	do.
Greenwich	19·4	Fair
Holborn	19·7	Middling
City of London	20·0	Bad
Paddington	23·6	Rather rank
Kensington (salt butter)	23·7	Wretched
Whitechapel	24·9	Very bad

Butters are commonly divided into two classes, viz., fresh and salt; but a small proportion of salt, say half an ounce to the pound, is added to the best fresh butter. When there is nearly one ounce of salt to the pound the butter should be classed as salt, and in that proportion exceedingly good butter is imported into this country. It is, however, usual to find a larger proportion in the salt butters of the shops, and two ounces in the pound is not very uncommon. In the report of Mr. Rowsell just referred to that quantity was never exceeded; but it was approached at Whitechapel, where there was 10·7 per cent. of salt in the pound.

Besides these two adulterations, it is very common to find cheap animal fats, as mutton fat and horse grease, added to butters, the former to even good but the latter to bad butters only. It is also said that palm-tree butter, which is used for candles and railway carriage wheels, is sometimes added with the like purpose.

The addition of water or other fluid is determined by liquefying the butter, and expelling the fluid by heat, when the loss of weight is the required measure of the fluid. When the fluid is whey, or other nitrogenous substance, an analysis of the butter will be required, if the flavour does not sufficiently indicate its character. Salt is removed by washing the butter in pure water, when the taste of the water will reveal its presence, and the loss of weight of the butter indicate its quantity. Care must, however, be taken that no appreciable quantity of butter remains suspended as oil globules in the water, or the weight will be unduly reduced.

Pure meat fat when added to butter may be distinguished by the taste, which may indicate the absence of the flavour of butter and the presence of another flavour not belonging to it; but the latter is usually avoided by adding a tasteless fat, as that of mutton, assisted by the more penetrating flavour of butter. Analysis can, however, detect the admixture. These fats are, moreover, harder than butter, and cannot be perfectly mixed together, so as not to be detected by a magnifier, without some care.

Hence, in determining the chemical composition of butter, we must regard it as a fat, and it will be the same as that of other fats in their natural state. Thus, butter and fats consist, in their ultimate elements, of—

No. 43.

C 77   H 12   O 11

The proximate elements in 100 parts of an average sample of butter are as follows:—

No. 44.

Water 15·0   Fat 83·0   Salts 2·0

The fat consists of 68 per cent. of margarine, or solid fat; 30 per cent. of oleine, or fluid fat; and 2 per cent. of fats, which yield butyric, caproic, caprylic, capric, and other acids, according to the animal from whose milk the butter was made.

Fresh butter has the same amount of carbon per pound as suet.

Ten grains of butter, when burnt in the body, produce heat sufficient to raise 18·68 lbs. of water 1° F., which is equal to raising 14,421 lbs. one foot high. The effect of butter on the respiratory process was very small, as shown in Diagram No. 95 (frontispiece).

There are but few instances in which the flavour of butter is purposely changed, but the Bedouins in the *Hedjaz* boil herbs with it, with a view to improve it.

Ghee.

Ghee is a clarified butter, prepared chiefly from bison's milk, and used very largely by the natives of India.

The milk is first boiled, and on being allowed to cool, a little sour milk, called *Dhye*, is added, which causes coagulation. The mass is then churned, and hot water added, and in about an hour butter is produced. After a few days the butter becomes rancid, when it is clarified by being boiled with the addition of *Dhye* and salt, or betel-leaf, and is then kept in closed pots for use. This form of butter is less free from nitrogenous matter than that made in this country, and it has a peculiar flavour, which is distasteful to Europeans.

The Fat of Flesh.

The fat of animals is derived from two sources, viz., their structures and their secretions, and both are of the utmost value to mankind.

We have already pointed out the proportion in which fat is found in animals, and have shown that in a well-fed pig or over-fed sheep and ox, it may amount to one-third of the whole carcass-weight (page 43). This is not restricted to one part of their body only, for there is none which does not contain some fat, but it is chiefly found in masses within the loins, which are almost separated from the flesh, and in layers upon the outside of the flesh.

The former is commonly cut off, and sold separately from the joint under the name of suet, and is unusually solid, so that it is especially fitted for, and is largely and universally used in, culinary operations with farinaceous substances. This kind of fat has a flavour which is usually less marked than any other in the animal, and may be eaten by invalids when ordinary meat fat would be rejected, so that it is accounted one of the purest and most useful fats in food. The latter is usually sold with the joint, and is cooked with it, but when the quantity is excessive a portion is cut off, and is not used as food for man. Besides these, there is fat in the juices of flesh, much of which is extracted by cooking, and also in bones and other offal, as has been already described.

When it is found in connection with flesh, or in a separate form, it is always enclosed in cells (No. 45), and in order to extract it quickly, and with as low a degree of heat as possible, the mass should be cut into small portions; but when it is met with in the juices, and in secretions generally, it is unenclosed, and appears as globules of various sizes.

The value of fat in the animal economy is exceedingly great, both chemically and physically. Chemically, it supplies the heat-forming elements of food in their most compendious form, and is much more rapidly transformed than starch under the influence of exertion, when a very large elimination of carbonic acid takes place. We have elsewhere intimated the close relationship which exists between muscular exercise and the elimination of carbonic acid on the one hand, and the necessity for an increased supply of the hydro-carbons on the other, to meet the waste; but have deferred a fuller discussion to the work on Dietaries. There can, however, be no doubt that, whether we refer to the intimate molecular actions in nutrition, to the supply of an essential element in growth, or to the daily use of the body, it is essential that there should be a full supply of fat in some of its forms.

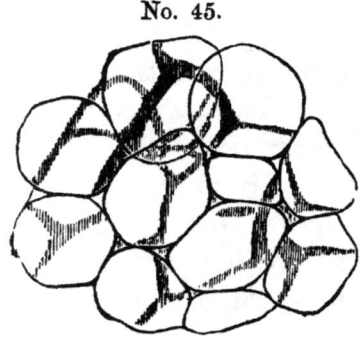

No. 45.

Fat-cells.

Physically, its action may be less important, but it is most desirable as an addition to bread and farinaceous foods generally. It supplies an agreeable flavour, without which they could not be readily eaten, and lubricates the passages through which the masticated food is the more readily conveyed. It is also very probable that it exerts an influence in the passage of refuse matter through the bowel; so that, with some excess of fat, the bowels will act more readily than where the dietary is deficient in that lubricating substance.

There is also another physical action, by which it

may act *indirectly* as a food, although it may not enter the body. Thus, when the skin is perspiring too actively, either for the wants of the body, or the degree in which loss of heat by that process can be borne, free inunction into the skin diminishes greatly the loss of heat, and thereby the necessity for food.

We have used this method with remarkable success in persons who had almost ceased to eat, and in whom it was with the utmost difficulty that the heat of the body and the circulation of the blood could be maintained. Moreover, without entering into a theological question, it may not be improper to point to this as a rational explanation of the use of anointing the sick, which has been practised, not as a religious but as a sanitary measure, by many Eastern nations, and of the habit of anointing the body among many savage nations.

The composition of suet is tolerably uniform, but there is a larger quantity of stearine, and less oleine in mutton than in beef suet. There are 4,760 grains of carbon in a pound; and if the free hydrogen be reckoned as carbon, the quantity will be increased to 6,720 grains per pound.

Lard.

This important culinary substance is derived from the loose fat of the pig, and is a very pure fat. It was used in Saxon and early Norman times, and as late as the fourteenth century, when butter was comparatively unknown, it was almost the only fat for cooking. Thus *Capon in grease* was a well-known dish. It is not entirely without flavour, but is nearly so, and this, with absence of colour, renders it particularly fitted for the preparation of any kind of pastry, or as a medium in which substances may be fried. It is rarely eaten with bread, since it is soon absorbed and disappears from sight, and has not sufficient flavour to sharpen the appetite.

The rendering of it from the leaf fat requires care lest it should be burnt, or obtain a flavour from the cooked membranes in which it is enclosed, and it is not necessary to prolong the process so far as to drive off all the water with which it is naturally associated. It is also desirable that the cooked membranes should not be so much pressed as to exclude all the liquid fat, and it is better that the fat should be allowed to drain off with no other help than the pressure of the spoon or ladle by the hand. When prepared in this way it is usual to mix a little salt with it, with a view to its preservation, but the quantity need not exceed half-an-ounce in the pound.

It is very frequently adulterated with fats of inferior value, as, for example, mutton fat, which is cut off the joints before they are sold, and is not one-half the value of good lard. When the fat used in adulteration is of greater consistence than lard it must be used sparingly, and, from whatever source the adulterating fat may be obtained, it must have little or no colour, (or be decolorised) and be nearly devoid of taste. Adulterated lard is, however, generally slightly coloured, and has not the flavour of the genuine fat. Moreover, it is not uncommon to mix starch with adulterated lard, in order to hide any colour that may have been imported into it. It is also said to be customary in Canada for dealers to add two to five per cent. of milk of lime, by which the colour is improved, and the lard made to absorb as much as 25 per cent. of water.

There are no means of discovering adulteration by other pure fats, but starch is readily ascertained by the aid of the microscope (No. 46), p. 147. An excessive quantity of salt may be proved by washing and subsequent weighing; the quantity of water may be ascertained by loss of weight on evaporation. Lard, whether pure or adulterated, is very commonly used to adulterate butter, and particularly when the colour of butter is so high as to bear lowering without exciting suspicion.

The composition of lard is that of a pure fat, and there are 5,320 grains of carbon; or if the free hydrogen be reckoned as carbon 8,237, in one pound.

DRIPPING.

Dripping, which is eaten as a fat, or used in cooking, is almost invariably obtained from the process of roasting flesh, and, as it should have very little water in its composition, it is one of the most nutritious kinds of fat.

Prepared in this manner it has also the flavour of meat, by which it is more agreeable than when made from the fat of animals alone; and that derived from camels' flesh is especially accounted a luxury by the Arabs. Its flavour will, however, be affected by the degree to which the flesh is roasted, and its subsequent use for culinary purposes.

Except in the care and cleanliness with which it is prepared, it differs little from the finest kinds of tallow which are still prepared in Russia, and those made in Australia and South America, by the boiling-down of the whole carcass of the ox or sheep for the fat alone. Yet, notwithstanding this relationship, it is one of the fats most highly valued by the poor for its flavour, nutritive quality, and cheapness. Wealthy families benevolently give it to their poorer neighbours, or sell it to them at a merely nominal price; but the more general practice of allowing cooks to have it as a perquisite leads to many and serious evils. Thus the cook is tempted to purchase more fat or fatter meat than is desirable, and to put the excess direct into her store; also to overcook the meat, so as to increase the dripping; to purloin butter, lard, and any other good fat to which she has no claim; to prevent its proper use in culinary operations and the preparation of pastry, and to substitute other fats for it; to bring her in connection, when selling it, with a class of people who will steal, and encourage her to rob her master, and, generally, to make a thief of an honest woman. There is little doubt that this absurd custom has led to great waste and extravagance in our kitchens, and to the dishonesty of our cooks; and every employer, of whatever rank, should prohibit it, and punish any infraction of his regulations.

When it is skimmed from broth, or similar fluids, it is less valuable, since it contains a larger portion of water, and has acquired foreign flavours, so that it is seldom used for culinary purposes.

Dripping properly made contains 5,320 grains of carbon in the pound, whilst butter and suet have only 4,760 grains. When the free hydrogen is reckoned as carbon, the quantity of carbon becomes 7,511 grains per pound.

Oils.

There are various animal oils which are used in the adulteration of food, but extremely few which are eaten in temperate and hot climates. Whale oil is eaten largely by the inhabitants of exceedingly cold climates; but even there it is consumed rather as blubber than after its extraction from the cells in which it is enclosed, and mixed with more solid fat. Seal oil, in the same manner, may be eaten as food; and cod-liver oil, which is now so largely used medicinally, probably acts chiefly as a food.

The effect of fats and oils upon the respiratory process is shown in Diagram No. 95 (frontispiece).

SECTION II.—VEGETABLE FOODS.

a. Nitrogenous.

CHAPTER XIX.

ANALOGY OF ANIMAL AND VEGETABLE FOODS, AND GENERAL CONSIDERATIONS ON SEEDS.

VEGETABLE FOODS.

It has already been shown that the same nutritive elements exist in both vegetable and animal foods, and that, within certain limits, the two classes of food are interchangeable. Also, that both are divisible into two sub-classes, viz.:—nitrogenous and non-nitrogenous, or flesh-formers and heat-givers, the former being the larger. The nitrogenous consists of all seeds and vegetable tissues; whilst starch and sugar are in vegetables that which fat is in animals, viz., the especial representatives of the non-nitrogenous. Hence, flesh in animal foods is represented by seeds in vegetables, and fat by starch and sugar; and, to continue the analogy, it may be added that seeds when digested will produce flesh, and starch when transformed in the body may produce fat.

Moreover, every other element, whether mineral or organic, which is required for nutrition is found in the vegetable kingdom; as for example—salts of potash, soda, lime, magnesia, iron, and manganese; substances analogous to fibrin, albumen, and gelatin; gum, pectin, and sugar; phosphoric, acetic, sulphuric, hydrochloric, and fluoric acids; besides many acids peculiar to vegetables.

It may then be asked, if there are the same elements in both, and if either will suffice to maintain life for a considerable time, what is the practical difference between them, for the purposes of nutrition?

It is probable that this depends upon the habits of men, for whilst the majority of people require and can digest a moderate quantity of both, there are some who from early use live chiefly upon one kind and eat many pounds of flesh or vegetables at a meal. It is, however, a general rule that, whilst flesh presents the elements of nutrition in a form the most compendious and easy of digestion, seeds are composed of substances which must not only be digested but thoroughly transformed before they can be used for the reparation of the body. The cooking of flesh is doubtless desirable, although it is not necessary to its digestion; but the cooking of seeds is still more so, in order to enable the stomach to dissolve and perfectly transform them. A good test is the amount of matter which leaves the bowel after the consumption of vegetable and animal foods, and if quantities supplying an equal amount of nutriment be taken, the refuse from the former will be twice as much as from the latter. It is commonly assumed that the digestion of vegetable is easier than that of animal food, and that the process is more quickly performed, but the experiments of Dr. Beaumont have shown that mutton will be digested more quickly than bread, and an egg earlier than a potato.

To this must be added the fact, that a greater bulk of vegetable than of animal food is required to provide a given amount of nutriment, and hence those who live chiefly on the former must be large eaters; but if it were possible to live on either alone, the difference in this respect would not be great.

It will be inferred from the above statement, that although the vital actions may be sustained by both kinds, they are more slowly moved by vegetable than by animal foods; and this is true, whether we regard the respiration, pulsation, or heat-production. When, therefore, we compare them it may be stated generally, that vegetable food must be eaten in larger volume, and be better cooked, than animal food, and that it requires a longer period for, and greater power of digestion, whilst it excites the vital processes more slowly, and in a lower degree.

We will now proceed to offer some general remarks on Seeds, as the chief representatives of the large and important class of nitrogenous vegetable foods.

These structures have much in common in reference to their organisation and nutritive elements. There are two essential parts in all seeds, apart from the seed-vessel, viz., the external rind or skin and the contained kernel. The rind differs much in thickness and appearance, as in the cocoa-nut and rice, but in all edible seeds it is comparatively thin, and when properly prepared is sometimes itself edible. It consists of woody fibre or lignine, through which the circulation is carried on in the seed, and by which form and strength are given to the skin, as also of a small proportion of starch, besides elements peculiar to each kind of seed which determine its flavour and properties. In many, as wheat, it is coated with a shining layer of silica or flint, which protects the underlying structures from the action of the atmosphere; and, unless its continuity be broken, the gastric juices cannot act upon the starch. Hence, grain when kept dry will remain sound for thousands of years, as has been proved in the Egyptian tombs, whilst oats which are unbruised by the teeth, or other instrument, will be found almost unchanged in the dung of the horse.

The interior, or kernel of the seed, consists almost exclusively of cells, which are filled with starch, and albuminous, glutinous, or mucilaginous matter. The former may be obtained from the finely-ground seed by continued washing, as in the preparation of commercial starch, whilst the latter are the substances remaining after this process has been completed.

The starchy material was known many centuries ago, for a substance, then called *Amidon*,[1] was employed to thicken broths. Cotgrave says that it was made from ' fine wheat-flour steeped in water, strained and let stand to settle, then drained and dried in the sun.' It is a loose material, which on decomposition affords a sour odour, whilst gluten is so adhesive and tenacious, as to have been used even by the ancients for taking birds, and called bird-lime, and on decomposition emits highly offensive odours. The grains or granules of starch differ much in form and size, and are readily distinguished under the microscope, although, as they all consist of the same material, they may not be distinguishable from each other by any chemical process.

The following are illustrations of the best known forms as observed under a microscope of moderate power, such as is usually in the hands of non-scientific persons, and are a very interesting and useful subject of enquiry. They may be seen by both transmitted and reflected light, but with the former they should be examined in a drop of water; and the power should be varied with the size of the granules.

[1] Still called *amidon* by the French, and the starch from roots *fécule*.

No. 46.—Starch Cells.
A.

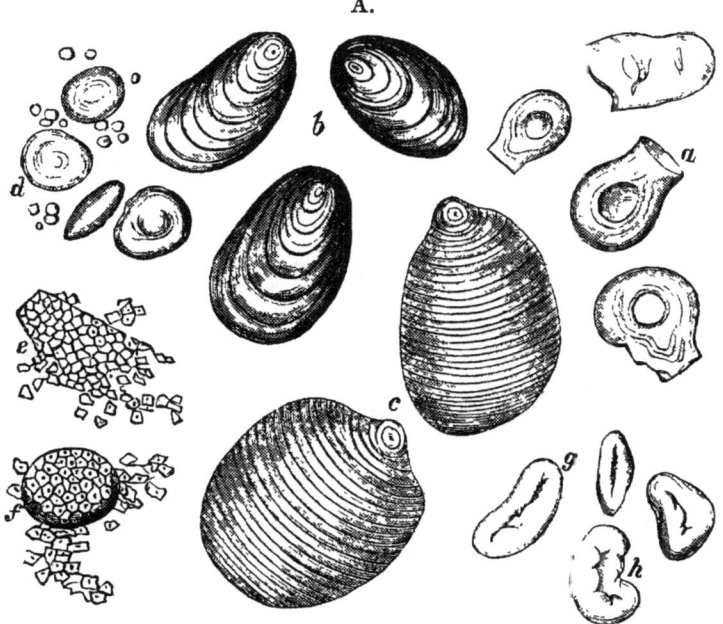

The best known forms of Starch Cells.—*a.* Sago starch, highly magnified. *b.* Potato starch. *c. Tous les mois* starch. *d.* Wheaten starch. *e.* Rice starch. *f.* Oat starch. *g.* Pea starch. *h.* Bean starch. (300 diameters.)

When the cells are heated, as in the process of boiling, they absorb water, and swell until the cell-wall bursts, and the contents escape. This is the true effect of cooking, as represented in the drawing, on the following page, and enables the gastric juices to act upon it immediately it is admitted into the stomach. The same result is no doubt effected when the seed is placed in its raw state in the stomach, for it is then

No. 47.—Starch Cells B.

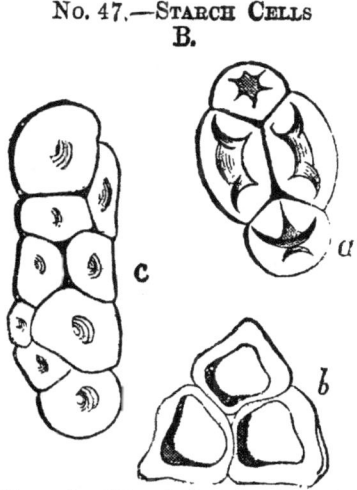

Shows in different degrees the central cavity and folding of the cell-wall. *a.* Colchicum autumnale. *b.* Iris. *c.* Arum maculatum.

exposed to heat and moisture; but the digestion and

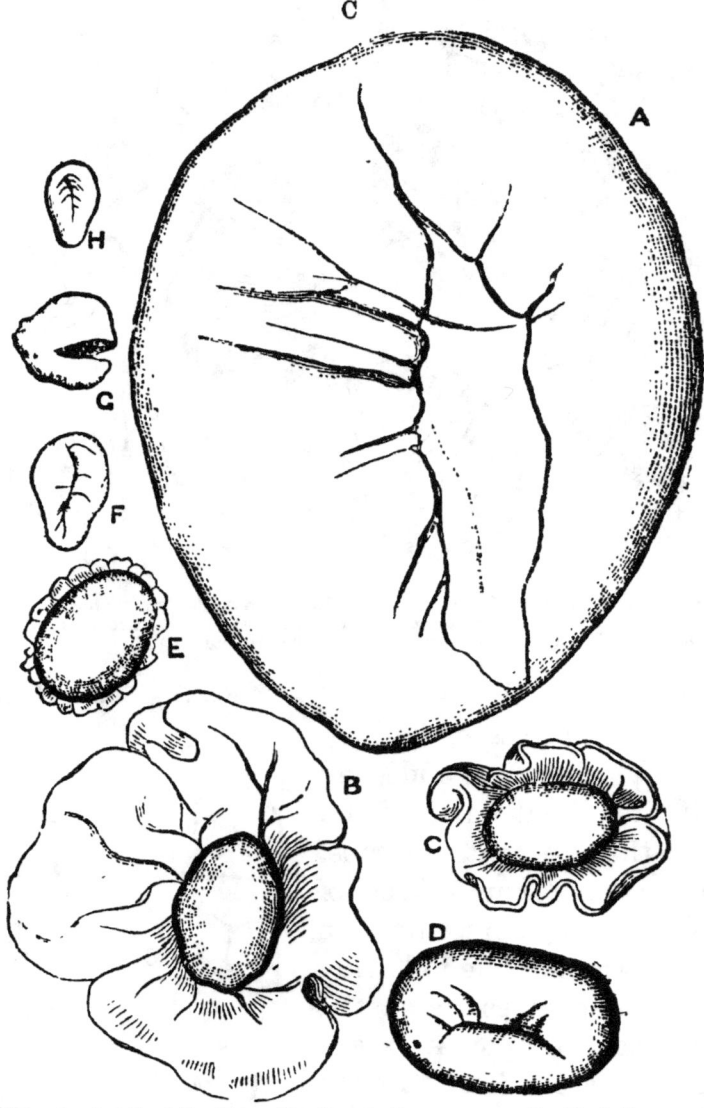

No. 48.—Cooked Starch Cells.

The starch-cell of the Horse-Chestnut in its progress under the influence of moisture and heat, from its dry form (H), to its cooked state (A). The progress is marked by the inverted order of the letters H to A, and the beautiful fringes in E. C and B snow the unfolding of the cell-wall until it is perfectly expanded and ruptured in A. The size is proportionate at each step of the process.

transformation of it are so greatly retarded, that it may

not yield the required nutriment when the body needs it, or be removed from the stomach before another supply is required.

Hence the two parts of the process are— (1) the minute division of the seed, so that all the parts or cells may be quickly brought under the action of the heated water; and (2) the continuance of the cooking process until all the cells shall have been distended and ruptured. When the whole seed is exposed to the action of heat, a longer period will be required for this process, and the result will be more uncertain.

In order that the seed may be ground into a fine powder, it should be previously dried. Grinding by millstones is a far more perfect process than by the teeth, for not only may the former be used for a prolonged period at a time, but the action of the teeth is to cut and press as well as to grind, and is more analogous to the action of the coffee-mill than of a pair of revolving stones. When, therefore, seeds are eaten raw and ground by the teeth, less nutriment is derived from them, the stomach is more taxed to digest them, the process of digestion is slower, the refuse emitted from the bowel greater, and the production of indigestion much more likely to occur than when they have been well cooked and ground.

This is not the case, or only in a slight degree, with flesh, for the teeth can sufficiently tear it asunder, the juices of the stomach can more readily act upon a mass, and when uncooked, it is still comparatively easy of digestion.

But whilst these processes are necessary in reference to all seeds, they are more easily performed on some than on others, and on the kernel than on the rind or skin. Hence, however finely peas and rice may be ground, the former will require a longer period for cooking than the latter, and so tough is the skin of the pea that if it be cooked whole, it may be boiled for many hours and still retain its form. The perfect cooking of the kernel is, however, always practicable, and of known duration, but that of the rind is not possible by ordinary methods, unless it be first ground into a fine powder.

This is a very important consideration in reference to the more expensive grains, as wheat, since it is desirable that every portion of it should be used as food, and the same would hold good of all edible seeds at periods of severe privation.

But however perfectly the skin or rind may be prepared for cooking, and however perfectly it may be cooked, the siliceous or shining layer before mentioned is insoluble in the gastric juices, and is therefore indi-

No. 49.—Siliceous Cuticle.

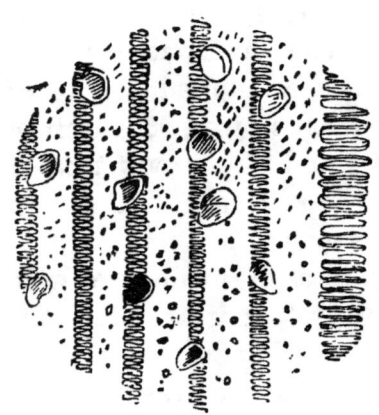 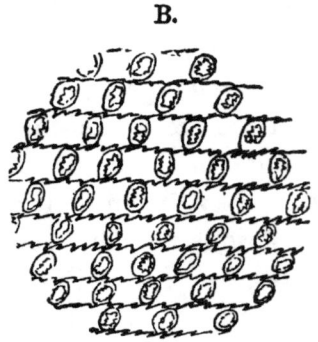

A. Siliceous cuticle of the Wheat (*Triticum vulgare*), showing cups for the insertion of hairs and a spiral vessel.

B. The same of the Meadow Grass (*Festuca pratensis*).

gestible. When the bran of wheat is eaten by an animal, whether man or horse, it will be found in the excrement, having this layer still perfect, but the more nutritious part which is covered by the silica will have been digested in proportion as it was reduced to a fine powder and allowed to remain sufficiently long in the stomach and bowel.

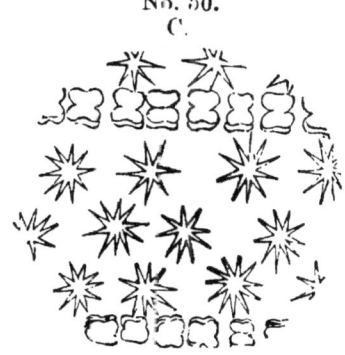

C. This coating is sometimes very beautiful, as in the star-like flinty hairs of the *Deutzia scabra* and the *Pharus ciliatus*.

This leads us to the question of the action of the skin or rind upon the bowel, and it is of common observation that in proportion as the piece of rind is large and indigestible, so is it a stimulant or irritant of the bowel. So far, therefore, however rich the rind may be in nutritive elements, it is more likely to prevent than to sustain nutrition, since it will lead to the quicker removal from the body not only of itself but of other and perhaps more nutritious matter. When, however, it is ground into a fine powder it does not produce this effect, and although the siliceous part may still be undigested, the lignine and starch which it covers may be partially digested and promote nutrition.

The chemical composition of lignine and starch is as follows; and whilst starch may be readily detected by the aid of the microscope, it may be even more quickly determined by the dark blue colour which it produces when iodine is added to a solution of it in water.

Lignine or cellulose, and starch, per cent. :—

No. 51.

O. 49·4   H. 6·2   C. 44·4

The proportion in which starch is found in some of the principal articles of vegetable food may be conveniently stated here :—

No. 52.

Arrowroot . . . . . 82·0 per cent.
Rice . . . . . . 79·1   ,,

Rye meal . . . . . 69·5 per cent.
Barley flour . . . . . 69·4 ,,
Wheaten flour . . . . 66·3 ,,
Indian corn meal . . . . 64·7 ,,
Oatmeal . . . . . 58·4 ,,
Peas . . . . . . 55·4 ,,
Wheaten bread . . . . 47·4 ,,
Potatoes . . . . . 18·8 ,,
Parsnips . . . . . 9·6 ,,
Carrots . . . . . . 8·4 ,,
Turnips . . . . . . 5·1 ,,

CHAPTER XX.

PEAS, BEANS, LENTILS, &c.

THE most highly nitrogenous seeds are peas (*Pisum sativum* and *arvense*), beans (*Faba vulgaris*), lentils (*Ervum Lens*), and analogous seeds from leguminous plants, and although the members of this class differ in appearance and flavour, they have very similar chemical and nutritive qualities. The following is the ultimate chemical composition, per cent., of peas, besides oxygen and hydrogen :—

No. 53.

C. 39　　　　N. 3·65

The proximate elements in 100 parts of peas, millet and lentils are :—

No. 54.

Peas . . { Water 15　Nitrogenous 23　Starch 55·4　Sugar 2
　　　　　 { Fat　 2·1　Salts　　　 2·5

Millet . . { Water 13　Nitrogenous 9　Starch 74
　　　　　 { Fat　 2·6　Salts　　　 2·3

Lentils. .　　　　　　 Nitrogenous 25

There is no temperate or hot country which does not produce some member of this family, or where the inhabitants do not eat them, but in none are they the sole staple article of vegetable food. This arises from three well known causes, *viz.*, their cost, a strong flavour which does not please all persons at any time nor any person constantly, and the comparative difficulty with which they are digested. We are familiar in this country with peas and beans, and in the fourteenth century our forefathers used vetches or lentils, then called *Chych*, whilst on the continent of Europe and in Egypt and Asia various kinds of lentils are still eaten.

Several species of beans are eaten daily and in great quantity by the inhabitants of Mexico and Central America, and are called *frijoles*; they are eaten when cooked with pork and fat, or with dry beef called *tasajo*, as in Nicaragua. Haricot beans are also consumed very largely in France, Italy, and the South of Europe, as well as in Egypt and in India, where the various members of this class of foods are known by the term *Dhal*.

As prepared in Mexico they are a very savoury as well as strong food. They are first boiled slowly and for a long time in soft water, or water softened with a little alkali, until they become perfectly tender, and then are simmered with a little lard and crushed capsicum pods and flavoured with onions or garlic.

The chick pea (*Cicer arietinum*) is called *Gram* in India, and is very largely cultivated there as well as in the south of Europe. It is of too delicate a nature to be grown in this climate with advantage, and is not so prolific as our common pea. The seeds are parched for use on journeys, and made into cakes and puddings, and in India they are made into sweetmeats with sesame oil and sugar. The pods are hairy, and exude oxalic acid, which irritates the hands and is injurious even to the boots of persons walking through a gram field, but it is collected by the natives as a medicine.

The skin or rind of the members of this class generally, but of the larger kinds in particular, should be rejected, and the kernel alone eaten, and both should be thoroughly cooked. This was well understood in early times, for in a recipe of the fourteenth century we find the direction to dry the beans in a kiln or oven and to shell them well, and winnow the shells (or skins) away, before using the beans to make broth, or to be eaten with bacon.

But it was not always so, as is seen in the following recipe for frying beans:—

'BENES YFRYED.

'Take benes and seethe hē almost til they bersten, take and wryng out the wāt clene. do thereto oynons ysode and ymynced (minced) and garlec 'pw^t, fry hem ī oile, or ī grece and do therto powdō douce (pepper) and sūe it forth.'

The importance of this direction varies with the maturity and dryness of the seed. Green peas and dried peas are very different foods, and whilst the former may be eaten with the skin, and be easily cooked and digested, the latter must be soaked for a long time in order to soften the skin; and after the kernel has been well cooked, the skin should be rejected. The same remark applies to the bean, but by proper selection and cultivation the haricot bean has a skin which is thinner and more easily detached from the kernel, whilst the flavour of the kernel is more delicate than that of the common bean. Hence this kind in its mature state is far more frequently used as an article of food than the dried pea or bean. Moreover, the pods of certain leguminous plants may be eaten when young, as those of the haricot (*Phaseolus vulgaris*), and scarlet-runner, of which both the pod and the immature bean are cut up into thin slices and boiled.

It may well be believed that the increased cultivation and greater consumption of the haricot bean by the working classes of this country would cause an improvement in their dietary, and in some degree supply the deficiency of animal food. The cost is, however, too great, and the want of acquaintance with the food too marked, to justify the hope that its use will ever be greatly increased in our time. On the contrary, there can be no doubt that the use of this class, as represented by the dried pea, has greatly diminished within this century, notwithstanding the general belief in its nutritive value, so that it is now more frequently found in compulsory than in voluntary dietaries. This is owing to the special qualities of the pea, as above stated, greater refinement of taste and increase in the means of the working classes, as well as to the cheapness and more agreeable flavour of such grains as wheat; and, in my opinion, it is not desirable to return to former habits so long as the conditions remain unchanged.

This class of seeds is not grown in extremely cold countries, as in the Arctic and Antarctic zones, where, indeed, all vegetable productions, except the lichen, are rare, and although so well fitted to support nutrition elsewhere are there supplanted by animal foods.

It might almost be said that peas alone are nowhere made into bread, although they are frequently added to farinaceous foods for that purpose; but in the southeast of Scotland very thick cakes, called peas-bannocks, are still made of pea meal.

The following quantities of carbon and nitrogen are found in 1 lb. of peas:—viz., carbon 2,683 grains, and nitrogen 252 grains.

The time required to digest beans when boiled is $2\frac{1}{2}$ hours and upwards.

According to Frankland's experiments, 10 grains of pea meal, when thoroughly consumed in the body, yield heat sufficient to raise 9·57 lbs. of water 1° F., which is equal to lifting 7,487 lbs. one foot high.

Pea meal, mixed with a proportion of refined fat, is made into small cakes by Messrs. Craig, for the use of soldiers, sailors, travellers, and housewifes, in the preparation of pea-soup. It requires to be boiled for half-an-hour in water, meat-liquor, or bacon-liquor, and with the addition of meat or bacon, will be found a very luxuriant way of preparing that excellent and nutritious food. It has been adopted experimentally for the use of the Volunteers. Prepared in a peculiar manner, it was also largely eaten by the Prussians in their late war with France, and supplied with meat in the form of sausages. Perhaps a better form still is the pea soup prepared by MM. Kopf, which is inclosed in a metal envelope in order to protect it from damp.

CHAPTER XXI.

THE CEREALS, MAIZE, OR INDIAN CORN, MILLET, &c.

This large class of farinaceous seeds comprehends those which are most frequently used by man in different parts of the world, and while affording much nutriment, are agreeable to the palate, capable of being perfectly cooked, and may be produced in an unlimited quantity in all but the coldest climates. They are inferior to leguminous seeds in chemical constituents, but, being more agreeable, are more willingly eaten, and at the same time, are more readily cooked and digested. The greatest attention has, moreover, been given to their growth and preparation, and they are so prolific that a given extent of land will yield a larger quantity of this than of any other food fitted for man.

There are many well-known members of this class, and each is a staple food in different climates, but in no one climate can all be equally well produced. Thus wheat is essentially a product of temperate and is not grown in very hot countries, whilst maize is a chief food in many temperate and hot climates, and rice is essentially the food of hot climates only. Oats are the product of the northern portion of temperate climates, and the general use of them does not extend beyond that region. None of these productions are grown in extreme northern latitudes, and are, therefore, not a principal part of the food of the inhabitants of cold countries.

This class of seeds is also used very largely in the production of ardent spirits, the feeding of animals, and the preparation of starch for commercial purposes, and thus by so much is the portion used by man as food lessened and its cost increased.

Maize (*Zea Mays*) may be placed first on our list, whether we regard its nutritive value or the immense regions of the globe where it is produced and consumed by man. The production extends throughout North and South America, the continent of Europe, and a very large part of the continents of Asia and Africa, besides numerous islands of the Pacific Ocean; but it is not grown in England, and but little in the English colonies of Australia. Hence it is, no doubt, a product of sunny rather than of temperate climes, and cannot in the latter compete with wheat, whether in flavour or production. It yields, however, the largest crop of any of the class in the countries where it is usually grown, and is by far the cheapest food.

The composition of this grain is as follows, in 100 parts:—

No. 55.

Water 14   Nitrogenous 11·1   Starch 64·7   Sugar 0·4   Fat 8·1   Salts 1·7

The quantity of carbon and nitrogen in 1 lb. is 2,800

and 121·6 grains. The time required for the digestion of Indian corn-cake or bread is from three to three and a half hours.

When eaten green its flavour is delicate, and boiled in milk, or roasted and eaten with other foods, it is accounted a luxury; but the proportion thus eaten is infinitely small as compared with that which is allowed to mature, and to become dry for storing and grinding. The plant is the most handsome of the class; and, growing to the height of six to ten feet, with several cobs upon each stem, gives an appearance of abundance unsurpassed by any product.

The flavour of the mature and dried seed is rough and harsh, so that one must be trained to like it, and it is the most preferred by those who have eaten it from childhood. It may be eaten when cooked with water only, but it is much improved by the addition of milk, eggs, sugar, and other substances of a softer and more agreeable flavour.

The whole mature grain is used by man, after having been parched, by which it is rendered more friable and digestible, and a very convenient food to travellers in India and many Eastern countries. With this exception, it is always ground, the skin and kernel together, for the skin is thin, and when well ground is digested, with the exception of the siliceous coating. The meal thus produced is much coarser than that of wheaten flour, and as it is not readily cooked, it cannot be made into large loaves, lest the central part should not be perfectly baked, but is prepared in cakes and baked before the fire or in an oven. Hence it is cooked very soon before it is eaten, (unless it be dried in very thin layers, like a cake); and it is commonly prepared fresh every day, and often at every meal. The grinding of the grain is often roughly done in the Western

States of America and other newly-peopled countries, and the cooking is too hasty and insufficient to make the food very digestible. Cakes baked on a board before the fire, or on an iron plate in the oven, in the Western States of America, are called 'Johnny cakes;' and in Nicaragua, Mexico, and other places, 'tortilla.' They are eaten hot, and usually with milk, but not unfrequently with butter, treacle, or other savoury food; and although very acceptable, the inhabitants would gladly exchange them for cakes of wheaten flour, if the latter were within their reach. Sometimes two cakes or tortillas are made into a sandwich, with some intervening food, as meat, or a sauce of tomatoes and Chili pepper, and eaten by travellers.

The tortilla is a food which has been eaten by the Mexicans from remote antiquity, and is prepared now with the same kind of materials and in the same mode as in former ages. It is really a cake, made of ground maize, of the size and thickness of a pudding-plate, and resembles our oat-cake, but is better cooked and rendered more digestible. It is prepared by boiling the whole corn in water with a little soda or mixed alkali, until it is completely softened but not broken up, and afterwards it is kneaded or rolled upon a stone, so as to become thoroughly homogeneous in texture. The paste is divided into portions, rolled or clapped between the hands into a cake, and then turned upon a hot plate, to be browned on both sides, until it is ready for the table.

But besides the use of maize as bread, it is very commonly made into pudding, with or without other and more agreeable foods, and eaten either alone or with meat. It is now known in Ireland as Stirrabout, and in Italy as Polenta, whilst it is eaten with pork in America under the designation of Hominy.

Maize has recently supplanted the millet in the food of the Kaffirs. The corn, termed *umbila*, is placed in a covered earthenware cooking-pot, with a little water, over a wood fire until it is partially softened. It is then pounded and rolled upon a flat stone until it is made into a soft paste or porridge (*isicaba*), in which state it is eaten with a wooden spoon, and, as in Ireland, with sour milk. Being rich in nitrogen, it is also known as a strong food, and numerous preparations of its starch have been introduced into this and other countries under the name of corn-flour, for the preparation of blanc-mange and puddings, for which it is perhaps better fitted than the flour of wheat. It is thus not only a food of very high nutritive value, but is consumed in a great variety of forms, and whilst a necessary to hundreds of millions of people in a limited sense, is a luxury and a very agreeable food to selected classes of persons.

The large proportion of nitrogen which it contains renders it a more stimulating food than wheat, so that when it is the constant food of horses, even in the countries where both the food and the horse are indigenous or acclimatised, it is liable to produce disease.

The consumption of this grain in Great Britain is no doubt greater than it was thirty years ago, and an impetus was given to its importation during the Irish famine of 1847, but it has not taken root as the ordinary food of any class. As a food for man it is known in England almost exclusively in gaol and Irish poor law dietaries; and although it was welcomed by the Irish people when the potato failed, and they were exposed to starvation, it is not preferred to oatmeal, much less to wheat or potato, and its use is receding before the claims of the two latter foods. So long as wheat can be obtained here in sufficient quantity, maize will never be generally accepted as an ordinary food for man, but at the same time its use is rapidly increasing for horses and other of the lower animals.

It may be doubted whether Indian meal is ever greatly adulterated; but for some years past a preparation, known as 'Jonathan,' has been sometimes added, which is indigestible and useless, if not injurious. It consists of calcined and ground oat-chaff, as was admitted in an enquiry before the magistrates at Guisborough, in May 1872, to which the following extract from a report refers :—

'It was found difficult, however, to prove what the article "Jonathan" really was. A witness for the prosecution stated that it was not fit for the food of man or beast, and a witness for the defence stated that it was meal. The case was adjourned for a proper analysis to be made. Mr. Frederick Wm. Rock, analytical chemist, of London, attended on Tuesday, and stated, after careful chemical examination, he found the article consisted entirely of fibre, generally resembling oat husks which had been calcined and ground; he could not positively say whether there was any wood in it. He found nothing but a trace—one-half part in a thousand parts—of nutriment. This article was worse than wood when eaten by either animal or man, because the husks would irritate the intestines and bring on inflammation. He could positively state that the article was not meal, and that oatmeal or the kernel of any grain would be adulterated if mixed with it. For the defence, it was contended that the article was not sawdust, as stated in the summons, but the husks of oats, and not a foreign substance within the meaning of the Act of Parliament. It was also objected, that a conviction could not be sustained on the ground, that the allegation of the "article supposed to be sawdust" in the summons was not proved by the evidence.'

Millet (*Panicum miliaceum*) is a small grain scarcely larger than a large pin's head. The *Sorghums* or larger millets grow very extensively in the south and east, as

China, India, Egypt, and other parts of Africa; one is described by Lord Elgin on his visit to China as a stiff, reedy stem, some twelve or fourteen feet high, with a tuft at the top. The grain is found in this tuft. There are numerous varieties, and it is a chief article of food over very large districts, but usually it shares that position with maize or rice.

This and similar seeds, under a variety of names, as cheena, warree, are parched in India, and form a very convenient food for travellers. It is used in Tartary for the preparation of a fermented beverage, called *bouza*, by pouring hot water over a portion of fermented seed and imbibing the infusion through a reed. The flavour of the beverage is said to resemble sour sherry and water.

Millet consists, in 100 parts, of—

No. 56.

Water 12·1   Nitrogenous 9   Carbonaceous 74   Fat 2·6   Salts 2·3

Jowaree, a species of millet (*Sorghum vulgare*), is used as food in many parts of India.

CHAPTER XXII.

RICE.

It is perhaps worthy of observation that rice, both whole and ground, was used in English cookery in the fourteenth century, seeing that it was not grown in England, and must have been imported, at considerable cost, from some foreign country. The following is a recipe of that period:—

'Re Smolle.

'Take almānd blanched and draw hem up with wat (water) and alye (thicken) it with flo of rys and do pto (thereto) powdo of gyng (ginger) sug and salt and loke it be not standying (thin or diluted) messe it and sūe it forth.'

In other recipes, the rice was used with dates, white roses and spices, or with meats; or made into a compote, with apples, saffron, honey, almonds, pepper, and salt; or eaten with salmon, Cyprus wine, and condiments. A preparation called *blank mang* was described in the following manner. The rice which had been soaked in water for a night, and on the morrow washed clean, was put upon a strong fire until it burst, but not too much. Then brawn of capon or hen was taken and drawn small. Milk of almonds was then mixed with the rice and boiled, after which the brawn was added to it, and thickened with it. When it was stiff enough, sugar and almonds were added, and it was fried in lard, and served. This is not the *blanc-mange* of our day, but rather a curry with the curry powder omitted.

Rice (*Oryza sativa*), upon which hundreds of millions of people chiefly subsist, is by no means equal to wheat in its nutritive properties, since it consists almost exclusively of starch, and is relatively deficient in nitrogenous elements. Hence rice-starch is a familiar article of commerce, and rice is nowhere regarded as a strong food.

There are two principal divisions of this grain, viz., upland and sea-level rice, but there are an almost infinite number of varieties in the different countries where it grows. It varies very much in size, colour, and general appearance, and in its properties when cooked, so that in one kind the grains remain quite distinct, whilst in others they are broken up, and furnish a mucilaginous material, which makes a very agreeable food. As a specimen of the former, we may cite that which is grown in India, and used in the preparation of curry, and of the latter the South Carolina rice, so generally used in puddings. The latter is the finest quality of rice known in commerce, due chiefly, if not entirely, to care in cultivation and husking.

Rice is cut with the sickle, made into shocks, stacked, threshed, and winnowed, like wheat, when it is called *paddy* in India, and *rough rice* in America. It is then sent to the cleaning or husking-mill, by which the outer yellow siliceous coating is removed, and the inner white grain separated. The former is of no use as food, but is of value as a material in which to pack fragile goods, very much as oat-chaff is used in this country. After the grain has been received from the mill it is winnowed, to drive off a part of the husk, and sifted into five parts, viz., the chaff, chaff and rice-flour, broken fragments of rice, middlings and smaller grains, and prime. The last is the best rice to be obtained from the crop, so far as its market value may be a test, but it is not sweeter, or, in a nutritive sense, better, than the broken rice, when the latter is free from grit and chaff.

The broken rice, being well dried, may be ground into flour of different degrees of fineness; and when white rice is selected for this purpose, it is not unfrequently used to adulterate fine wheaten flour, in order to increase the whiteness of the whole, and, the cost of broken rice flour not being so great as that of fine wheat flour, the operation is a profitable one.

New rice is regarded as inferior in quality to old, inasmuch as it is much less digestible, and likely to produce indigestion, diarrhœa and rheumatism. It should not be eaten for at least six months after it has been gathered, and some Indian authorities interdict its use for three years.

Whole rice cooked by boiling in water until it is softened throughout is known in almost all parts of the world, but particularly in India, where it is called *bhát*; but it is more usual in this country to bake it with milk, eggs, and other foods, as rice pudding.

A food, called *Julpaun*, is made with parched rice and other grain in India; and parched rice alone is very commonly eaten by travellers and labourers, who are unable to cook food when it is required.

Ground rice is not necessarily of the best quality, but its value is determined by its colour and thickening properties. It is ground somewhat roughly, so that the grains are perceptible and impart a rough and dry taste to the food. When mixed with wheaten flour and made into bread it gives a dry taste, whilst it may increase the whiteness of the food.

Rice cakes are prepared as small loaves or biscuits, and are easily masticated and digested, and being usually mixed with sugar and other agreeable substances, are luxurious foods.

Rice flour is often adulterated with the flour of other grains, as maize and wheat, and although the nutritive value may be increased thereby, it is desirable to be able readily to detect the adulteration. So small an admixture as two per cent. may be determined by adding a saturated solution of picric acid to equal parts of the flour and water, when a precipitate occurs in about an hour. There is no precipitate when rice flour only is present.

The chemical composition of an average sample of rice is as follows, per cent., besides hydrogen and oxygen:—

No. 57.

C. 39   N. 1

The proximate elements in 100 parts are—

No. 58.

Water 13   Nitrogenous 6·3   Starch 79·1   Sugar 0·4   Fat 0·7   Salts 0·5

According to Frankland's experiments 10 grains of ground rice when thoroughly burnt in the body produce heat sufficient to raise 9·66 lbs. of water 1° F., which is equal to lifting 7,454 lbs. 1 foot high. The effect in my experiments of eating 4 ozs. well cooked, was to give a maximum increase of carbonic acid in expiration of 1·9, 1·67, and 1·54 grain per minute in different persons, and to increase the quantity of air inspired by 54 and 96 cubic inches. The rate of pulsation was increased, but not that of respiration, and the effect was very enduring (No. 95). The addition of 1 ounce of fresh butter did not add to the respiratory effect of the rice, but the rate of pulsation was increased 9 per minute.

The time required for the digestion of boiled rice is somewhat over an hour, so that it is a very digestible food.

CHAPTER XXIII.

OATS.

THIS grain (*Avena sativa* or *orientalis*) is grown in comparatively low temperatures, whether in reference to latitude or elevation, as in the Peak of Derbyshire, the northern districts of England and Scotland, and generally through the north of Europe, where wheat cannot be produced as a remunerative crop. The grain is larger and its nutritive qualities greater in the climate of Scotland than in England, and in elevated rather than in lowland districts generally.

The husk is particularly hard, so that it is not only somewhat difficult to break, but unless it be broken the gastric juice cannot act upon the kernel, and it will pass through the stomach and bowels undigested. It is also furnished with long sharp spikes, which are apt to become fixed in the folds of the intestines and to accumulate in large masses in the bowels of the horse. Hence it is desirable for proper nutrition that it should be entirely removed or be so divided that it may not prevent the action of the gastric juice and the solution of the kernel. This is well known to horse-keepers, and every prudent man bruises, if not grinds, the oats before feeding the horse.

The proportion of this part of the grain to the kernel is one of the measures of the value of the corn, since inferior grain has a small kernel and much husk. It is also said to be very nutritious, because it is richer in nitrogen than the kernel; but that is a fallacy, since the husk can neither be thoroughly masticated nor digested, and hence cannot yield up all its nitrogen. This is important when we consider the nutritive value of the ground grain, for whatever proportion of the husk may remain in the meal the nutritive quality of the food is scarcely, if at all, increased. Notwithstanding the greater proportion of nitrogen, as there is no known method by which the husk may be ground into a very fine powder, it should be removed from the meal in the dietaries of men, for not only does its weight not represent nutriment, but it acts as an irritant to the bowel, and may thereby prevent nutrition. It is, however, very possible to remove the husk as a whole and leave the kernel alone, as anyone may prove for himself.

The husks are used both in Scotland and Wales for the preparation of a kind of porridge, by being steeped in water for one or two days until they begin to ferment, and the mass skimmed and boiled. This is called Sowans in Scotland, and Succan or Llymru in Wales, and when cold it assumes a gelatinous or *blanc-mange* appearance.

The kernel is really the digestible and nutritive part of the oat, and is an exceedingly valuable article of food.

The flavour although sweet, is rough, and to be thoroughly approved must be eaten in early life, but notwithstanding the alleged preference for this food, a Scotchman so trained is very apt to exchange oatmeal for white wheaten flour on removing to England, and his children brought up in England eventually discard it as a daily article of diet.

Oatmeal is known as a strong food and one that requires much cooking in order to break its starch cells, but when it is well cooked it thickens milk or water more than the same weight of wheaten flour. It also yields a jelly or blanc-mange of a firmer quality than that derived from wheaten flour, and is doubtless the stronger and better food.

When it is ground in the ordinary way a proportion of the husk is left in the meal, but less in meal made by millstones than from crushing corn mills. The meal is ground in two forms, namely, in somewhat large grains as in the Scotch oatmeal, and a fine powder as in Derbyshire oatmeal, but either may be obtained from the same grain. The Scotch always prefer the rough grain, and boil it for a long time; by which they obtain a thicker and sweeter porridge than the English, who use a finer meal and boil it for a shorter time. The longer it is boiled the more digestible is the food produced.

It is worthy of remark that the cost of this food has increased of late years, while that of other farinaceous foods has decreased, not because there is a greater demand for it, but a less supply. It has, in fact, changed from a necessary food of a coarse character for the use of the poor to a luxury eaten in small quantities by all classes, and produced on so small a scale that it is not only not a cheap food for the poor man, but being dearer than wheaten flour is in many places too costly to be obtained by him.

Groats are the whole kernel of the oat when freed from the husk, and in like manner they are regarded as a luxury and not a necessary food, since the expense of preparation and the selection of the best and largest grain renders them more expensive than their nutritive value warrants. They are not eaten in the form of bread or cake, but are boiled in water or milk in the preparation of gruel; and if the rough-ground meal requires much boiling the whole grain demands still more. When thoroughly cooked with milk, they make a very nutritious pudding, but as the flavour is far less delicate than that of rice they are rarely used for that purpose.

Ground oatmeal is cooked in two principal forms, namely, as porridge and cakes. The word 'porridge' in Scotland and the north of England means oatmeal boiled well in water, in which state it is known in England as hasty pudding, but it is more usual in England to boil a smaller portion of it with milk or milk and water in the preparation of milk porridge. The former kind is eaten as a thick pudding with cold milk, into which it is thrown, or it is sweetened with treacle, or sugar and butter, as in eating hominy.

When oatmeal is prepared by simply stirring it in boiling water it is called brose, and is still used both in the highlands and lowlands of Scotland, but particularly in the bothies. It is not unusual to find this the sole food of the bothy men, with the addition of milk, and being prepared so easily is a very convenient if not an easily digested food.

When oatmeal is steeped in water from twenty-four to thirty-six hours until it begins to ferment, and is then skimmed and boiled to the consistence of gruel, it is called budram or mwdran or brwchan in Wales. Another Welsh preparation of oatmeal is termed bargout.

Oat-cakes are made of two degrees of thickness—one very thin, as in the Passover cake of the Jews, and the other about a quarter of an inch thick. It is not fermented, but soda or carbonate of ammonia are sometimes added to levigate it, and it must be well dried to preserve it for subsequent use. The Passover cake, being very thin, may be thoroughly dried, and will resist the action of the atmosphere for an indefinite time, or so long as it remains dry; but the ordinary oat-cake, being thicker, is rarely dried throughout, and is very apt to become so sour that only those who are accustomed to its use can eat it.

No attempt has been made to bake it in large loaves, like wheaten bread, and in this it again resembles Indian corn, the explanation being the great difficulty of thoroughly expanding and rupturing the starch cells, and therefore of cooking the meal in that form. If it were desirable to make the attempt, it would be better to first heat it to a temperature of 200°, taking care to stir it frequently, and not to allow it to become too dry until most of the cells have been ruptured. This process would, if carefully carried out, cause the meal to be better cooked in water, and would make it more digestible, but it is useless to recommend it wherever fine wheaten flour can be obtained. The ultimate analysis of oatmeal shows the following constituents, besides oxygen and hydrogen, per cent. :—

No. 59.

C. 40   N. 2

There are 140 grains of nitrogen and 2,768 grains of carbon per pound.

The proximate elements in 100 parts of oatmeal are :—

OATS. 171

No. 60.

Water 15   Nitrogenous 12·6   Starch 58·4   Sugar 5·4   Fat 5·6   Salts 3·0

The time required for the digestion of oatmeal is somewhat greater than that of wheaten flour when cooked in the same manner.

Ten grains of oatmeal, when thoroughly burnt in the body, produce sufficient heat to raise 10·1 lbs. of water 1° F., which is equal to lifting 7,800 lbs. one foot high.

In my experiments the effect upon the vital functions of eating oatmeal was considerable, and very similar to that following the use of rice and wheaten bread. Four ounces of good Scotch oatmeal, when well cooked as porridge with water, gave maxima of increase of 1·63 grain and 1·32 grain of carbonic acid per minute in different persons. The volume of air inspired per minute was increased by 55 cubic inches, and the frequency of respiration was lessened. The whole effect was sustained and very enduring. (No. 95, frontispiece.)

The chaff of oats is used for the adulteration of Indian meal and barley meal, as described at page 161, under the term 'Jonathan.'

CHAPTER XXIV.

WHEATEN FLOUR AND BREAD.

WE now proceed to consider the most important vegetable production of temperate climates (*Triticum vulgare, turgidum* and *durum*)—that upon which the life of man in these regions mainly depends. Its importance rests upon several properties, by which it is an acceptable and good food for all ages and classes of the people. It is produced abundantly, and cheaply; is easily ground and refined, is readily and thoroughly cooked, has a mild flavour which is universally agreeable, and contains nearly all the essential elements of nutrition. It is preferable to any of the other great vegetable products on which men chiefly live, since it is a far more agreeable food than maize, and a more nutritious food than rice. It is probable that the health and mental and bodily vigour of the inhabitants of temperate climes are more attributable to this food than to any other single cause.

Wheat is of two principal kinds, known as white and red wheat; but there are numerous varieties of the plant which do not affect the colour of the grain. The red is the stronger food, and the grain is usually smaller and harder; whilst the white is a large grain, and particularly adapted to the production of fine white flour, and to mix with red wheat for the same purpose. The red variety is the most widely grown, and in nutriment is to be preferred.

The quality varies not only with the selection of seed and cultivation, but with season and climate, so that a hot summer and a sunny clime produce grain with the least proportion of water and the greatest of nitrogen. Hence wheat from Southern Europe, the shores of the Black Sea, and the Steppes of Asia and the Caucasus, is preferred, as is also that of any temperate clime in which the heat of the sun is great during the summer months, as in the interior of America and Russia. It does not, however, flourish under a tropical sun, or in a high northern latitude. It is grown in India chiefly in the upper provinces and on lands at a considerable elevation. The effect of season is practically as great as that of climate, and the product of a hot season is harder and more nitrogenous than that of a wet or cold season.

Hence the art of the miller consists not less in properly mixing the kinds of grain to produce the best flour, than in well grinding and preparing it for food, and offers scope for intelligence and knowledge.

By the same art, wheat flour is adulterated with rice flour, potato starch, plaster of Paris, pea flour, alum, sulphate of copper, and other materials which cost less than flour, or add to its weight at a cheaper rate.

Rice, potato starch, and pea flour are readily ascertained under the microscope by the form of the granules (No. 46); plaster of Paris, by being insoluble when the flour has been washed in water and separated for analysis; alum, by dipping a slice of bread into a watery solution of logwood, when a claret colour will be produced if alum be present; and sulphate of copper, by the colour of chocolate brown, which is produced when prussiate of potash is added to a solution of the bread in water.

Perhaps a more convenient form of applying the logwood test for alum is to macerate $4\frac{1}{2}$ oz. of logwood chips in 8 oz. of spirit for twenty-four hours, and filter. A few drops, added to moistened bread or bread-crumb or flour in a little water, will show the dark red colour, if alum be present. Such a preparation can be made and kept ready for use in every house.

Rice flour, possessing much less nitrogen than wheaten flour, lessens the nutritive value of the mixture, and is used only when it can be obtained at a cost less than that of wheat, or when it is desired to prepare flour and bread of extreme whiteness, as French flour, or of great dryness and friability, as in rice-cakes.

Pea flour is added to inferior wheat flour, to give strength, by its greater quantity of nitrogen, and is not an injurious addition. Plaster of Paris increases the weight of the flour at a cheap rate, and although it is not a poisonous substance, it is not useful to the body, and is liable to interfere with the proper action of the bowels, so that its use is very reprehensible. Alum is added to give strength to the flour, not in the sense already mentioned by supplying nitrogen as an important food, but to enable the flour to absorb a larger quantity of water, and to produce a greater weight of bread from a given weight. It is extremely likely to produce indigestion; but if employed in very small proportions, it might be useful when the flour is of inferior quality, as the result of a cold and wet season, or of sprouting, and in that proportion might not be injurious to health. The usual extent of adulteration varies from ten to thirty or forty grains in the 4 lb. loaf. A mixture of potato starch and boiled rice is added to enable the flour to take up more water, (so that five additional loaves may be sometimes made from a sack of flour,) and the bread remains moist for a longer time. Sulphate of copper is said to be used very frequently to give increased whiteness to the bread.

The bran of wheat is of value in the nourishment of both man and the inferior animals, and its real merits are better appreciated now than at any previous period. It cannot be detached entire, but in the process of grinding it is broken up into scales of various sizes and qualities. The inner part is called cerealin, and acts like diastase in the conversion of starchy food into sugar, and is therefore an aid to digestion. The several layers become thinner and whiter as they approach the kernel, and when broken up in the process of grinding receive different names, and obtain different prices in the market. These are as follows, besides the fine and second flour:—

1. Tails or tippings, in the proportion of about 2 per cent.
2. Sharps, in the proportion of about 3 per cent., selling at 2s. a bushel.
3. Fine pollards, in the proportion of about 3 per cent., selling at 1s. a bushel.

4. Coarse pollards, in the proportion of about 4 to 6 per cent., selling at 10*d.* a bushel.
5. Bran, in the proportion of about 5 to 10 per cent., selling at 9*d.* a bushel.

The market price is, in a rough way, a measure of the nutritive value of these layers, for their properties have been well tested in feeding the inferior animals, and it increases as we proceed from without inwards.

The outside layer, or coarse bran, is the least nutritious, and as the exterior is covered with a layer of silica, it is so far indigestible, and remains as a foreign body in the bowel, setting up irritation or diarrhœa. Hence, its nutritive value in this form is limited to the starch and gluten which lie on its inner side; but if it irritate the bowel, it may be removed before these have been digested, and in its removal carry away other nutritive material, and rather lessen than increase nutrition. This laxative quality may be medicinal, but it is not nutritious, and may be more useful in one form than in another, and at one time than at another. That it can add directly to nutrition is impossible; and whilst it may be very useful to those who are well fed and need a laxative, it may be worse than useless to the ill fed who need nourishment.

Years ago in England, as now in India, this part of the skin was left in the flour (thence called brown or batch flour), so as to enable the whole to be sold at a lower rate than fine flour, for the use of the poor; but the disadvantage of its use was insisted upon by me, in 1864, at the Society of Arts, when treating on the dietaries of the poor and the compulsory dietaries of prisoners and paupers, and it was shown that the discarding of it by the poor in favour of a finer flour was based on sound experience of its nutritive value. It was also shown, in my Report to the Privy Council in 1863, on the dietary of low-fed populations, that as bread was more agreeable without it, the children of the poor were content to eat it dry when they would have been disgusted with dry brown bread, or would have required treacle, or some savoury and expensive addition, to induce them to eat it. Thus, it was shown that white bread is now the poor man's food; and it may be repeated, in illustration of the contradictory course of events, that brown bread has become a luxury—the luxury of the rich man, and too dear for the use of the poor.

These facts having been strongly insisted upon, attempts have been made to remove the objection to use it. This has been effected by grinding it very finely, and we believe it has been effectual, not only in preventing the mechanical irritation of the bowels; but in so exposing the particles to the action of the gastric juices that the stomach appropriates nearly all the nutritive matter.

There are two reasons for regarding this change with favour—the lower price of the bran as compared with that of the kernel, and the greater proportion of nitrogen in the bran. The value of the former will depend upon the usefulness of the latter, and it is therefore very desirable to prove that the bran in this finely divided state is readily digested. The proof is still wanting, and I think we are only entitled to affirm that a much larger portion of it can be digested than of the whole bran; but the woody fibre and the silica will probably remain undigested. Hence, this partial advantage may be counterbalanced by the increased cost of grinding, and also by the increased value which would be given to the bran in the market.

There is but little of the inner scales which cannot be employed in nourishing the body; and, as they contain a larger proportion of gluten, in relation to the starch, than is found in the kernel, as well as the valuable ferment, called cerealin, they should be desir-

WHEATEN FLOUR AND BREAD. 177

able additions to the flour, so far as their nitrogenous element is concerned. The flavour, however, of these layers is far inferior to that of the kernel, and diminishes rather than enhances the value of the flour.

It is now very unusual to eat the grain of wheat in its entire form, but it was long employed in the preparation of *frumity*, or furmenty, either for daily use, or to celebrate the harvest-home. It is prepared by steeping the new wheat in water, and placing the pan in an oven, where it may be kept at a temperature of 100° to 120° F. for eighteen to thirty-six hours, when the grain will have swollen and ruptured its skin, and at the same time, the kernel will be softened, and the saccharine process will have commenced. It is then ready to be boiled with milk, and, when spiced and sweetened with sugar, is a preparation of delicious flavour, but it is indigestible, so far as regards the whole wheat and the husk, which will ultimately pass off by the bowel, and it could not be used constantly as an article of food.

In the middle ages this old English food was called frumity, frumetye, furmenty, furmenti, and formenty; and, in the fourteenth century, was prepared according to the following recipe:—

'Nym (take) clene Wete and bray it in a morter wel that the holys (hulls) gon al of & sethe yt til it breste, and nym yt up and lat it Kele (cool), and nym fayre fresh broth and Swete Mylk of Almandys or Swete Mylk of Kyne & temper yt al, and nym the Yolkys of eyryn (eggs). boyle it a lityl, and set yt ad ōn and Messe yt forthe wyth fat Venyson & fresh Moton.'

In other recipes it is prepared with porpoise, instead of venison and mutton.

The Arabs of Syria boil it with leaven, and afterwards dry it in the sun, and eat it with butter or oil.

They call it *Burgoul,* and when thus dried it may be kept good for years. At other times they grind the wheat very coarsely and boil it in water, after which it is eaten with butter or milk.

In the grinding of wheat by the miller the grain is passed over a series of sieves, by which the parts of it which are large are ultimately removed from those which are more finely ground. The finest white flour is derived chiefly from the central parts of the kernel, and is known as the finest biscuit flour, whilst all the succeeding portions vary in colour and fineness according to the degree of admixture of the different layers or coats of the skin. Hence the cost of the finest flour must be greatly enhanced, both by the small proportion of it, and the skill and material required to prepare it. It has therefore a value in its very whiteness and so-called purity and rarity; but let us further enquire into the true nutritive value of this and other qualities of flour.

We will assume that all are made from the same grain, and therefore that there is one standard of quality common to all. The finest flour has lost, as we may suppose, every trace of the skin, and is truly the kernel only. It is therefore composed of starch in the starch cells, and of the glutinous matters in the cell walls and intervening structures with certain mineral matters which are associated with them. So far as starchy or heat-giving matter is concerned, it is superior to any other quality of flour. The other qualities vary only in the proportion of the layers of the skin which they contain; and as they possess a larger proportion of gluten and other nitrogenous materials with phosphoric acid and various salts, by so much does the flour contain less starch and more of the flesh-forming and other nutrient principles. Hence *primâ facie* the latter qualities are preferable to the first if the nutritive quality and not the flavour be the test. Subject to the removal of so much of the skin of the wheat as would, if present, colour or deteriorate the flavour of the flour in a marked degree, the so-called inferior qualities are to be preferred by the poor and by those who would obtain nourishment at the least proportionate cost.

It may be convenient to state, in a summary manner, the nitrogenous and mineral matters which are found in the whole grain of wheat and in its several parts.

Thus the gluten or chief nitrogenous principle varies from 10·5 in English to 11·5 in Virginia, 13·5 in New Orleans and Dantzig, and even to 15 per cent. in Black Sea wheat in good seasons, which is nearly one-half more than that produced in our climate.

The increased proportion of gluten and salts in the husk, as compared with the kernel, is very striking. Thus, whilst the nitrogen is 1·7 and salts 0·7 per cent. in fine flour, the following proportions are found in the different layers of the husk:—

No. 61.

	Nitrogen	Salts
Bran	2·3 per cent.	7 per cent.
Coarse pollard	2·4 ,,	6·5 ,,
Fine pollard	2·4 ,,	5·5 ,,
Coarse sharps	2·5 ,,	3·8 ,,
Fine sharps	2·2 ,,	1·9 ,,

Hence I have no doubt that a good seconds flour is the cheapest and most nutritious, if not the most digestible of the series, and that when thirds are produced, or even when the bran is ground to a fine powder and added to the flour, the actual amount of nutritive matter which the stomach can extract is probably less than with the seconds, whilst its flavour does not recommend it to either the poor or rich.

There is, however, a mode of grinding the whole grain, by Mr. Hart's process, so as to leave it in small masses, instead of reducing it to an impalpable powder, which gives a peculiarity, if not advantage, to the bread produced from it, and by which it may be among wheat flours that which Scotch oatmeal is in the kinds of oatmeal. I have eaten it with enjoyment, but the flavour is rough and somewhat coarse, and it would not be eaten, except as a luxury by the wealthier classes. Ladies generally object to it, until they have by habit acquired a new taste.

The importance of the subject has stimulated inventors to give more variety, sweetness, or economy to this product. Mr. McDougall has, with the same view, prepared an addition of phosphates, by which he assumes that he has added an important element. This was suggested by the presence of that material in the bran of wheat, as well as by the known value of the phosphates in supplying material against brain-waste, and I think it very probable that it is based on truth.

Comparisons of flour prepared by different processes are open to great fallacy, since different parcels of the grain vary considerably in the proportion of nitrogen, whilst flour varies in the proportion of husk which is allowed to remain in it, and both circumstances may occur quite irrespective of the particular process employed.

The following are said to be analyses by Dr. Crace Calvert of the kinds of London flour, namely, the ordinary seconds flour and Hart's whole-meal flour:—

No. 62.

	Seconds	Hart's Meal
Water	14·73 per cent.	14·33 per cent.
Starch	65·48 ,,	48·27 ,,
Gluten and albumen	10·31 ,,	19·87 ,,
Sugar, gum and oil	7·01 ,,	4·01 ,,
Vegetable fibre	1·43 ,,	11·86 ,,
Mineral oils	1·04 ,,	1·66 ,,

In Hart's meal there is the indigestible vegetable fibre of the husk to the extent of nearly 12 per cent., which is nearly 10½ per cent. more than in the seconds flour, and replaces the more nutritious starch of the latter. There is a much greater proportion of nitrogenous elements in the former than in the latter, due to the presence of the inner layers of the husk and to the better quality of the flour submitted to analysis.

Mr. Hart sees an advantage in the presence of the indigestible fibre on the ground that it gives volume to the fæces, but if it be so, 'Jonathan' would be cheaper and as effectual.

The ultimate composition of good seconds flour is as follows, besides oxygen and hydrogen, per cent. :—

No. 63.
C. 38   N. 1·72

The proximate composition is in 100 parts :—

Water 15   Albuminous and allied substances 10·8   Starch 66·3
Sugar 4·2   Fat 2   Salts 1·7

The quantity of carbon and nitrogen which is contained in 1 lb. of flour is 2,656 and 120 grains.

Ten grains of wheaten flour when thoroughly burnt in the body produce sufficient heat to raise 9·87 lbs. of water 1° F., which is equal to lifting 7,623 lbs. 1 foot high.

CHAPTER XXV.

WHEATEN BREAD, BISCUITS, AND PUDDINGS.

THE most important use of wheat flour is in the manufacture of bread.

The mode of preparation of bread is essentially the same everywhere. It consists in the gradual but thorough admixture of flour, water, and salt in proper proportions so as to produce dough, which is subsequently placed in a heated oven until sufficiently baked.

The water employed is somewhat warmed, and, in preparing some fancy kinds of bread, milk is wholly or partially substituted for it. The use of the water is to cause the expansion of the starch cells and to give a convenient and agreeable consistence to the bread after it has been baked. Salt is used as a condiment, and supplies the hydrochloric acid and soda which are insufficiently provided by natural foods. Heat is required to cause the rupture of the starch cells and to promote the rapid evaporation of superfluous water. Hence the action of these agencies on the flour is physical and not chemical, or, if chemical, only in a slight degree.

When loaves of bread are required it is needful to lighten the mass of dough by the introduction of air or other gas, so that it may assume a spongy character, and be readily broken up by the teeth when baked. This has usually been effected by fermentation set up in the dough on the introduction of yeast, when the starch being partially converted into sugar or diastase, carbonic acid gas is eliminated and dispersed through the mass. When the admixture of yeast and flour has been perfectly accomplished, the action of fermentation is set up equally in all parts, and the mass is permeated throughout by small bubbles of a tolerably uniform size; but when the yeast has not been well mixed with the flour, there is an accumulation of gas in the parts exposed to its action, which disfigures the bread by large vacuities, whilst there is little or no action elsewhere. The process requires a certain time for its completion, but it is stimulated by heat, so that it is customary to

WHEATEN BREAD, BISCUITS, AND PUDDINGS. 183

place the mixture in a warm place and to add the lukewarm water containing the yeast little by little to the flour. Hence this action is chemical and causes a loss of a portion of the nutritive parts of the flour.

The introduction of the German yeast has been of great service in the preparation of home-made bread, because it may be purchased daily and rarely fails to produce good fermentation.

Brewers' yeast is attainable in country villages only at irregular periods and will not keep good for more than a day or two; but it has recently been ascertained that if it be made into a thick syrup with sugar, it may be kept for weeks or months. The same process may be carried on with even better effect with German yeast, since it contains much less water.

A successful effort has been made to obviate the waste of nutriment by pumping carbonic acid gas into the dough and thus to make it as light as when acted upon by fermentation. This is known as Dr. Dauglish's process, and the method has been so perfected that the whole dough is evenly permeated by the air in the course of a few minutes, and a very uniform aspect given to the bread.

The gas is supplied in the water to the dough thus:—There are two strong iron vessels each capable of sustaining a pressure of 120 lbs. on the square inch, of which one is the mixer and the other the water vessel containing water to be charged by compression with the gas. The mixer receives the flour, and being furnished with arms is ready to mix it with the water. Water containing a proportion of common salt having been introduced into the water vessel is charged with the gas under great pressure until a proper quantity of gas has been absorbed. The communication is then opened between the two vessels, and the charged water is admitted to the mixer whilst at the same time the arms of the mixer are set in motion and the water and flour are rapidly mixed together. During this part of the process the gas not actually in solution becomes liberated from the water and permeates the bread, whilst the prepared dough is squirted out of the mixer and made into loaves for the oven.

This process is, I think, the greatest improvement of our time in the manufacture of bread, for it leaves the chemical constituents of the flour intact, and there is not only a greater proportion of bread produced by it, but none of that which remains has been deteriorated by chemical decomposition. It is of course very desirable that the gas thus introduced should be pure.

The same effect is produced by yet another method, viz., the introduction of a gas-generating mixture into the dough. Thus, if carbonate of soda be well mixed with the flour in its dry state, and tartaric acid or weak hydrochloric acid added to the water, the incorporation of the two will be accompanied by the union of the two chemical substances and evolution of gas. There are, however, two disadvantages in this procedure which prevent its general use, viz., the uncertainty as to the uniform distribution of the chemicals and the retention of the tartrate of soda in the bread. The former may lead to an uneven levigation of the bread, by which parts of the mass would be comparatively sodden or tough, or there may be lumps of the chemicals in various parts, which give a slightly yellow colour and a disagreeable flavour. The tartrate of soda is not usually injurious to health, but it is not required by the system, and does not improve the flavour of bread. Moreover, if the mass thus levigated be not baked with sufficient rapidity, or if it be over-baked, the gas will escape, and the bread will be less light than is desirable.

When it is desired to add anything to the bread, as, for example, the phosphates of soda, in McDougall's process before mentioned, it is mixed with the flour or dissolved in the water, or sea water may be used.

The method of mixing the ingredients to make the dough was formerly by hand in small bakings, and by treading the mass with the feet in large ones, but the invention of Mr. Stevens has provided an apparatus by which the dough is mixed by curved iron levers, which move the mass in a concentric manner when the axle is rotated. This process is free from objection on the score of uncleanliness and is very expeditious.

Thus the process of bread making is complete when the starch cells have been ruptured and the mass rendered capable of easy and agreeable mastication; and it is not essential that there should be any chemical change in the flour itself. The degree, however, in which these results occur may vary with the wish of the operator. To produce a very light bread the mass must be more completely permeated by the air or gas, and if this be effected by fermentation, a much larger proportion of the starch will be destroyed in the operations; or the bread may be highly baked, and both these conditions are more commonly found in the long bread of the Continental than in the thick loaves of the English people. The economical aspect of this question is evident, but the process may not materially interfere with the digestibility of the bread, for in either case it may be capable of ready mastication.

If, however, the bread contain too much water, it may not only be less agreeable to the taste, but less capable of being masticated and digested; or if it be over baked it may have lost much of its nutritive value, and also be less capable of ready mastication. English bread is moister than French, but it should be always so dry as to crumble between the fingers, and it is usually more friable than the French bread. The art of baking consists both in the due preparation of the dough and the degree of moisture which should remain in the bread.

These conditions materially affect the weight of bread to be obtained from a given weight of flour, but, it must be added, that according to the quality of the flour will be the weight of water which it will take up and retain in the process. Flour produced from wheat of the finest quality, and in hot summers, or in hot countries, takes up much water, and is known as strong flour, the strength being due to the larger amount of nitrogenous elements which it possesses; but sprouted corn, or the produce of cold climates and cold summers, yields flour of the contrary tendency.

Moreover, of the different kinds of flour produced from the same wheat, the finest will absorb the greatest quantity of water, and produce the greatest weight of bread. This increase is, however, due to the water, and so far the economy of the operation is only apparent; but if the wheat itself be of the finest quality, there will be also a substantial advantage.

When good seconds flour is used, as is ordinarily the case in England, and the bread is baked in the usual degree, 14 lbs. of flour will produce 19 lbs. or 19½ lbs. of bread, or a sack of 250 lbs. produces 95 four-pound loaves, or even more; but if with the same degree of baking the finest flour derived from the finest wheat be employed, the weight of bread will be 1 lb. per peck or stone more. When the baking is carried to the usual degree observed in French bread, the quantity from 14 lbs. of the finest flour will be not more, but frequently less, than that from the seconds flour when made into English bread; but whether 18 lbs. of French bread will contain as much nutriment as 20 lbs. of English bread will also depend upon the degree to which the process of fermentation was carried. Where neither real nor apparent economy is of essential moment, it is certainly desirable to select the finest flour and to bake the bread well; but when bread is baked for the market, or when a loaf must be cut up into a given number of rations, it will be more profitable to use seconds flour of moderate quality, and to rather under than over bake it.

When bread is baked in very small cakes, it is usual to levigate it by adding carbonate of soda or ammonia, but although this will lighten the mass it will not improve the flavour of the cake. The varieties of bread made as cakes are almost infinite, and not unfrequently milk and other foods are added which increase its flavour and richness.

There is an impression that newly-made bread is less digestible and healthful than that which has been kept for a certain time, and when this is so, it is not due to any difference in composition, but to a degree of toughness which renders the bread less capable of mastication, whilst after bread has been kept for a short time it has lost some of its water and is more friable. It varies, however, with the quality of the wheat and the mode of preparation. Bread made from unsound wheat is always more difficult to masticate than from good wheat, and the better the flour the more easily may the bread be masticated. When bread is made with yeast, and particularly if the process of fermentation have been insufficient, the difficulty of mastication will be greater than when it is aërated or made 'short' by the addition of milk or baking powder.

The mode of manufacture has no doubt caused a change in this respect, for formerly country and home-made bread was improved by being kept a few days, whilst the baker's bread of the present day need not be kept longer than one day.

It is probable that new and hot bread is eaten rapidly and with less mastication than old bread, and is consequently swallowed in lumps, so that the saliva, which is the transforming agent, has not acted upon the whole mass. In such a case the bread will lie in the stomach unchanged for an unusual time.

The preparation of biscuits is now a very important trade, and the variety of the articles renders the manufacture a very interesting process. Huntley and Palmer's manufactory at Reading may be quoted as an example of variety, and well repays a visit; whilst the great ship biscuit factories in our dockyards show the simple process on a very large scale.

These are for the most part unleavened, and must be baked in thin masses, and highly dried, so that they may be broken by the teeth in mastication, and be fitted to resist atmospheric changes for years if kept dry. When, however, they are exposed to damp air or to the action of water, they lose their brittleness and become mouldy, so that it is usual to carefully enclose them in tins or well-made casks for use in distant countries.

The process of preparation is yet more simple than that of bread, since it is needful only to well mix the flour with the water or milk, and to add salt, butter, sugar, and any other flavouring or colouring matter that may be desired, until a dough of sufficient consistency shall have been produced. This is collected in receivers and put through a series of rollers to be reduced to the proper thickness, after which it is cut by machines into the desired sizes and forms, and is either baked in tins or passed through ovens on an endless band until sufficiently dried.

The fancy kinds of biscuits require special manipulation, in order to add the decorations, and this is done with great rapidity upon the cut paste before being passed through the oven.

The various kinds having been thus prepared, are mixed in certain proportions, or retained separately, to be packed in tins or casks.

The fancy kinds are eaten as luxuries in health, but, in the feebleness of disease, are often of great value, and may be strictly necessaries. The ordinary kinds are very convenient foods, as to carriage and keeping, and contain a much larger proportion of nutriment than the same weight of bread; but they are more difficult to masticate, and, by reason of their dryness, more difficult to digest when eaten. It is desirable that the process of mastication should be perfect, and that there should be plenty of fluid in the stomach to ensure their rapid solution. Hence, notwithstanding their nutritive value, plain biscuits are not preferred to good bread, and although they may be necessaries they are not luxuries.

The other chief uses of wheaten flour are in the preparation of puddings and pastry. When flour is sprinkled into boiling water, and boiled for a time, the cells of starch become ruptured, and an agreeable hasty-pudding is made. This is usually eaten hot, with butter and sugar, or milk; but it should be well boiled and well stirred during the process, both to complete the rupture of the cells and to prevent the burning of the substance.

It is customary in Somersetshire to make dumplings of unleavened dough of thinner consistence than for the manufacture of bread; and, strange to say, the people like them. In other parts of the country the dough is leavened, and if the dumpling be boiled carefully, it remains light. In neither case is the food very easy of mastication and digestion, for it is tough and not easily broken up by the teeth; but the latter is to be preferred.

It is, however, far more common to mix the flour with fat when making puddings, pie crusts, or pastry, either with or without the addition of soda or ammonia. Where suet is added, it is usually cut into very small pieces, and lightly mixed with the flour, so that the two are rather mixed than incorporated, and such a composition is agreeable, and readily masticated and digested. When lard, butter, or dripping is added in the preparation of pastry, it is usually closely incorporated, not only by kneading but by rolling the dough, and ultimately there are many layers in one crust. Such a crust, when baked, is easily broken to pieces, but not easily pulverised or ground into a pulpy mass by the teeth, and the eater very commonly swallows flakes of a comparatively large size. Thus there is a mechanical cause for the indigestibility of pastry; but there is also a chemical cause, viz., the imperfect application of the saliva to the starchy matter when enveloped in fat, by which the first step in the conversion of starch into sugar is prevented. From these two proceed the evil results so frequently experienced. Hence, very rich crusts are not desirable, and when they are eaten should be very patiently and thoroughly masticated.

The preparation of macaroni, from fine wheat flour, requires a notice. It is now, as it has been for ages, a chief food of the inhabitants of Naples and Southern Italy, and is prepared simply by selecting the finest flour, making it into a very thin paste, and gently drying or baking it, so that it may keep good for future use. Various *pâtes* used in soup are made from the same kind of paste.

It appears that in the fourteenth century we had acquired a knowledge of this food, as is shown by the following recipe for *macaroni au gratin* :—

'MACCOROS.

'Take and make a thynne foyle of dowh, and Kerve it on peces and cast hem on boillyng wat and Seethe it wele. take chese & grate it and butt caste bynethen and above as losyns, and sūe forth.'

The ultimate chemical composition of English bread, made from good seconds flour, is, besides oxygen and hydrogen, per cent. :—

No. 64.
C. 28·5   N. 1·29

The proximate elements in 100 parts are :—

Water 37   Albuminous and allied substances 8·1   Starch 47·4
Sugar 3·6   Fat. 1·6   Salts 2·3

The quantity of nitrogen and carbon in 1 lb. of bread baked in the ordinary English manner is 92 and 1,968 grains.

The time required for the digestion of bread is $3\frac{1}{2}$ to 4 hours. Ten grains of dryish bread, when burnt in the body, produce sufficient heat to raise 5·52 lbs. of water 1° F., which is equal to lifting 4,263 lbs. one foot high.

The effect of eating 4 oz. of white home-made wheaten bread was to give maxima of increase in the quantity of carbonic acid evolved of 1·48 grain and 2·4 grains per minute on different persons. The quantity of air inspired was increased by 60 cubic inches per minute. The influence was very enduring. (No. 95, frontispiece.)

The law of this country inflicts penalties for the adulteration of bread, and prescribes that when ordinary bread is sold over the counter it shall be sold by weight, so that a quartern loaf shall weigh 4 lbs., and half a quartern loaf 2 lbs.; and the price of the loaf varies only with the price of flour.

This subject has always attracted the attention of Governments, so that in the reign of Edward IV. we find it stated that the Lord Mayor 'did sharpe correction upon bakers for making bread otherwise than of floure and light of weight, and caused divers of them to be put in the pillory.'

At a later period the assize of bread by the Court of Mayor and Aldermen of London prescribed the weight and price of bread, as shown in the following precept, copied from a Bill of Mortality, No. 28, London, Tuesday, the 27th day of June, 1775:—

'WILKES, *Mayor.*

'London, to wit.

The Assize of Bread, set forth this 13th day of June, 1775, by order of the Court of Mayor and Aldermen of the said city, to commence and take place on Thursday next, and to be observed and kept until the further order of the Lord Mayor of the said city, or the said Court of Mayor and Aldermen, by all persons who shall make, or bake for sale, any bread within the jurisdiction of the said Court of Mayor and Aldermen, that is to say,

		lb.	oz.	dr.
The penny loaf, or two half-penny loaves, to weigh	Wheaten	0	8	11
	Household	0	11	9
The two-penny loaf	Wheaten	1	1	6
	Household	1	7	3
The three-penny loaf	Wheaten	1	10	1
	Household	2	2	12

To be sold for:—

	lb.	oz.	dr.		s.	d.	f.
The peck loaf, to weigh	17	6	0	Wheaten	2	8	0
				Household	2	0	0
The half-peck loaf	8	11	0	Wheaten	1	4	0
				Household	1	0	0
The quartern loaf	4	5	8	Wheaten	0	8	0
				Household	0	6	0

NOTE.—All loaves, if complained of, must be weighed before a magistrate within twenty-four hours after baking, or exposing thereof to sale, and must be according to the respective weights in the above table.

Sixteen drachms make an ounce, and sixteen ounces a pound.

ITEM.—It is hereby ordered and appointed that no person within the jurisdiction aforesaid shall, after Wednesday next, until the further order of the Lord Mayor, or of the said Court of Mayor and Aldermen, make, or bake for sale, or sell or expose to or for sale within the jurisdiction aforesaid, any half-quartern loaves.

And the better to distinguish and ascertain the two sorts of bread hereby ordered to be made, one from the other, there is to be imprinted and marked on every loaf of bread which shall be made, sold, carried out, or exposed to or for sale within the jurisdiction aforesaid as wheaten bread, a large Roman W; and on every loaf of bread which shall be made, sold, carried out, or exposed to or for sale within the jurisdiction aforesaid, as household bread, a large Roman H. And the penalty for every omission is twenty shillings.

R. I. X.

The price of salt, set by order of the Court of Lord Mayor and Aldermen, dated the 21st of October, 1735, is five shillings the bushel, 56 lbs. to the bushel, and so in proportion for any lesser quantity; and whosoever shall sell at a higher price, or shall refuse to sell at the price aforesaid, forfeits five pounds.'

CHAPTER XXVI.

BARLEY, MASLIN AND RYE—SCOTCH AND PEARL BARLEY—GLUTEN.

BARLEY (*Hordeum*) is now rarely used in this country for the preparation of solid food, but is chiefly devoted to the preparation of malt and the manufacture of ales and spirits. It is, however, eaten as bread in certain parts of South Wales and in the Northern Counties, and is sometimes added to wheat flour in making brown bread. It is also largely eaten, and is still the ordinary farinaceous food of the peasantry and soldiers in certain parts of the Continent, and in parts of India and other Eastern countries.

When used whole as food it is first parched, as in many districts of India, and was so given by Boaz to Ruth, and referred to at a yet earlier period. Mrs. Finn writes from Palestine:—'It is still usual for reapers, during barley harvest, to take bunches of the half ripe wheat, and singe or parch it over a fire of thorns. The milk being still in the grain it is very sweet, and is considered a delicacy.' When baked into bread it has a brown colour, and although rough it has a sweetness and a moist consistence which are not disagreeable. It is sometimes adulterated with 'Jonathan.'

The composition of barley-meal is as follows, in 100 parts:—

No. 55.

Water 15   Albuminous and allied substances 6·3   Starch 69·4
Sugar 4·9   Fat 2·4   Salts 2·0

The quantity of carbon and nitrogen in 1 lb. of barley-meal is carbon 2,500 grains, nitrogen 68 grains.

Hence it is not equal in nutritive value to wheaten bread, but its cost being less it is possible to obtain more nutriment from that source for a given sum than from wheaten bread. As, however, there is good reason to believe that a larger proportion passes off by the bowel undigested, it may be really a less economical food. When it is added in small quantity to the whole-meal bread made of wheat, it certainly keeps the bread moist, and improves the flavour.

The special quality which it possesses is its ready

conversion into saccharine matter in the process of malting, as will be described in the chapter on beer, while it is also well fitted to produce alcoholic spirit and other elements of ale.

The German *Weiss-bier* is made from a mixture of wheat and barley.

A mixture of barley meal and wheat flour is in very general use among the poorer (not the poorest) classes in India, the proportion of each being equal, as at Gonda, or one-quarter barley meal and three-quarters wheat flour as at Bareilly, or one-seventh barley meal and six-sevenths wheat flour as in Meerut, and oftentimes the husks of both grains remain in the meal.

Scotch and pearl barley appear to be very different foods from ordinary barley, and in fact they are used rather as rice than barley. They are never made into meal or bread, but are used whole in puddings, and seem from their flavour and thickening properties to occupy a position between wheat and rice. They are, however, prepared from one of the species of barley by the removal of the husk, as in the instance of the Scotch barley, and by further polishing and rounding, as in the pearl barley. They are much approved as a farinaceous and mucilaginous ingredient in puddings, and when prepared with new milk and flavoured with sugar, if not spices, produce a delicate and delicious food.

The ultimate chemical composition is as follows, besides oxygen and hydrogen, per cent. :—

No. 66.
C. 38   N. 1·3

The quantity of carbon and nitrogen in one pound is 2,656 and 91 grains.

The time required for their digestion when boiled is twice that of rice, viz., two hours, but less than that of wheat.

Maslin is a mixed product grown in Yorkshire and other northern districts of England, and is composed of two or three parts of wheat and one part of rye; but it is not very extensively used at the present time. Neither rye nor barley is preferred to wheat in the production of bread, for the colour of the bread is less white and the flavour is rougher than that of white wheaten bread. Moreover, as rye is inferior to wheat in nutritive elements the mixed product is less valuable than wheaten flour alone, so that it is only as an inferior food selling at a lower price than wheat that it is accepted. At the same time the admixture of rye with wheat is not disagreeable as a change of food, and bread prepared from maslin is more agreeable, if not more nutritious, than that from thirds or fourths wheaten flour.

Rye (*Secale cereale*) is grown on light soils in various parts of the country; but chiefly in the north of England. It is, however, a product of less importance now than formerly, since bread is rarely if ever made in England from rye flour alone, and when it is used it is mixed with wheaten flour. It is inferior to wheaten flour in nutritive properties, flavour and digestibility, but it is still used in some parts of the Continent and particularly in Russia.

The Russians also use this grain for the preparation of a fermented liquor called *Quass*.

The following is the composition of rye meal in 100 parts:—

No. 67.

Water 15·0   Nitrogenous 8·0   Starch 69·5   Sugar 3·7   Fat 2·0
Salts 1·8

There are 2,660 grains of carbon and 86 grains of nitrogen in 1 lb. of the meal.

A diseased state of this grain is the well-known and

highly poisonous ergot of rye (*Secale cornutum*), by which many persons have been poisoned. It is a valuable medicine, but when taken incautiously produces convulsions and gangrene.

GLUTEN.

This is a most important part of food, and represents the nitrogenous element of grain to an extent varying from nine to sixteen per cent. It may be obtained by washing the flour of wheat or any other grain until all the starch has been removed, when it is a tenacious colourless substance, known in ancient times as birdlime and used for catching birds. It can scarcely be regarded in this state as a food, since it is not agreeable, neither can it be entirely masticated. As, however, it is so important a principle of food, it was desirable to determine its influence, and after eating all the gluten which could be washed out of four ounces of wheaten flour, we found it produce a maximum increase of 0·84 grain of carbonic acid and of 11 cubic inches of air per minute in the respiration. It may be added, that in this effect it is very similar to albumen, gelatin, and other animal compounds, having the same chemical constitution. Hence gluten is an excitor of the respiration in a degree beyond the respiratory material which it supplies.

CHAPTER XXVII.

SUCCULENT VEGETABLES.

THE class of succulent vegetables is a very large one, and is valuable both for its nutritive and saline constituents, by which it is at once a food and a medicine. It will not be possible for us to enter into very large detail, but it will suffice to group the various members of the class in three divisions: one of which will be represented by the Potato, the second by Cabbage, and the third by Mushrooms and Lichens.

A.—*Potato and similar Foods.*

Of all kinds of fresh vegetables grown in temperate climates, none is so valuable as the potato (*Solanum tuberosum*), when we have regard to its agreeable flavour and nutritive and medicinal qualities, so that it is eaten daily by almost every family, and with the addition of butter-milk or skim-milk, was until recently nearly the sole support of a whole nation.

It was introduced into this country from Chili in the sixteenth century; but in hot climates it is very generally supplanted by the sweet potato, artichoke, and similar vegetables; and in the Arctic regions, oil and other animal foods are substituted for it. Hence it may be said to be practically excluded from the hottest and coldest parts of our globe, whilst in reference to more moderate temperatures it may be remarked that it is influenced by comparatively slight conditions of soil and climate, for it grows of very different qualities in various parts of this country, and is a better crop in Tasmania than in the neighbouring continent of Australia.

The best known varieties of potato do not differ very greatly in chemical composition, and are selected less on that ground than according to the taste of the individual as to the hardness or mealiness of the tuber after it has been cooked; yet there is some difference in flavour which influences the selection. Thus some are called waxy or watery, as, for example, young potatoes, and those grown in the bog lands of Ireland and Scotland, whilst others are mealy, as the York Regents, grown in Yorkshire and other parts of this country.

The varieties which are brought into the market are too numerous to warrant a reference to them here, but it may be added that but few vegetables have received and rewarded better selection and cultivation.

The proximate elements necessarily vary much with the season, kind, ripeness, and soil, so that the analyses differ much among themselves, but a good average composition per cent. may be stated as follows :—

No. 68.

Water 75　　Nitrogenous 2·1　　Starch 18·8　　Sugar 3·2
　　　　　　　Fat 0·2　　Salts 0·7

The ultimate composition, besides oxygen and hydrogen, may be stated to be, per cent. :—

No. 69.
C. 11　　N. 0·35

The relative value of potatoes is determined in a general manner by their specific gravity, just as a similar estimation is made of an apple or an orange by weighing it in the hand, for the heavier in relation to bulk is any given potato, so is the greater amount of starch. If several potatoes be thrown into a strong solution of salt, and water be added until some of them sink and others swim, the specific gravity of the saline solution will be that of the potatoes as a whole.

J. J. Pohl found the following relation between the specific gravity and the proportion of starch :—

No. 70.

Sp. Gr.	Starch in 100 parts
1·090	16·38
1·093	17·11
1·099	18·43
1·101	18·98
1·107	20·45
1·110	21·32
1·123	24·14

The quantity of ash or saline materials present in

100 parts of the fresh potato, varies with the kind, as shown by T. J. Herepath. Thus:—

No. 71.

Fortyfold 0·88 Prince's Beauty 1·06 Maggie 1·09
Axbridge Kidney 1·27 White apple 1·30

The proportion of potash in the salts of potato is very large, and does not vary much in the several specimens examined, the limits being 53·03 per cent. in the Fortyfold, and 55·73 per cent. in the Maggie.

Potatoes are deficient in mineral matter, so that they are unfit to be a sole food, but that defect is supplied by the addition of hard water, milk, and other elements of food.

There are 760 grains of carbon and 24 grains of nitrogen in 1 lb.; so that more than $2\frac{1}{4}$ lbs. of potato are required to equal 1 lb. of bread in carbon, and more than $3\frac{1}{2}$ lbs. in nitrogen. Hence when potatoes are 1d. per lb., and bread $1\frac{1}{2}d$. per lb., the former are two to three times dearer than the latter, but the former are valuable for their juices in addition to their nutritive elements.

New and waxy are said to be less digestible than old and mealy potatoes, but the time required for digestion may be regarded as nearly the same as for bread, viz., from $2\frac{1}{2}$ to $3\frac{1}{2}$ hours.

The potato disease, which produced so much distress in Ireland and elsewhere, was accompanied by the production of a fungus and the destruction of starch in the starch cells, as is represented in the following drawings.

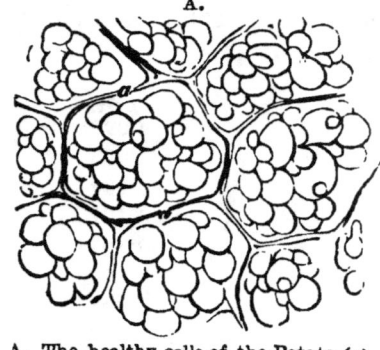

No. 72.—POTATO.
A.

A. The healthy cells of the Potato (*a*) filled with starch cells (*b*).

SUCCULENT VEGETABLES. 201

No. 72.—Diseased Potato.

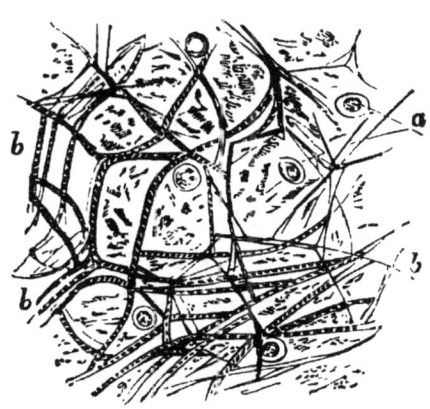

B. The same after the process of germination or in a state of disease, when the starch cells have been removed, except at (*a*).

C. The fungus in the diseased potato is shown at (*b*), and the healthy starch cells (if any) at (*a*).

The kinds of potatoes which were found in Cambridgeshire to be the least subject to the disease were Myatt's Ash-leaf Kidney, the River Royal Kidney, the Nonpareil Prolific, and the Rock.

10 grains of potato consumed in the body produce heat sufficient to raise 2·6 lbs. of water 1° F., or to lift 1,977 lbs. one foot high.

Potatoes and similar vegetables should be well cooked, with a considerable degree of heat. If it be intended to boil them, they should be placed at once in hot water; and if to be roasted, the oven should be moderately hot, and care taken lest they should burn. When peeled and soaked in cold water a larger proportion of the fecula and salts will be extracted than is desirable, and with a slow oven the peel will be hardened and thickened.

Potatoes are best boiled in their skins, by which means both their flavour and the salts of vegetable acids are preserved. If peeled before cooking, they should be stewed, as in Irish stew and the like.

In my experiments, the effect of eating good potatoes, whether new or old, was less than that of rice.

In two experiments 8 ounces increased the carbonic acid evolved in the respiration by 1·27 grain per minute, and the quantity of air inspired by 52 cubic inches per minute. The rate of pulsation and respiration was slightly lowered. (*Phil. Trans.* 1859.)

There are many vegetable productions which somewhat resemble potatoes in flavour or quality, and are not very inferior to them as food, to which we will now briefly refer.

The sweet potato (*Batatas edulis*) is rarely used, although it is not unknown in this country; but is in daily use in Central America, the West Indies, and in many other hot countries where the common potato does not grow or cannot be obtained.

The cassava (*Manihot utilissima*), from which meal is obtained, is not a bad substitute for potato, and is in common use in Central America and India.

The red yam (*Dioscorea alata*) is a similar food, and grows over a very extended area. It is particularly appreciated by the negroes, but Europeans do not prefer it to the potato.

The *Dioscorea batatas,* or Chinese yam, has the following composition, per cent., according to Frémy:—

No. 73.

Water 79·3   Starch 16.   Nitrogenous 2·5   Fat, &c. 1·1   Salts 1·1

The common yam (*Dioscorea sativa*) is said by Suersen to yield 23 per cent. of starch.

The common artichoke (*Cynara Scolymus*) and the Jerusalem artichoke (*Helianthus tuberosus*) (perhaps so called as a corruption of the Italian name Girasole or Sunturner) are well known foods, both in temperate and hot climates, and are highly nutritious, whether cooked alone or made into soup. They contain a large amount of salts, of which, 55·9 per cent. is potash, in

SUCCULENT VEGETABLES.

the Jerusalem artichoke, whilst there are potash, soda, lime and magnesia, with sulphuric and phosphoric acid, with silica and salts of iron. The leaves are used as fodder for cattle in the United States of America, and contain salts which yield 40 per cent. of lime.

There are numerous plants in Chili and the neighbouring States, which produce edible tubers somewhat resembling the common potato, such as the *Ullucus tuberosus*, *Oxalis crenata*, *O. tuberosa*, and *Arracacha esculenta*. In North America the tubers of the *Apios tuberosa* and the prairie turnip, which are twining plants of the leguminous order, are eaten as food.

The great value of such tubers as food is well shown by the contrast in the chemical composition of the apios and the potato as made by Payen:—

No. 74.

	Apios	Potato
Nitrogenous matter	4·5	1·7
Fatty matter	0·8	0·1
Starch, sugar, pectose, &c.	33·55	21·2
Cellulose and epidermis	1·3	1·5
Inorganic matter	2·25	1·1
Water	57·8	74·4

Turnips vary in nutritive qualities and flavour, so that the Swede variety is far more nutritive than the white, but the flavour of the latter kind is preferred as the food of man. They are also much less valuable as occasional foods than carrots and parsnips. The following is the ultimate composition of turnips and carrots:—

No. 75.

	Turnips		Carrots
	Swede	White	
Carbon	4·5	3·2	5·5 per cent.
Nitrogen	0·22	0·18	0·2 ,,

The following is the quantity of carbon and **nitrogen** (in grains) contained in 1 lb.:—

No. 76.

	Turnips		Carrots
	Swede	White	
Carbon	304	173	384
Nitrogen	15·3	11·2	14

The proximate elements in 100 parts are:—

No. 77.

Turnips	{	Water 91	Nitrogenous 1·2	Starch 5·1
	{	Sugar 2·1	Fat	Salts 0·6
Carrots	{	Water 83	Nitrogenous 1·3	Starch 8·4
	{	Sugar 6·1	Fat 0·2	Salts 1·0
Parsnips	{	Water 82	Nitrogenous 1·1	Starch 9·6
	{	Sugar 5·8	Fat 0·5	Salts 1·0

The time required for the digestion of parsnips and carrots varies from 2½ to 3½ hours.

Ten grains of carrot, when consumed in the body, produce heat sufficient to raise 1·36 lb. of water 1° F., which is equal to lifting 1,031 lbs. one foot high.

BEETROOT.

This valuable food (*Beta vulgaris* and *B. Cicla*) must be considered apart from other vegetables, on account of the large proportion of saccharine matter which it contains, a quantity so large as to enable our French neighbours, who lack sugar-producing colonies, to prepare nearly all their sugar from it, and even to export sugar to this country in competition with the cane sugar which we import. It is also grown extensively in Germany and Russia for this purpose, as well as for the distillation of alcohol. Its production for sugar has been tried unsuccessfully in this country, although sugar factories and distilleries were established here.

It has been shown that carrots and parsnips possess an unusual amount of sugar, but beetroot exceeds them in this respect. Thus the amount of sugar in parsnips, carrots, and beetroot is 5·8, 6·1, and 8·0 to 10·0 per cent. in their order.

The proximate composition in 100 parts is, according to Payen:—

No. 78.

Water 83·5   Sugar 10·5   Pectose, &c. 0·8   Nitrogenous 1·5
Salts and pecten 3·7

There are 350 grains of carbon, and $17\frac{1}{2}$ grains of nitrogen in each pound. (Lawes and Gilbert.)

The time required for its digestion is greater than that of similar sugar-yielding foods, and even greater than that of bread, viz.: $3\frac{3}{4}$ hours.

There are other plants used as food which are not tuberous, and may be called fruits or vegetables indifferently, but as they are not eaten uncooked, we shall refer them to the latter, and mention a few of them in this place.

The bread-fruit tree (*Artocarpus incisa*), growing to the size of the largest oak, provides a farinaceous fruit which has so much the appearance and dietetic qualities of food as to resemble fine white wheaten bread, and to be called bread. It grows freely in the Indian Archipelago and the islands of the Pacific, and is a chief article of food to the inhabitants. It produces a fruit singly, or in clusters, about the size of a child's head, which in its early stage contains a thickish milk, whilst in the second stage it is fibrous and pulpy, and in the third juicy and rotten. It is edible at all its stages of development, but is usually cut for food in the second stage.

This fruit is usually cooked before being eaten, and the process is carried on in a hole of the earth. It is cut into several pieces and the core removed, after which it is placed on heated stones for half an hour in layers, alternating with layers of leaves. When ready for eating, it is brownish on the outside with a yellowish pulpy substance in the inside, resembling somewhat the inside of a white wheaten loaf.

The following is its chemical composition per cent.:—

No. 79.

Water 63   Starch 14   Albumen 3   Gluten and woody fibre 19

The plantain (*Musa paradisiaca*) bears a long fruit, which is cooked like other vegetables. The tree will produce 133 times more food in weight than the same space of ground could if cultivated with wheat, and 44 times more than if planted with potatoes, but the nutriment is not equal to that of wheat.

The juices contain very valuable salts as shown by the following analysis, per cent.:—

No. 80.

Potash 25·27   Soda 9·52   Lime 15·85   Magnesia 5·0
Alumina 0·87   Chlorine 6·3   Sulphuric anhydride 0·96
Phosphoric anhydride 0·87   Silica 0·81 and Carbonic anhydride 34·17

The horse plantain grows in Ethiopia and South Africa, and bears a fruit a foot in length filled with hard black seeds, which are fried in butter and eaten.

The kemmaye is described by Burckhardt, in his 'Notes on the Bedouins and Wahabys,' as a favourite dish which forms the chief vegetable food whilst it lasts of those people, as also those of Damascus and Eastern Syria. It lies about four inches under the ground, and at length by growth throws up a little mound of earth, over which the camels may stumble. It is dug up and boiled in milk or water until it forms a paste over which butter is poured, or it is eaten when washed with butter. There are three species, distinguished by their colour, viz., black, white, and red, but all alike resemble the truffle in shape and appearance

B.—*Cabbage and similar Green Succulent Vegetables.*

Numerous green vegetables as borage, radish, turnips, parsley, cabbage, mustard, hyssop, mint, rue, spinach, onions, coriander, rosemary, fennel, chervil, and also mushrooms were in common use in this country among our Saxon ancestors; and we find the following recipe of the fourteenth century for the preparation of a salad, which in this point of view may not be without interest:—

'*Salat.*

'Take Psel, Sawge, garlee, chibol'i, oynons, leek, borage, myn't poneet, fenel, and ton tressis (cresses), rew, rosemarye, purslarye, laue, and waische hem clene, pike hem, pluk hē small wip phu (thine) hand and myng hem wel with rawe oile, lay on vyneg^r and salt and sue it forth.'

This represents the least nutritious class of vegetable foods, and is perhaps less valuable for its direct nutritive elements than for its indirect and medicinal saline juices. With so great a variety in common use as the common cabbage (*Brassica*); Brussels sprouts (*Brassica oleracea*); broccoli (*B. oleracea italica*); turnip tops (*B. rapa*); cauliflower; spinach (*Spinacia oleracea*); water-cress (*Nasturtium officinale*), &c., there will be some difference in chemical composition, but for all practical purposes it will suffice if we take cabbage as a type of class in which the leaves are used as food.

Anderson found the following difference in composition in the leaves of the common cabbage per cent.:

No. 81.

	Young Plant	Ripe Outer Leaves	Ripe Heart Leaves
Water	91·8	91·1	94·4
Nitrogenous	2·1	1·6	0·9
Woody fibre, gum and sugar	4·5	5·0	4·1
Ash or salts . . .	1·6	2·2	0·6

It may be stated generally that there are 420 grains of carbon and 14 grains of nitrogen in 1 lb. of cabbage and mixed vegetables.

It would not be possible to sustain life for a lengthened period upon such substances alone, since the quantity required to supply the requisite food would exceed the rate of digestion and the limits of the human stomach, but they are nevertheless most agreeable and useful adjuncts in their season. They should be well cooked, and indeed it is said that they cannot be too much cooked, but even then they are not so quickly digested as certain kinds of animal food, but require from $2\frac{1}{2}$ to 4 hours.

10 grains of cabbage when consumed in the body produce heat sufficient to raise 1·08 lb. of water 1° F., which is equal to lifting 834 lbs. one foot high.

The nettle (*Urtica dioica*), a well-known weed, is sometimes used in this country and more frequently on the Continent as a table vegetable in the early spring when it is yet young and tender, but probably less for its nutritive qualities than for the medicinal virtue of its juices. It is boiled with a little soda, and served like spinach, or it is perhaps more frequently boiled in water or milk and the juice alone taken.

The charlock (*Sinapis arvensis*) and the dandelion (*Taraxacum dens leonis*) may also be eaten as green vegetables.

The goatsbeard (*Tragopogon orientalis*) grows near the line of perpetual snow, and forms a principal article of fresh vegetable food in the dietary of Kurdistan. The stem is peeled and eaten raw, and makes an agreeable salad in the European manner.

Salsify (*Tragopogon porrifolius*) is eaten cooked in Europe.

Fennel (*Foeniculum vulgare*) was eaten in England many centuries ago, and is still used as an aromatic

SUCCULENT VEGETABLES.

vegetable in the cooking of fish, but in various Eastern countries the stem is eaten like rhubarb stalks, and offers an agreeable acidity.

Tomatos (*Solanum lycopersicum*) may be eaten raw, or cooked, and made into an agreeable sub-acid sauce.

Vegetable marrows (*Cucurbita ovifera*) are grown very rapidly, and having a mild agreeable flavour are a favourite food.

Cucumbers (*Cucumis sativus*) are a favourite vegetable, which some eat in lumps with or without the skin, whilst others carefully peel it, cut it up into thin slices and plentifully flavour it with vinegar and pepper. There is much difference of opinion as to the digestibility and general wholesomeness of this food, but whilst it may be admitted that the flavour sometimes repeats itself in the mouth for some time after it has been eaten, and that if made acid with vinegar and eaten in large quantity, it is very apt to cause pain at the stomach, less evil will follow its use when cut and eaten like any other vegetable. It is also said that an unwholesome property resides in the skin, and in the fluid which drains from the pulp, but without good grounds. The seeds, which are numerous, should be removed, and the pulp should be well masticated.

The pumpkin (*Cucurbita Pepo*) grows to an enormous size in both temperate and hot climates, but is not highly appreciated in this country as a culinary vegetable. It somewhat lacks flavour, and has a tendency to toughness, and certainly cannot compare favourably with the vegetable marrow and analogous plants. Gourds were in common use in England in the fourteenth century.

Sea kale (*Crambe maritima*) is a delicate and even delicious vegetable, when grown in perfection and well cooked.

Rhubarb (*Rheum*), is a vegetable growing rapidly and yielding an agreeable sub-acid juice. It is to some tastes the most agreeable of its class, quite equal in flavour to gooseberries, apples, or other fruit which are usually made into pies, and it should be ranked with fruits rather than with ordinary garden vegetables. The juice yields a considerable quantity of saccharine matter, and makes a good home-made wine. The well-known medicinal quality of rhubarb resides in the root, but the English cannot compare with Turkey rhubarb as a medicine. It is carefully cultivated, and a number of choice varieties have been produced. Those grown in the market-gardens near London produce an early crop of leaves, and have a fine flavour.

C.—*Mushrooms and Lichens.*

MUSHROOMS.

Much difference of opinion exists as to the varieties of mushroom (*Fungi*) which may be eaten with safety, since some allow but few, whilst others believe that nearly all may be eaten with impunity. In the first view they may be regarded as luxuries or as condiments and agreeable adjuncts to food, whilst the supporters of the latter look forward to the time when they will become important and generally used foods. With every desire to see substances used as food, which may be so rapidly and cheaply grown, it is impossible to ignore the cases of poisoning from mushrooms, which have been recorded. The discussion would occupy too much of our space, and we shall therefore leave it to be dealt with by an able writer in the present Series.

Mushrooms generally consist of about 90 per cent. of water, and 10 per cent. of cellulose, with salts and

SUCCULENT VEGETABLES.

acids. Schlossberger and Döpping give the following elements in 100 parts of *dried* mushrooms:—

No. 82.

	Nitrogen	Ash
Agaricus Cantharellus	3·22	11·2
,, Russula	4·25	9·5
,, glutinosus	4·61	4·8
,, deliciosus	4·68	6·9
,, muscarius	6·34	9·0
,, arvensis	7·26	19·82
Boletus aureus	4·7	6·80
Dædalea quercina	3·19	3·1
Polyporus fomentarius	4·46	3·0
Lycoperdon echinatum	6·16	5·2

If we estimate the composition on the *dry* matter, we must allow that these substances are highly nutritious, ranking indeed with the most nutritious of vegetable products; but such would scarcely be a practical view, for the bulk of *fresh* mushrooms to obtain a given quantity of nourishment would be too great to allow them to be used as a necessary food. About seven pounds weight of mushrooms would supply the minimum requirements of a day in nitrogen, assuming that all the material is digestible.

It is said, that mushrooms of the edible kinds which grow under trees are not always harmless, and that they should be gathered in the open field and when of comparatively small size; but they may be grown readily in dark cellars with plenty of manure, without poisonous qualities. Some persons eat them when raw, after peeling off the outer skin; but it is more common to roast or fry them with a little butter, pepper and salt, or to add them to sauces when served with meat. Perhaps they are at present the most useful in the preparation of ketchup, which is one of our most agreeable and harmless condiments.

Truffles are now a favourite food and belong to this

class of vegetables. The common variety (*Tuber cibarium*), called also black truffles, grows in many parts of England, as the downs of Hampshire, Wiltshire, and Kent, but more abundantly in the southern parts of Europe, and varies in size from a nut to a potato. It grows about a foot under the ground, whence it is dug up, and it is said to be found by dogs which have been trained to the search. A white variety (*Rhizopogon albus*) also grows in this country and in Germany, and appears partly above ground.

They are very costly in relation to the nutriment supplied; but are aromatic, agreeable, and fashionable foods.

SEAWEEDS AND LICHENS.

There is very little doubt that there are valuable foods in this class of vegetables, but they have not yet been generally eaten. The inhabitants of the sea-coast, and particularly in the Orkneys and other isolated localities, where the people are frequently on the brink of starvation, have long had recourse to them, and in inland districts the inhabitants are aware of the existence of Iceland and Carrageen moss, and of laver as an agreeable adjunct to roasted mutton.

In chemical composition they appear to rank high, since they are said to contain 10 to 15 per cent. of nitrogenous and 60 to 70 per cent. of carbonaceous matter; but the flavour is not such as to lead to their general use. When steeped in water with a little soda, they lose the adherent salt and much of their bitterness, and may then be prepared as a mucilaginous food by long stewing in water, milk or soup. When eaten with meat, it is usual to allow the bitter flavour to remain.

Professor Stenberg found in the lichens of the Northern regions 29 to 33 per cent. of starch in *Evernia jubata*, and 27 to 31 per cent. in *Cetraria Islandica*, or Iceland moss, quantities considerably exceeding those in the potato, and there is reason to believe that the bitter flavour of the *Evernia* may be extracted. It may be interesting to add, that the reindeer moss (*Cladonia rangiferina*) contains but 1 per cent. of starch.

Laver is the name given to a preparation of seaweeds (*Porphyra laciniata* and *Ulva latissima*), and is in somewhat frequent use in London as an accompaniment to roasted mutton, and has an agreeably acid flavour. It may be eaten on bread, after thoroughly warming it with a few tablespoonsful of hock in a saucepan, and flavouring it with a little lemon-juice. It should be immediately removed, and whilst hot eaten on buttered toast.

CHAPTER XXVIII.

FRUITS.

A.—*Succulent.*

This very large class of vegetable products comprehends representatives from every hot and temperate climate, and offers the greatest variety of flavours, and those of the most agreeable character, of all vegetable and animal foods. Many of them are too well known to render it needful to mention them, and the whole number is so large that it would be impossible to cite them; but the English-speaking race, with their Asiatic and African experience, are probably better acquainted with them than all other races of the civilised world.

The true position of these foods is less that of nutrients than of agreeable luxuries, for nowhere are they a food on which the life of man depends, whilst all races, however diverse in habits, eagerly seek after them as additions to their dietary. It is true that in hot climates they occupy a higher position as foods than in temperate regions, not, however, by reason of any superiority in their chemical composition, but from the necessity of obtaining food which does not greatly produce heat, and juices of an agreeable flavour to moisten the mouth, and to stimulate the sense of taste. In these respects they are but imperfectly appreciated in temperate and cold climates; and, it may be added, that the flavours of such tropical products as reach us are not by any means equal to those of the fruits immediately they are gathered.

It must not, however, be inferred that we suffer from lack of fruits in this country which may not favourably compare with those of any clime, for our strawberry is probably unsurpassed in delicacy and lusciousness of flavour; and the pine-apples, melons, grapes, peaches, apricots, nectarines, and pears of our hot-houses excel those of any country. This is owing, however, not to our temperate climate, but to care and skill of cultivation and to the artificial climate which we provide for them.

But with all this, the Englishman who has returned from India, Ceylon, or the West Indies longs for a fresh orange or cocoa-nut instead of the old and stale samples, which alone are within his reach, just as when in India he would have been delighted with an English pear, apple, or bunch of currants, could he have gathered them from the trees. There is probably nothing in which Nature has been so bountiful to man, in whatever temperate or hot climate he may be found.

It is a characteristic of all fruits that when ripe they may be eaten in their raw state, and of many, that they may be eaten cooked or raw. They consist essentially of two parts, viz., the juices, and the cellular structures in which the juices are contained; and it is necessary to add that, whilst the juices may be readily transformed, the cells are not easily digested, and when possible are thrown away. This is readily seen in such a food as the orange and apple when not of good quality or not quite ripe. In such fruits as the strawberry, the pine-apple, the grape, and even the banana the cell-wall is very thin, and is easily broken up, so that its presence is not perceptible, and the digestion of it cannot be difficult.

As a general expression, it may be stated of any fruit that the variety which yields the richest juices in the greatest quantity, whilst the cellular frame-work is the least perceptible on mastication, is the most preferred, and the most digestible.

It is not necessary in a work of this class to enter at great length into the chemical composition of different fruits; but it is of interest to remark that chemical science has become so practical that the delicate flavour of the pine-apple, pear, strawberry, and other fruits are produced by the chemist in his laboratory, and that under the term 'essences' we may enjoy the agreeable delusion of believing that we eat the juices of choice fruits when the fruits themselves cannot be obtained.

All fruits agree, however, in containing much fluid in relation to the solid matter, and in supplying sugar, acids, salts, and the various volatile essences on which their flavour depends. As an illustration, we may cite the following as the composition per cent. of the ripe grape:—

No. 83.

Soluble		Insoluble	
Grape sugar	13·8	Skin, stones, &c.	2·6
Tartaric acid	1·12	Pectose	·9
Nitrogenous matter	·8	Mineral matter	·12
Gum, fat, &c.	·5		
Salts	·36		
Water	70·8		

The quantity of sugar varies with the different vineyards, and as examples the following Rhine grapes may be cited:—

No. 84.

	Per cent.			Per cent.
Kleinberger, quite ripe	10·59	Oppenheim, ripe	.	13·52
White Austrian . .	13·78	,, over ripe		15·14
Red Asmannshäuser, ripe	17·28	Johannisberg . .		19·24

The dried grape or raisin and dried figs have been long in repute, and enter largely into the recipes of the fourteenth century. There are many kinds in use at the present day. Thus the muscatels, which are so much prized for dessert, are dried whilst hanging on the tree, and are thence called raisins of the sun. All the vine leaves around them are first cut away, and the foot-stalk is half cut off, after which the bunches are left undisturbed and the fruit unbroken until they are carefully removed and packed in boxes. These are the largest and sweetest raisins in the market, and sell at a very high price. The sultanas are the smallest which are sold under the name of raisins, and their flavour is very delicious. The former grow in Southern Europe, and the latter in Turkey. The ordinary raisins are produced from the ripe fruit, and are imported from France and Southern Europe, as well as from Asia Minor. They have greatly improved in quality and appearance within the last twenty years, and probably from the use of lye in which they are dipped before being dried.

Currants from the Ionian Islands, which have been recently restored to Greece, are small seedless grapes, having less juice, and containing a less proportion of saccharine matter than the muscatels.

Raisins and currants are used chiefly in the preparation of puddings, but may produce good wine.

The date is the fruit of a palm (*Phœnix dactylifera*), which grows extensively in the Sahara, Persia, and generally throughout Asia and Africa, and bears bunches of fruit weighing 25 lbs., with perhaps 200 dates. It contains a very large quantity of saccharine matter, and is eaten when either fresh or dried; and so valuable is the tree that it is the chief source of wealth of the inhabitants. The fruit is largely imported into this country, and is highly esteemed, whilst where it grows it is nearly a necessary food. It contains, according to Reinsch, 58 per cent. of sugar, and 9 per cent. of pectin, besides other soluble substances, and water.

The edible fig (*Ficus Carica*) attains perfection in Southern France, Turkey, and the Mediterranean coasts, but it grows and is appreciated throughout Central Europe, Asia, and Africa, and ripens on the chalk soils of our own country. It has a more delicate and less satisfying flavour than the date whether fresh or dried.

The mulberry (*Morus*), both white and black, is an agreeable sub-acid fruit, well known in this climate even in Saxon times, and ripening in the southern parts of the island. It contains more sugar than blackberries or bilberries, as shown by the following analysis of Fresenius (per cent.):—

No. 85.

	Mulberries	Bilberries	Blackberries
Water	84·707	77·552	86·406
Sugar	9·192	5·780	4·444
Free acid	1·860	1·341	1·188
Nitrogenous	0·394	0·794	0·510
Salts	0·566	0·858	0·414

Cherries were grown in England in Saxon times, and used in the 14th century in the preparation of broth. They were called chyryse, and there is a Saxon word nearly resembling it.

The composition of cherries, according to Fresenius, is as follows, per cent. :—

No. 86.

	Water	Sugar	Free acid	Nitrogenous and Pectinous	Salts
Sour . . .	80·494	8·772	1·277	2·656	0·565
Rather sour .	81·998	8·568	0·961	3·529	0·835
Sweet black .	79·700	10·700	0·560	1·680	0·600
Sweet light red.	75·370	13·110	0·351	3·189	0·600

The apple and pear were known in this country before the Conquest, probably even before the Saxon invasion. Apple fritters, milk fritters and herb fritters were eaten in the middle ages, and in the 14th century were usually fried in oil or lard. They were eaten with honey or sugar, and much resembled the pancakes of our day.

An apple tart was made as follows :—'Tak gode applys and gode spycis and figys and reysons and perys (pears), and wan they are well ybrayed colourd wyth safroñ wel, and do yt in a cofyn, and do yt forth to bake wel.'

Fresenius's analysis of certain kinds of apples and pears is as follows, per cent. :—

No. 87.

Apples	Water	Sugar	Free acid	Nitrogenous and Pectinous	Salts
English rennets . .	82·04	6·83	0·85	7·92	0·36
White dessert . .	85·04	7·58	1·04	2·94	0·44
English golden pippin	81·87	10·36	0·48	5·11	
Pears					
Sweet red . . .	85·007	7·940	trace	4·646	0·284

The custard apple obtained from *Anona reticulata*, growing in South America, Africa, the East Indies, the Caribbean Islands, &c., is a fruit with a juicy pulp, which, when ripe, has the flavour of clotted cream and sugar, and is highly prized by rich and poor. The fruit varies

SUCCULENT FRUITS. 219

in different species of the order, from the size of an artichoke to that of a melon, and in the latter case it has a reticulated appearance somewhat like a pine-apple.

The alligator or Avocado pear is the succulent fruit of *Persea gratissima*.

The guava (*Psidium*) is both pear-shaped and apple-shaped, and grows generally throughout the East. It is aromatic, but not so agreeable as some other Eastern fruits, yet it is highly prized, stewed in wine, and for the excellent jelly which is obtained from it.

The mangosteen, alligator, or Avocado pear, as well as the mammee apple, are favourite tropical fruits, whilst the prickly pear, which grows in Majorca and over extensive tracts of country of the Old and New World, has long been known, but is not a fruit much sought after by the English people.

The pomegranate (*Punica Granatum*) grows extensively in the East and in Spain and Portugal. Its rind is a mild astringent, and its juice sub-acid.

The banana (*Musa sapientum*) is the prolific fruit of an allied species to the plantain, growing in Ceylon, Africa, and very generally throughout the East and West Indies. It is of the form and size of a large finger, and grows in clusters, two or three feet long, containing 100 to 200 specimens. Its flavour is luscious and delicious, but a little cloying to the appetite, so that a great number could not be eaten at one time. It consists of a thick skin and an enclosed pulp, and the latter alone is eaten. The analysis by Corenwinder is as follows, in 100 parts: Water 73·9, cane sugar and grape sugar, &c. 19·66, nitrogenous 4·82, cellulose 0·2, fat 0·63, with phosphoric anhydrid, lime, alkalis, and iron.

There are two kinds of melon (*Cucumis Melo* and *Citrullus*) sold in our markets, both of which are delicious

—viz. the water melon (*C. Citrullus*) of Spain and France, and the musk-melon of our English glasshouses, but the latter is the more agreeable.

The cannon-ball tree (*Couroupita Guianensis*) has a fruit as large as an infant's head, with a pulp of a vinous and pleasant flavour when fresh.

The peach palm (*Gulielma speciosa*) grows in South America, and bears clusters of fruit with a flavour like chestnuts and cheese, and Mr. Bates reports a belief that it is more nutritious than fish.

A large number of delicious fruits have been introduced into our South African possessions, amongst which may be mentioned the granadilla, yellow peach, white mulberry, Cape gooseberry, and the loquat. The latter is a delicious small apple, obtained from the Eastern Archipelago, with a juicy pulp and flavour between that of a gooseberry and peach, far excelling the English apple. It is the fruit of the *Eriobotrya Japonica*.

Souari nuts are the seed of the *Caryocar butyrosum*, a very large tree growing in Guiana and other parts of South America. They are like the Brazil nut in flavour.

Before quitting this part of the subject, it may be interesting to add, that when oranges are freshly gathered from the trees in the West Indies, the aromatic essential oil in the skin is so pungent as to blister the lips and disorder the stomach.

Strawberries, which trail on the ground in Great Britain, abound in many countries, and in parts of North America are so abundant in the woods as to feed a large detachment of soldiers. They are of delicious flavour, and grow on bushes three or four feet high.

Fresenius's analysis of strawberries and raspberries is as follows, per cent. :—

No. 88.

	Strawberries		Raspberries	
	Wild	Cultivated, quite ripe	Wild red	Cultivated red
Water	87·019	87·474	83·860	86·557
Sugar	4·550	7·575	3·597	4·708
Free acid	1·332	1·133	1·980	1·356
Nitrogenous	0·567	0·359	0·546	0·544
Pectous	0·049	0·119	1·107	1·746
Salts	0·603	0·480	0·270	0·481

The pine-apple plant (*Ananassa sativa*) grows in every hedge-row in Zanzibar, the East and West Indies, and very widely within the tropics, and has a flavour far more delicious than the same fruit when imported into this country. The flavour is well imitated by a solution of pure butyricite of ethyl in eight or ten times its weight of alcohol, and is sold as pine-apple essence.

A large portion of this class of fruits may be preserved in an edible state for a lengthened period, after they have been detached from the tree, either by allowing them to dry, as in the case of raisins, Normandy pippins, figs, dates, apricots, and tamarinds, or by simply keeping them in a suitable manner, as apples are kept, so that they may be used at nearly all seasons of the year, or transported to other climates and distant countries.

The preservation of fruit, whether produced in this country or prepared in other countries for exportation, is an art of great interest and importance, and has led to a very large commerce between nations. When not effected by drying, as already mentioned, it is chiefly by being immersed in a preserving liquid, after being cooked. In reference to all saccharine fruits, the preserving material is sugar dissolved and boiled in the juices of the fruit, as in the preservation of gooseberries, currants, apricots, and strawberries; whilst oleaginous fruits, as olives, are preserved in salt and water. Much care is required.

Plums, apricots, and peaches, were eaten by the Saxons in England, but probably not of the fine quality of our own day. They are too well known to require a description, and we shall simply indicate their composition, per cent., according to Fresenius :—

No. 89.

Plums	Water	Sugar	Free Acid	Nitrogenous	Salts
Greengages, yellow	80·84	2·96	0·96	0·47	0·318
,, large and sweet	79·72	3·40	0·87	0·40	0·398
Mirabelle, common yellow	82·25	3·58	0·58	0·19	0·570
Apricots					
Large fine-flavoured	82·115	1·53	0·76	0·38	0·75
Small	83·55	2·73	1·60	0·41	0·72
Peaches					
Large Dutch	84·99	1·58	0·612	0·46	0·42

Currants and gooseberries are amongst our most useful fruits and preserves, and grow as well in our Australian colonies and in America as in England, but are not suited to very hot climates.

The following is the analysis of Fresenius, per cent.—

No. 90.

Gooseberries	Water	Sugar	Free Acid	Nitrogenous	Salts
Small red	84·831	8·23	1·58	0·35	0·50
Yellow	85·364	7·50	1·33	0·36	0·27
Large red	85·565	8·06	1·35	0·441	0·31
Currants					
White	83·42	7·12	2·53	0·68	0·70
Red	85·27	6·44	1·84	0·49	0·57
Very large red	85·355	5·647	1·69	0·35	0·62

SUCCULENT FRUITS.

But however valuable chemical analyses may be, they afford but a very imperfect idea of the differences which exist in the flavours and other qualities of these delicious foods; and although such investigations are carried farther than is here indicated, it would almost have sufficed for all questions of food, to cite one example for all of them.

The elephant or wood apple (*Feronia elephantum*) of Bengal makes a preserve of an unusually soft and creamy flavour. The fruit is about the size of our apple, and the fleshy part which is adapted to this purpose is found underneath a very hard woody rind.

The mango, which grows generally in the tropics, is not only a luscious fruit, but makes a delicious preserve. It grows to the size of a large Jersey pear, and encloses a very large stony seed.

The fruit, known as Amatungula or Natal plum, and the Kei apple or Dingaan's apricot, are African fruits, which make excellent preserves. The former varies in size from an olive to a plum, and possesses a milky juice due to Caoutchouc, and of an agreeable sub-acid flavour. The latter is a berry which in its fresh state is so acid as to be used alone as a pickle, but when preserved with abundance of sugar is most agreeable.

The *Carissa Carandas* furnishes a good substitute for red currant jelly.

The quince (*Cydonia*) makes better marmalade than the orange.

Preserved nutmegs (*Myristica*) are perhaps the most aromatic and delicious of all conserves, but they must be used when quite young before the seed has begun to harden. The preserve is but little known in this country.

Preserved ginger (*Zingiber officinale*) when young and tender is prepared with sugar as a preserve in the East Indies and China, and large quantities are sent to England. It is a most agreeable addition to the dessert or to the tea-table.

It is not necessary to refer to the fruits from which preserves and jelly are made in this country, which are not excelled by the fruits of any climate; but it may be added that for this purpose the fruit should be fresh and dry when gathered, and well boiled with an equal weight of fine loaf sugar.

A review of the qualities of these productions enables us to place them in the first rank of subsidiary or luxurious foods, since they supply an agreeable and refreshing material when taken alone or with other foods which in health is desirable, and in disease almost necessary to life. They may be taken by the sick when nothing else is desired, and by acting upon the sense of taste may ultimately induce the invalid to eat food of a more nutritive character.

Moreover, certain of them approach nearer to foods than others, so that, for example, the olive as a fruit is less luscious than the more succulent fruits, but it has the two advantages of offering a true food in the form of oil, and of stimulating the appetite when pickled: so that in the former it is the servant of the poor and ill fed, and in the latter of the rich and over fed, but in its fresh state it is eaten daily by poor and rich alike, both as an agreeable fruit and a true food.

In their indirect action as the producers of wines and alcohols, they exert a totally different influence which, whether it be analogous to that of food or not, is of almost universal use, and exercises an influence for good or evil, second only to that of necessary foods in the maintenance of life. The readiness with which fruits assume the fermentation process, and the large quantity of sugar which they possess fit them for the production of alcoholic liquors, and give rise to that great trade of nations connected with the production and consumption of wines and spirits. This is particularly the case with the grape, since in addition to its special qualities it grows in temperate climates very abundantly, and offers a sufficient variety of flavour to repay the greatest devotion to its cultivation and the preparation of wine. But there is not a member of the class which may not be employed for the like purpose, and a thousand other liquors of the same nature are prepared in different parts of the world.

There is, however, a limit on the score of expense to the production of spirits from these substances in this country, for alcohol can be produced much more cheaply from grain, and it is only the finest qualities of spirits, as Cognac, which are usually produced from the grape. But in Eastern nations spirits are prepared readily and economically from sugar-bearing fruits as well as from the sugar-bearing juices of the locality.

The cost of production has moreover been lessened by diminishing the duration of the process through the distribution of air in the mass of fluid, as has been recently effected in California, so that the fermentation may now be completed in a few days instead of several weeks.

B.—*Albuminous.*

The class of fruits now to be considered are really seeds, and might have been referred to under that head, but as they are eaten when uncooked, and have the character of luxuries rather than of necessaries, it seems more fitting that they should be considered with fruits than with grain. Such are nuts, as the cocoa nut, hazel nut, or filbert; the almond, walnuts, hickory nuts, and many similar products.

The nut itself consists of the envelope or pericarp, which is almost always inedible, and the contained edible and nutritive substance. The latter consists of a solid material, which in composition resembles animal albumen or casein, and is therefore very much more nutritious than succulent fruits. These substances may be regarded as foods of high nutritive value if we consider their chemical value only, but they fail in that they could not be eaten in quantities sufficient to constitute a chief part of a dietary except for a short period when proper food might be unattainable. This is a familiar fact, as it respects nuts in use in this country, but in countries where the cocoa palm abounds, and great poverty exists, the cocoa nut is perhaps a not unimportant part of their daily food. It is also a familiar fact that, owing to difficulty of mastication, this kind of food is not easily digested, for any one eating a piece of cocoa nut is aware of the great length of time and careful attention necessary to reduce it to a pulp, but it is true of this class of food, as of cheese, that whilst difficult of digestion they promote the digestion of other foods. Hence, like cheese, they should be eaten with great moderation.

Of the whole class the cocoa nut (*Cocos nucifera*) is by far the most important to mankind, whether, considered as a delicious and nutritious food, or as supplying valuable oil and many other articles, useful in social life. Each tree yields from 80 to 100 nuts yearly, and will continue to bear during two generations of men. The edible part when ripe is composed of fat or oil to the extent of about 70 per cent., so that a quart is obtained from six or eight nuts. The quality of the oil or butter is very good, and might readily be used as food, but as it quickly becomes rancid, it is more commonly employed in the manufacture of fine soap and candles. It has also an albuminous or nitrogenous substance and various volatile oils and salts.

When green and unripe the nuts are lined with a creamy substance which subsequently becomes the hard white solid albumen which we eat in this country, and filled with water so cold as to render it desirable to add a little brandy to it. They are readily obtained by cutting the end of the nut across with a cutlass, and holding it upright. The enjoyment of this fruit by one living in a hot climate cannot be appreciated by those in temperate regions.

It is an Arab saying, that the cocoa nut and the date cannot exist together.

The cocoa palm supplies the nut as meat, milk, wines, spirits, vinegar, sugar and syrup; besides mats, cords, sails, strainers, tinder, firewood, houses, boats and fencing, and is undoubtedly the most useful tree in the world.

Almonds (*Amygdalus communis*) are both sweet and bitter, and are produced abundantly in Southern Europe and Africa. The sweet variety has been in use in this country since, if not before, the 14th century, in the preparation of almond milk by trituration with water, but in our day it is eaten more as a nut, with or without raisins. The bitter almond contains a proportion of prussic acid, but the quantity is not sufficiently large to render the nut itself dangerous.

Like other nuts, they contain fixed and volatile oils. The former which is obtained by compression, is bland and agreeable, and although not used as a food is esteemed for other purposes, and particularly by the hairdresser. The volatile oil is separated by distillation, and that from the bitter almond is highly poisonous until after the prussic acid has been expelled by heat.

The hazel nut was used in England during the middle ages in the preparation of milk, after the manner of almond milk for the cook. It grows in most temperate climates, but attains its greatest perfection in Spain and Southern Europe. The filbert, however, which is a congener, has by careful cultivation been produced in its greatest perfection in this country.

The earth nut or ground nut (*Arachis hypogœa*) is a vetch-like plant, which buries its pods in the earth, and produces oily seeds of excellent flavour when roasted.

The Brazil nut (*Bertholletia excelsa*), or Peru almonds as they were called three centuries ago, is enclosed in a capsule as large as a 24 lb. shot, very hard, and full of closely packed seeds. When falling in the forests they are a source of great danger to travellers.

The edible chestnut is doubtless an important article of food to the poor in Southern Europe, and possesses many excellent qualities. Its flavour is agreeable, it is very nutritious, it may be easily cooked, is much more digestible than other nuts, and may be eaten with bread alone. It is unlikely that this food will occupy the highest place among foods anywhere, and least of all in northern climates where it cannot be produced; yet it must be admitted that its use has largely increased of late years, both among the rich and poor, and that it is only the expense which prevents its more general extension among the latter class. At present it is regarded as a luxury.

The first step to a great extension of its use would be to make the ordinary horse chestnut a safe and agreeable food, since it grows in our climate, and could be obtained in large quantities. It has an acrid bitter principle, which does not exist in the Spanish chestnut, and makes it distasteful and injurious as a food. It is said that means might readily be adopted to remove this poison; but they could not be applied by the eater, and would therefore imply some preparation which would destroy the appearance of the nut and enhance its cost.

Kola or Guru nuts are the seeds of *Sterculia acuminata*, growing in Central Africa, which are bitter but highly prized, having the effect of causing water which is drunk to taste afterwards like white wine and sugar.

CHAPTER XXIX.

CONDIMENTS.

ALTHOUGH Condiments are rather adjuncts to food than foods, and rather medicines than foods, they are of extreme value in rendering food more palatable, stimulating a jaded appetite, supplying a necessary substance, and assisting in the preservation of food.

They are of three classes, which may be represented by salt, vinegar, and pepper, since, all compounds of the class contain some or all of these elements.

Salt, or chloride of sodium, has been in use from the earliest times, and is eagerly sought after as food by both man and animals. So necessary is it for man, that when revenue was the most urgent a tax was placed upon salt, since, nearly everything would be sacrificed to obtain a portion of that material, and the saline earths called *salt licks* are the greatest attraction to the wild animals of the prairie or the desert. Both the chlorine and the sodium, of which it is composed, are, moreover, part of the elements of the body, and are not yielded in sufficient quantity by the foods which we eat, and hence we crave for a further supply.

The immediate use appears to be to stimulate the sense of taste and to increase the flow of saliva, but its preserving action is due to its power to attract moisture, by which it tends to harden whatever moist substance is brought into contact with it, and when it has obtained moisture it becomes soft, and loses its flavour. There is no other compound of chlorine which effects both of these purposes or could supplant common salt. It is met with in an unlimited quantity in sea water, and in certain salt springs, from which it is obtained by evaporation and crystallisation. Salt springs are preferred by salt manufacturers to sea water, since they do not contain iodine or other elements which would modify the flavour of the salt. Another source is the earth itself, as in the salt mines of Cheshire, Salzburg, Cracow, and the Punjab, whence it is extracted by digging or washing. Such salt is generally mixed with colouring matter, and is prepared for the table at greater cost than from salt springs.

Vinegar is an acid fluid which is prepared from many sources. Thus all saccharine materials may put on the acetous fermentation and produce vinegar. The finest is that prepared from the grape, but malt is well known as the largest source of vinegar in this country. This kind of fermentation is produced by a temperature of 75° to 80°, maintained for some days or weeks. At a lower temperature, vinegar is gradually produced from beers or wines which contain little spirit, and are exposed to the atmosphere. Such is more or less coloured, but a tolerably white vinegar may be produced from white wine, and perfectly white vinegar is the product of distillation of wood or pyroligneous acid.

The strength of the vinegar varies much, but may be increased by evaporation, and is not unfrequently fortified by the addition of a strong mineral acid or sulphuric acid. The proof strength is 5 per cent. of pure acid and has a sp. gr. of 1·019, but vinegars are usually lower than that standard.

Vinegar of very excellent quality and at a moderate price may now be obtained in every part of the country, so that it is no longer necessary to make it at home, but the following recipe will produce it: viz., 1 gallon of water, 1¼ lb. of raw sugar, and ¼ pint of yeast. At a temperature of 80° it will be sufficiently acid in three or four days to be drawn off, when an ounce of cut raisins and the like weight of cream of tartar should be added, and after a few weeks the sweet taste will have entirely disappeared, so that the fluid may be bottled. A more simple method for the production of a small quantity is to procure the vinegar plant (*Penicillium glaucum*) and place it in a weak solution of sugar in a warm place for a few days. The same plant will continue to increase and may be used again and again for the same purpose. The production of vinegar from any saccharine material is accompanied by a fungoid plant, so that vinegar produced in the purest manner from wine lees deposits a material called 'mother of vinegar,' which is a mycoderm (*Mycoderma Vini*) and when added to weak alcohol produces vinegar. It consists of cellulose and a nitrogenous principle. The process has, however, some connection with the extractive substance of plants, for when that of wine is lost by age the acetous fermentation is not easily produced.

The addition of beech shavings is valuable for the purpose of clarifying the vinegar and the acetous fermentation process is greatly expedited by the free admixture of air with the alcoholic liquor as it passes through the shavings.

The flavour is due in some degree to the acid itself, but much more to the substances which accompany it, such as various acetic ethers; and in proportion to its strength in acid, and fulness and delicacy of flavour is the vinegar esteemed. Distilled vinegar is weaker in acid if not in flavour than raw vinegar, and having less colour is generally preferred. As acetic acid is the true acid of vinegar, all others must be regarded as adulterations. Such are hydrochloric and sulphuric acids, the former of which may be ascertained by the white deposit which follows the addition of nitrate of silver, and the latter of chloride of barium.

Although acetic acid is transformed into carbonic acid, and so far is a food, it would be a refinement to say that vinegar supplies nutritive materials, and its sole use, as a food seems to be, to flavour food and stimulate the nerves of taste. It has also a powerful preserving action by which it hardens flesh and prevents the decomposition of both animal and vegetable substances.

Peppers and spices constitute a very large and important class of condiments, and have been in use in all ages and in nearly all climates. It is true that peppers are the product of hot climates solely, and if used in temperate or cold climates, must be imported, but they are light and compendious, and particularly fitted for the mode of conveyance which was in use in distant ages. But many other substances growing in temperate climates have similar properties, and are substituted as occasion may require.

Spices were always highly esteemed, and not only were a principal article of merchandise, but acceptable presents even to a king. Thus the Queen of Sheba presented spices in great abundance, and it is written, 'Neither was there any such spice as the Queen of Sheba gave King Solomon.'

So important were they even in our cold climate, that in our early history the spicery was a special department of the Court, and had its proper officers. Spices were necessarily rare and costly in the 14th century, since they were imported from the Levant, and were not then in general use. Chaucer and Wiclif mention cinnamon or canella, mace (or as then called macys), cloves (or clowe), galyngal, pepper, ginger, cubebs, grains of paradise (or de Parys), nutmegs, caraway and spykenard de Spayn, are specially noted in the recipes. Certain compounds of spices, as our allspice, were then used, as powder douce and powder fort. Of all these, only the carraway seed is now produced in England, but at that time it was an exotic, and imported from Caria. It was not unusual to flavour fluids with leaves as bay leaves, and with flowers, as the white and red rose and the hawthorn. The class is, however, much too large to allow us to devote the requisite space for detailed description, and we shall content ourselves with indicating certain typical specimens.

1. *As to Fruit.* The seeds or berries of plants are the chief source of this pungent material, including the peppers with which we are so well acquainted. The black and white pepper of commerce are produced from the same plant (*Piper nigrum*), a low shrub which grows in the East Indies and Ceylon, and indeed extensively throughout the East. The berry has a black or dark brown cuticle, and when the whole is ground, the product is black pepper, but when the skin is removed, white pepper is produced.

Long pepper is the fruit of several species of the

Chavica, and particularly of the *Chavica officinarum*, so called, because it has been used in medicine in almost all ages, and is more pungent than white pepper.

Jamaica pepper or Pimento is the dried berry of the *Eugenia Pimenta*, which grows in Jamaica to the height of 20 or 30 feet, and bears flowers of an aromatic odour. The berries are gathered when green, and most carefully dried, after which they are of a brown colour. The flavour is much less pungent and more aromatic than that of black pepper, and is so rich as to be called Allspice.

Cubebs is the dried fruit of the *Cubeba officinalis*, a climbing shrub, which grows abundantly in Java, Sumatra, the Maluccas, and other Eastern countries, where it is in common use as a condiment; but in this country it is now employed almost exclusively as a medicine, and *Cubeba Clusii* in Western Africa.

No. 91.

Showing the nutmeg with its foliaceous arillus or mace.

Coriander, cummin, and carraway seeds are produced in temperate as well as in hot climates, and have an agreeable aromatic pungency.

The nutmeg is the fruit of the *Myristica moschata*, and *fatua*, and other trees of the same genus, growing in the Straits Settlements, the West Indies, Maluccas, and many parts of Asia. It is enveloped in a fleshy sarcocarp, and when unripe, the whole fruit makes an excellent preserve. The seed of the ripe fruit is dried and sent to Europe whilst whole, but it is easily torn into small portions by the grater, which is in common use. It is very delicate and aromatic in its flavour, but not by any means so pungent as ordinary pepper. Hence it is more fitted for use with farinaceous foods, having a delicate flavour, which would be destroyed by the use of a strong pepper. Mace is a thin foliaceous aril, which is attached to, and envelopes the nutmeg. It has the character of the nutmeg, with perhaps greater delicacy of flavour, and it is used whole (No. 91).

Cardamom seeds are the fruit of species of *Elettaria* and *Amomum*, growing in Malabar, Travancore, and Siam, as well as in many other parts of Asia. They are said to be divided commercially into three classes, viz., shorts, short-longs, and long-longs, according to the length of the fruit. They are very aromatic.

Grains of paradise are also the product of the same genus, the *Amomum Melegueta*, growing in tropical West Africa, and are used in this country chiefly to give increased pungency, or false strength, to beer. They are not employed here in cookery.

Mustard seed (*Sinapis nigra* and other species) was used by our Saxon forefathers, and in the 14th century was mixed with honey, vinegar, and wine, as a condiment. It is produced largely in this country and Europe generally, and is perhaps the chief condiment of this class. It is never used whole, but is ground into an impalpable powder, and is commonly mixed with a proportion of flour to mitigate its natural pungency. Hence the mustard makers are also manufacturers of starch, and there is a tendency to use an undue proportion of flour by which the pungency of the powder is so much enfeebled, that capsicum is added to restore it.

The seed-pods of numerous plants are used as condiments. Thus the cayenne pepper is the roughly ground pod of the *Capsicum frutescens*, and other varieties, which grows in both temperate and hot climates. The same product is called chillies in Mexico and other parts of America, and is used largely when cooking beans.

The flower bud of the *Caryophyllus aromaticus* is the clove of commerce, and is so called from its resemblance to a nail (*clou*). The tree grows in the Maluccas, Mauritius, Sumatra, Zanzibar, and many other tropical countries. The flower is dried in the sun, or by artificial heat, and in the process becomes much duller in colour. It has a somewhat pungent aromatic flavour.

2. *The Bark of Plants.* This division is well represented by the cassia and cinnamon, which grow in Java, Ceylon, China, Japan, and many other Eastern countries. They are the product of the *Cinnamomum Cassia, Cinnamomum zeylanicum*, and other species, but of the two the cinnamon bark is much more aromatic and useful as a condiment. So highly was cinnamon esteemed, and so close the monopoly which the Dutch had in Ceylon, that the punishment of death was inflicted on those who injured the plant or illegally exported the bark or oil. It is used in pieces or it may be ground into a fine powder. The finest quality is obtained from the young shoots or suckers, since it has a less proportion of woody fibre and a more delicate aroma.

3. *Leaves.* It is necessary to premise that whilst there are leaves which are used as condiments, the same property is generally extended to the fruit, and very frequently to the flower and bark, if not to the wood, of the shrub or tree.

Bay leaves of various species of *Laurus*, spear mint, peppermint, penny royal, and all our well-known garden herbs, belong to this division. The powdered leaves of the *Baobab* are eaten as a condiment in Central

CONDIMENTS. 237

Africa, and the leaves of the betel-pepper (*Chavica Roxburghii*) are chewed by the Malays. It may also be added that the leaves of the tobacco plant (*Nicotiana Tabacum*) are chewed rather as a condiment than as a narcotic.

4. *The Root, Rhizome, or Bulb.* This division is well represented by the ginger plant (*Zinziber officinale*), growing in the East and West Indies. When fresh, it makes a gently stimulating preserve, and when dried may be used either whole or ground into powder. It enters into almost every compound of this class, and is one of the most useful and least injurious members.

The following drawing represents the *Zinziber officinale*, with the rhizome, which is preserved.

No. 92.

The Ginger Plant.

Turmeric is the dried tuber of the *Curcuma longa*, and grows abundantly in the East Indies, where it is one of the chief condiments used in the preparation of food, and particularly of the various seeds which are called *dhal*. It is ground into a fine powder before use; but in this country it is not used separately as a condiment, but as a yellow dye, and is so delicate that it is used as a test-paper for determining the presence of alkalies.

Garlic (*Allium sativum*) is eaten both in hot and temperate climates; what is termed the clove is the young bulb. The onion is an allied genus, and although it is eaten as a food, the small onions are used as a condiment in pickles.

In reference to nearly all the substances comprised in this division of Condiments, it may be remarked that their property is due to special volatile oils, and that such oils are obtained in the largest proportion from the fruit and leaves and sold separately.

From these various sources which may be termed elements, are obtained the materials from which the infinite varieties of sauces, pickles, and similar appetising compounds are made, and to which we may only slightly refer.

Curry powder is a mixture of various peppers and other condiments coloured with turmeric, which is in universal use in India and the East, but which varies in flavour with each manufacturer. Its use is less necessary and defensible in a temperate than in a hot climate, and it is rare for one in England to tolerate the quantity of capsicum which is relished in India. It is said that a curry mixture should contain all the following ingredients, viz.: greenginger, garlic, coriander and cumin seeds, onions, Chili pepper, turmeric, rasped coco nut pulp and lime or lemon.

A composition very similar to this was in use in the spicery of the middle ages.

Cassareep is produced in the West Indies and Guiana from the root of the tapioca plant (*Manihot*), and is extensively used both to flavour sauces and to preserve meat. As a condiment it is very valuable and is prepared by boiling the expressed juice of the roots until it is reduced to the consistence of syrup, when it is highly seasoned with pepper, cinnamon and mace. The scum is carefully removed during the process. It is so strong in flavour that a tablespoonful is sufficient for a tureen of soup, and is useful for every purpose of a condiment and as a menstruum in which to impregnate and conserve animal food.

The palaver sauce is prepared from the leaves of the Baobab, and flavoured with pungent spices in Central Africa.

Ketchup or catsup is obtained from mushrooms and spices.

β. *Non-Nitrogenous.*

CHAPTER XXX.

STARCHES, SAGO, ARROWROOT, TAPIOCA.

THE non-nitrogenous vegetable products are of two classes, viz., starch and fat, which have the same elements of composition, but differ in their proportions. They are very interesting and important foods, and although not constituting an essential part of any dietary, are amongst the more agreeable of ordinary foods, and their use could not be dispensed with without great regret.

The starchy foods to which we shall now refer are used as puddings, and are never made into loaves of bread

Such are sago, arrowroot, tapioca, cassava meal, manioc, semolina, and many other similar productions. They are rarely absolutely free from a trace of nitrogen, but the quantity is inappreciable when considered as a food. It may also be added, that nearly all the members of this class are particularly useful as food for children and the sick.

Sago and Arrowroot.

Sago is the product of various species of *Sagus* and other palms, and a little from the *Cycas revoluta* and other trees in India, Ceylon, and Eastern countries (No. 94). It is obtained by washing the pith with water and drying the sediment, after which the sago appears in round grains, somewhat varying in size, or in small granulated masses. The grains, although so hard when dry as not to be easily broken by the teeth, absorb water somewhat readily, and make a mucilaginous food when prepared as pudding or gruel. It possesses but little flavour, but is not in any degree disagreeable, so that with suitable adjuncts it becomes an agreeable food.

Its nutritive value is not very high, since it consists almost exclusively of starch, and is therefore inferior to rice, and much inferior to the farinaceous foods grown in our own climate. The ultimate elements present in sago must be considered as if it were so much starch, on the following percentage:—

No. 93.
C. 44·4   H. 6·2   O. 49·4

There are 2,555 grains of carbon and $1\frac{3}{4}$ grain of nitrogen in 1 lb.

When cooked it is digested in about one hour and three-quarters.

Arrowroot is a well-known food, derived from various

SAGO. 241

No. 94.—Sago Plant.

Represents a part of a plantation of a Sago-producing tree (*Cycas revoluta*).

sources, but chiefly from the plant (*Maranta arundinacea*), which grows in the Bermudas, the West India Islands, the East Indies, and various other parts of Asia as well as in Africa. It is prepared from the rhizomes or roots by washing and drying. The rhizomes are first very carefully peeled, so as to exclude a resinous substance, which would give an unpleasant flavour to the arrowroot, and the inner portion is rasped or beaten and broken to pieces by mills or deep wooden mortars. The arrowroot being insoluble subsides to the bottom of the vessel in which it is washed, and may then be collected on hair sieves, and separated from fibres and other foreign substances.

English arrowroot is obtained chiefly from potatoes, but may be profitably prepared from rice or maize. It resembles sago in being almost a pure starch, as well as in its nutritive and culinary properties; and those substances are preferred from which to prepare it that contain very little nitrogen, and whose cost is less than that of wheaten flour.

Bermuda arrowroot is the most esteemed, and is not unfrequently adulterated with less costly preparations of starch, as potato starch, sago meal, and Brazilian arrowroot; but the chemical properties are not thereby changed. When the adulteration is made with wheat flour, it may be detected by the foam which remains upon the surface of the liquid in which a portion of the arrowroot has been boiled and persistently stirred, for with pure starch, as arrowroot should be, there would be no froth. This test depends for its action upon the presence of gluten in wheat flour.

The proximate elements in 100 parts when pure are water 18·0, and starch 82·0; so that it is or should be free from nitrogen. There are 2,555 grains of carbon in 1 lb.

The time required for digestion is the same for sago and arrowroot, viz., from one hour and three-quarters to two hours.

Ten grains of arrowroot when thoroughly consumed in the body produce heat sufficient to raise 10·06 lbs. of water 1° F., which is equal to lifting 7,766 lbs. one foot high.

In my experiments on arrowroot when eaten alone and on an empty stomach it gave no sense of satisfaction, but, on the contrary, there was a sense of sinking and *malaise* in the stomach and bowels in about an hour. The effect of arrowroot upon the respiratory and other vital functions was very small. After eating 500 grains well cooked in water, the average increase in the emission of carbonic acid was only 0·154 grain per minute, whilst there was a subsidence of the rate of pulsation and respiration. When the same quantity was taken after a perfect fast of twenty-four hours, the effect was still very small, but it was greater than when eaten under ordinary circumstances. The maximum increase in the carbonic acid evolved was only 0·45 grain per minute (No. 95, frontispiece).

The addition of one ounce of fresh butter scarcely increased the effect of arrowroot alone on the respiration, for the total average increase was only 0·17 grain, and the maximum increase 0·4 grain per minute. The rate of respiration was somewhat lessened, whilst that of pulsation was increased four beats per second (No. 95, frontispiece).

When 250 grains of sugar were added to the arrowroot, as in the preparation of pudding, there was a further increase in the effect, for the maximum increase in the carbonic acid evolved was one grain per minute, and of the quantity of air inspired twenty cubic inches per minute. The rate of pulsation was somewhat increased, whilst that of respiration was lessened (No. 95; see frontispiece).

The sense of satisfaction was greatly increased by the addition of sugar or butter to the arrowroot.

When starch was obtained by washing wheat flour, or as commercial starch, the effect varied with its purity, that is to say, with the gluten which had not been removed by the washing, and with gluten there was a much greater increase in the carbonic acid evolved and in the rate of both respiration and pulsation. Hence there can be no doubt that starch to be used as a nutrient should not be entirely free from gluten or other nitrogenous matters.

Tapioca or Cassava.

This starchy food is prepared from numerous plants which grow in tropical America, Asia, and Africa; but particularly from the *Jatropha Manihot,* and other species of the Euphorbiaceous family, although some yield highly poisonous juices.

It is found in the rhizomes or roots, which resemble large turnips, and grow in clusters, and is prepared in a manner very like that of arrowroot. The fecula, or starch, which has been washed out of the bruised roots, is carefully dried on heated plates, by which many of the starch-cells are broken, and dextrin is produced. The heat thus applied is apt to lessen the nutritive value of the product; but is essential, in order to drive off the poisonous acid which is found in the juices. It is imported from Brazil, under the name of Brazilian arrowroot, and both are essentially the same in nature and nutritive value.

It is very largely used in South America in the preparation of cassava cakes, by the farina, or meal, being heated, but in nutritive value they are far inferior to oat-cakes. Slices of the root are also dried and eaten as cassava bread. There is also a cassava which possesses the same substance without the poisonous juices, and known as sweet cassava, or sweet yucca, the rhizome of which may be eaten raw, roasted, or boiled.

The chemical properties are identical with those of arrowroot, and almost identical with sago, and as foods they are all practically equal, the one to the other.

Captain Burton says that this food is full of gluten (in which he is doubtless mistaken), but by no means nutritious, and after a short time it produces an inordinate craving for meat; yet it is a common article of food for the poorer classes on the western coast of Africa, where there are fifty ways of cooking it. It tastes like parsnips or watery potatoes.

Semolina.

This favourite food, from which puddings are made, is perhaps the most valuable of all the substances so employed, since it is derived from the finest and hardest wheat, and as it contains a larger proportion of nitrogen than tapioca or arrowroot might have been described elsewhere.

It is that portion of the central part of the grain of wheat which is not reduced to powder in the process of grinding by stones, and is produced from the grains of very sunny climates, as Spain, Odessa, and some other parts of Russia and the south of Italy, where the grain becomes very dry and hard. Its production is, in a degree, a necessary result of the imperfection of the grinding process, since it lies in the furrows of the millstones, but the quantity may be increased at the pleasure of the miller.

It is used in Italy for the preparation of polenta, and in France is made into a favourite kind of bread, whilst in this country it is employed in the preparation of puddings alone. In Algeria it forms the national dish called *Couscousou*, with the addition of vegetables, butter and fowl, and is as good as its composition and cooking are complicated. A preparation very similar to it is made from millet and maize, but the flavour is inferior.

CHAPTER XXXI.

VEGETABLE FATS AND OILS.

WE have now to consider a class of substances which are used by man everywhere in temperate and hot climates, and which offer very great variety in their source of production and flavour.

Fat, whether in a solid or liquid state, is found in nearly all vegetable productions, and adds to their nutritive value and flavour, but the quantity is sometimes so small as to preclude the attempt to obtain it separately. It is more particularly stored up in the seeds, and can be obtained by expression, with or without the aid of heat. Thus, the cocoa-nut is now extensively employed in the East Indies and other countries for the production of oil only, and yields a pure and clear oil, which is valuable as a food.

Numerous other palm-trees of Eastern countries supply fats, which are called butter, and which, when properly treated, yield oils which are used as food. Hence, we have the butter-nut and the Shea butter, as well as palm-oil, which are bringing the African people into communion with civilised nations, and upon which our greatest hope of the abolition of slavery and the cultivation of legitimate commerce chiefly depends.

The Brazil nut, ground-nut, the hazel-nut and the walnut are amongst many which are known in this country as possessing oils, and from the two former the oil is expressed, and sold as articles of commerce.

Fruits, which are less albuminous in their nature than nuts, also yield oil largely, so that edible vegetable oils are best known in this country, and in Europe generally, by the olive-oil, which was also used instead of butter in England in the fourteenth century; but in hot climates, and particularly in India, a great variety of vegetable oils are obtained and used as food.

The olive tree (*Olea Europœa*), which produces olive-oil, grows in Syria, and other parts of Asia, as well as in the southern countries of Europe, and is everywhere highly esteemed. Its fruit varies in size, so that the Spanish is twice as large as the French olive, and has a fleshy coat or sarcocarp, which contains the oil. The olive is used as food by the poorer classes in the country where it is produced, and when eaten with bread is agreeable and nutritious. The drupe is, however, as commonly used for the preparation of olive oil; or, after pickling in salt-water or lime-water, it is exported to countries where it is eaten as a luxury, to stimulate the appetite or digestion.

The oil is obtained from the fruit by pressure, and that which comes over first is the purest, and is called virgin oil, whilst a further portion is expressed by means of heat and water, and is much less valuable. The oil thus obtained is not so clear as it subsequently appears in the market, until subjected to a purifying process. This consists simply in allowing it to settle in tuns for about twenty days at a temperature of 60° F., and then drawing it off clear.

When it is fresh and pure it has only a very slight yellowish green colour, and but little smell or flavour, so that it may not only be drunk by those who like oil, but is particularly fitted for cooking delicate foods. It cannot be doubted that it is one of the most easily digested fats in food, and its use might be properly extended in this country, notwithstanding our excellent animal fats.

So valuable a substance excites the cupidity of those who deal in it, and becomes adulterated with cheaper oils, such, for example, as the oil of the ground-nut of Africa (*Arachis hypogœa*), the oil of other seeds, and even animal products, as lard oil. Such additions or substitutions may not lessen the nutritive value of the oil, but they diminish its peculiar flavour, and, being cheaper, are fraudulent.

But the largest source of vegetable oils is the small seeds of plants, and some of them are used as food. The seed of the cotton-plant, mustard-seed, linseed, and rape-seed may be quoted as illustrations of these products, and oils of a very fine quality are procurable from most of them. The seeds of the common cucumber may be especially cited as yielding an edible oil of delicious and delicate flavour; and that of the large cucumber grown on the African coast far exceeds in flavour the finest olive-oil. Seed-oil is much more commonly eaten in India, and other hot countries, than in England. The seeds of the sunflower, yielding 40 per cent. of oil, and sesamum oils may be added to the list, as representing Indian oils, which are used in cooking gram and other vegetables; whilst in the gloomy forests of Central Africa, where milk and butter are rarely attainable, and in the great mangrove swamps, where the cassava, plantain, and yam are the chief foods, palm-oil and vegetable fats are almost necessaries of life.

There can be no doubt that we have in this product of the seeds of plants, which seem otherwise to be useless, a great storehouse of most valuable nutritive material; and if we know but little of them in this climate, it is because we have the olive-oil at hand and are bountifully supplied with many kinds of animal fats. It is, however, probable that the cheapness of some of these vegetable oils, in addition to the delicacy of their flavour, will, ere long, force themselves into notice, and obtain a place among our foods.

CHAPTER XXXII.

SUGAR, TREACLE, HONEY, AND MANNA.

THE modern researches of chemists have so extended our knowledge of sugar and saccharine compounds, that, were we to enter into detailed chemical descriptions, we should have to describe a class of substances rather than the one food which is popularly known as sugar; and as that would be incompatible with the scope of this work, we shall strictly limit our observations to the popular conception of the subject.

Sugar is not a product of Nature, which, like a fruit,

may be detached by mechanical means from the plant producing it, but a component part of the juices of plants, and is obtained by expression, evaporation, crystallisation, and purification. It is met with in the juices of nearly all edible plants, but particularly in fruits and in the sugar-canes, which are cultivated solely with a view to its production. It is also found very largely in the subterranean stems and rhizomes of many plants, as beet-root, carrot, and parsnip, and in the juices of certain forest trees, as the sugar maple of the New World, some palms, and in all edible seeds. It is found in the blood, flesh, and secretions of animals, particularly milk, and is a product both of health and disease.

Hence, the sources whence it might be derived are exceedingly numerous and widely distributed, and there is probably no part of the globe where the inhabitants do not obtain it, so that it would scarcely be beyond the truth to say that it is the most generally distributed of all proximate chemical elements.

Sugar was not introduced into England until the fourteenth century; and, as it was imported from countries so distant as the Indies, by Damascus and Aleppo to Venice and Pisa, or from Cyprus, it was an expensive luxury. It could not compete with honey, which was the native saccharine food, but was reserved for choice dishes and for the use of the wealthy classes. It was even refined at an early date, but the process was effected with wine, and probably did not proceed further than the removal of foreign matters.

It was, however, known to, but rarely used by, the ancients as far back as the fourth century before Christ. It is not mentioned in the Old Testament Scriptures; and in those records the quality of sweetness which we attach to sugar was connected with honey, so that the expression used was, not 'As sweet as sugar and the sugar-cane,' but, 'As sweet as honey and the honey-comb.'

Amongst an almost infinite variety of so-called sugars, as understood by chemists, there are three kinds which demand notice, viz., milk-sugar, cane-sugar, and grape or fruit-sugar. They have the same chemical elements, but differ much in sweetening power. Milk-sugar is so called because it is chiefly derived from the milk of animals, and cane and fruit sugars from having their chief sources in the sugar-canes and fruits. The former is less soluble and has less sweetening property than the latter, and its use is infinitesimally small.

The usual sources of ordinary crystallisable or cane-sugar are four, viz., the sugar-cane, maple-tree, palms, and beet-root, but of these the former far exceeds the others.

The sugar-cane grows, or may be grown, freely and almost universally, in hot or tropical countries, as the lowlands of the Mississippi, the West Indian Islands, Spain, the Mauritius, Java, the Brazils, parts of Africa, Australia, and the East Indies, and also over extensive tracts in Asia. It is not grown in the temperate regions of Europe, or in districts where the elevation is so great as to reduce the temperature to that of temperate climates, and everywhere the product varies as the degree of heat, other conditions being equal.

The cane grows to twelve or even twenty feet in height, and to one inch and upwards in width, so that a field of canes fully grown and in flower presents a very handsome appearance (No. 97).

The chemical composition of the cane is as follows, per cent. :—

No. 96.

Water 72·1 Sugar 18·0 Woody fibre and salts 9·9

252 NON-NITROGENOUS VEGETABLE FOODS.

There is a larger proportion of sugar in the juices at the lower part of the stem, and 100 lbs. of mixed cane yield about 70 lbs. of sugar.

No. 97.

The Sugar Cane (*Saccharum officinale*).

The machinery now employed in the manufacture of sugar from the sugar-cane is ingenious and expensive, but the process of manufacture is simple. The canes are passed through slowly revolving rollers, which press out the juice, and the juice is then mixed with a little quick-lime and passed through a series of heated pans. In this process it is heated more and more until it boils furiously, by which the impurities rise as a scum to the surface, whence they are removed, whilst the juice is reduced to the consistence of oil. The further evaporation and crystallisation takes place in vacuum-pans, where the crystallised is separated from the un-crystallised sugar or molasses, and is ready for the market, and much skill and care are exercised in the manufacture of the sugar.

The crystals of sugar when purified are colourless, but before decolorisation are usually tinged by the adhesion of a dark syrup, consisting of uncrystallised sugar and fruit sugar. The chief object in the process of refining is to get rid of these impurities, and this is effected by two processes, known as the clay process (from which such sugars are known as clayed sugars) and the centrifugal process.

In the former, a layer of very moist clay is placed over the sugar, and the water leaving the clay passes slowly through the mass of sugar, and carries off the syrup adhering to the crystals, whilst, at the same time, it dissolves and removes but a very small quantity of the crystallised sugar. In the latter, the sugar is caused to revolve at a great speed in the inside of a drum, which is covered by fine wire gauze, and the syrup is thrown through the meshes of the gauze, whilst the sugar is retained within.

Refined sugar is of two descriptions, viz., loaf sugar and pieces, the former produced in London, the latter at Bristol and in Scotland; but it is by no means easy for one ignorant of the trade to distinguish the one from the other.

In both alike the raw sugar is dissolved in water and boiled, and the impurities are removed either by the addition of bullocks' blood, which causes them to rise to the surface, or by passing the syrup through cotton bags. It is rendered colourless by filtering the clarified syrup through layers of charcoal, and it is again concentrated by placing it in the vacuum-pan. After a certain time crystals begin to form, and are enlarged by the addition of fresh syrup and a continuance of the boiling process; and when large enough the sugar is withdrawn from the vacuum-pan, and placed in 'forms' to complete the crystallisation.

When it is intended to make loaf-sugar, the uncrystallised syrup is removed from the crystals by the addition of a concentrated solution of pure sugar, which carries off the syrup and leaves the crystals clear; and the latter, having been dried in a stove and turned in a lathe, form the loaf-sugar of commerce. The drainings are, however, again concentrated by boiling, and yield a proportion of crystals, which become loaf-sugar of an inferior quality, whilst that part which cannot be crystallised is sold as treacle. When, however, loaf-sugar can no longer be produced, a moist sugar, containing some treacle, is made and sold as 'bastards'; and all that cannot be so prepared is disposed of as treacle.

In making 'pieces,' the separation of the uncrystallisable part is effected by the centrifugal machine, and the first production of crystals is called 'lumps,' whilst all others from the same syrup are termed 'pieces.' Ultimately all, or nearly all, the treacle or syrup is absorbed by a sort of bastard crystals, and sold as raw sugar.

It is a familiar fact that all kinds of sugar have not the same sweetening properties, but the degree of variation is not by any means well known. This is due, in part, to the quality of the source whence it is derived, for where there is much grape or fruit-sugar mixed with the cane-sugar, as, for example, in beet-root juice, the sweetening property is lessened in proportion to the amount of fruit-sugar. In other part it is due to the process of manufacture, for the heating process converts cane-sugar into fruit-sugar, and the more so the longer the process is continued. Hence, the first crystallisation contains the greatest proportion of cane-sugar, and is the sweetest, whilst the quality deteriorates at every succeeding crystallisation; and finally the treacle, or uncrystallisable syrup, consists largely of fruit sugar. These qualities may be in great part determined by the size of the crystals, for it lessens as the quality deteriorates, and at length is supplanted by granulation, or bastard crystallisation.

The aim of the refiner is, therefore, to select a raw sugar which contains the least quantity of impurity, or any flavour distinct from that of sugar, and which also yields the greatest proportion of cane-sugar, whilst in the process he seeks to produce the largest and whitest crystals, and to cause the least quantity of granules or crystals to absorb the greatest amount of uncrystallisable syrup; that is to say, to make the greatest quantity of sugar and the least of syrup.

That kind should be preferred for domestic use which consists of the largest crystals, and has the least proportion of moisture; and, as it respects loaf or lump sugar, considerable time should be allowed for the perfect solution of such crystals. A very moist raw sugar, which has an homogeneous appearance when rubbed between the thumb and finger, and a refined sugar, which is broken with comparative ease, and not of a pure white colour, should be avoided. It is also very useful to judge of sugar by the smell, for 'pieces' and raw sugar from beet-root have an unpleasant odour, whilst pure sugars are entirely, or almost entirely, free from it.

The amount of sugar in different specimens varies from 100 per cent. in pure refined sugar and sugar candy to 94, 88, 80, and 67 per cent. in the first, second, third, and fourth classes of unrefined sugar, in accordance with the classification adopted by the Sugar Convention of the different Governments.

The adulteration of cane sugar is effected with starch and glucose or grape sugar, all of which lessen its sweetening power without introducing any deleterious property. Sand and water are also added with a view to increase the weight, but it is believed that this mode of adulteration is now much less frequent than formerly. Starch may be readily detected by the iodine test, and sand by its insolubility and grittiness. Iron is not unfrequently found in the low class of sugars as an impurity, and renders the infusion of tea black by its action on the tannin of the tea leaf.

The Chinese sugar grass (*Sorghum saccharatum*), a kind of millet, is also largely used in China, America, and Europe, for the production of sugar, as the date, fig, and similar fruits are used in Asiatic countries for the same purpose. It is well known that the common barley of this country contains so large a proportion of saccharine matter that by the process of malting it yields the sugar contained in beer and porter, and is therefore an important source of that food.

The sugar maple (*Acer saccharinum*) abounds in North America, and resembles the sycamore tree. It is not generally grown purposely for the production of sugar, but its juices are very abundant in the spring and summer, and yield a large amount of it. One tree yields from two to six pounds of sugar yearly, and one pound of sugar is produced from four gallons of sap. The method adopted to procure sugar is very simple. Holes are bored obliquely upwards into the lower part of the trunk of the tree in the spring season, through which the juice exudes, and by a proper apparatus the juices may be collected in vessels, without waste and almost without labour. Thus obtained, they are of a brown colour, and on evaporation and condensation by artificial heat yield sugar in both a crystalline and non-crystalline form, which is run into moulds. It is not generally prepared for sale, and therefore is rarely purified or decolorised, and hence, although it sweetens well, it has other flavours besides those due to the sugar, and would not be preferred to the purified cane sugar of the market where that could be obtained at or about the same cost.

It is for the most part prepared by the settlers in America for their own use, and is therefore obtained at small cost.

There are also numerous palms which yield sugar very freely, and particularly in India, where the juice is expressly used for this purpose. It is obtained by cutting off the male flower when young, and allowing the juice to flow for several months, during which time it is collected, and then evaporated and purified. Such palms are the *Wild date* and the *Arenga saccharifera*.

Beet-root sugar was first made in 1747, and is now more largely produced in France than in any other country, not only for home consumption, but for export to England. This is, however, due less to the special superiority of this source of production than to the absence of French colonies, whence the sugar could be obtained from the sugar-cane and imported into France. Hence, in order to foster native industry, as well as to supply sugar economically, the French have given great attention to the cultivation of beet-root (both for this product and for the distillation of alcohol), and have succeeded in producing a very refined sugar, which looks extremely well, if it does not sweeten so well as the less refined sugar of the sugar-cane. It is, moreover, grown largely for the same purpose throughout the continent of Europe, and its production is increasing in Australia and Tasmania. In this country, the beet-root has been cultivated for the preparation of both sugar and alcohol, but looking at the advantages which we possess in the importation of cane sugar, it is not probable that this will ever become a profitable and important business.

The preparation of sugar from beet-root is carried on with the best skill and apparatus, whilst the mode is probably more simple than that of sugar from the sugar-cane. The beet-root is first detached from the green top, and then cleaned from dirt, and washed in a cylinder or other washing machine. The next process is to rasp or tear it into the smallest pieces, and thus to open the cells which contain the juices, and allow the latter to be readily removed. This is perhaps the most important part of the proceeding, for it has been found that pressure alone, however great, does not remove more than two-thirds of the amount which may be obtained by first rasping the root. It is then subjected to pressure, by which probably 80 per cent. of all the juices are extracted and afterwards mixed with lime. The processes of purification, concentration, and crystallisation, are carried on until the sugar is prepared for the market or for the further operations of the refiner.

It is needless to show how universally this substance enters into the dietaries of every class in every place. There are of course immense tracts of country where sugar in a separated form is not obtained, but even there it is eaten in milk, fruits and other animal and vegetable productions. Thus in the deserts of Arabia the tamarind and camel's milk are the chief sources of its production, whilst the fig, date, and innumerable luscious fruits yield it in tropical and other Asiatic countries.

The following table shows the quantity of sugar which is contained in ordinary foods:—

No. 98.

Sugar in various products (per cent.).

Raw sugar	95·0	Wheat flour		4·2
Treacle	77·0	Rye meal		3·7
Butter-milk	6·4	Wheaten bread		3·6
Carrots	6·1	Potatoes		3·2
Parsnips	5·8	Turnips		2·1
Oatmeal	5·4	Peas		2·0
Skim milk	5·4	Indian meal	}	0·4
New milk	5·2	Rice		
Barley meal	4·9			

It is impossible to estimate the effect of the entire withdrawal of an important article from the food of mankind; but it may be doubted whether the loss of any one element of food would be so keenly felt as that of sugar. So necessary is it, in fact, that starch, the other universal vegetable food, is transformed into sugar in its course of chemical change within the body.

The ultimate chemical composition of dried sugar is as follows, in 100 parts:—

No. 99.

	C.	H.	O.
Milk sugar	42·1	6·4	51·5
Cane sugar	40·0	6·6	53·3

There are 2,800 grains of carbon in 1 lb. of ordinary moist sugar, and there should not be any nitrogen.

Ten grains of lump sugar when burnt in the body produce heat sufficient to raise 8·61 lbs. of water 1° F., which is equal to lifting 6,649 lbs. one foot high.

It was indicated as the respiratory food *par excellence* in my *Mémoire* published in the *Annales de l'Académie des Sciences de Montpellier*, 1860; and is the most striking illustration of a respiratory food which we possess, for not only does it exert a great and rapid influence over the respiratory process, but it is itself entirely transformed into carbonic acid and water, both of which pass off from the body by the lungs.

Its action is very rapid, commencing within five to ten minutes after it is eaten in solution, attaining its maximum in about thirty minutes, and disappearing within two hours. My experiments on the different kinds of sugar have been very numerous, but it will suffice here to give a general indication of the results.

One ounce and a half of white sugar dissolved in water gave a maximum increase in the carbonic acid evolved of 2·18 grains per minute, and of the air inspired of 111 cubic inches per minute. The rate of respiration and of pulsation was lessened (No. 100).

In an experiment when the whole of the carbonic acid evolved was collected, it was proved that after taking 500 grains of white sugar in cold water, there was an increase of 1·57 grain per minute during the first half-hour, and of ·58 grain per minute during the second half-hour, whilst after one hour and a-half the quantity was still increased by 0·12 grain per minute. The increase in the quantity of air inspired was 19·3 cubic inches per minute during the first half-hour, of 30·6 cubic inches per minute during the second half-hour, and of 8·3 cubic inches per minute during the third half-hour. When the sugar was eaten dry, the effect was less, but it was rapidly increased on drinking water (No. 100).

The addition of 6 drachms of good vinegar to 750 grains of sugar was to increase the effect of the sugar. The maximum increase of carbonic acid was 3·3 grains in 20 minutes, and the whole average increase through-

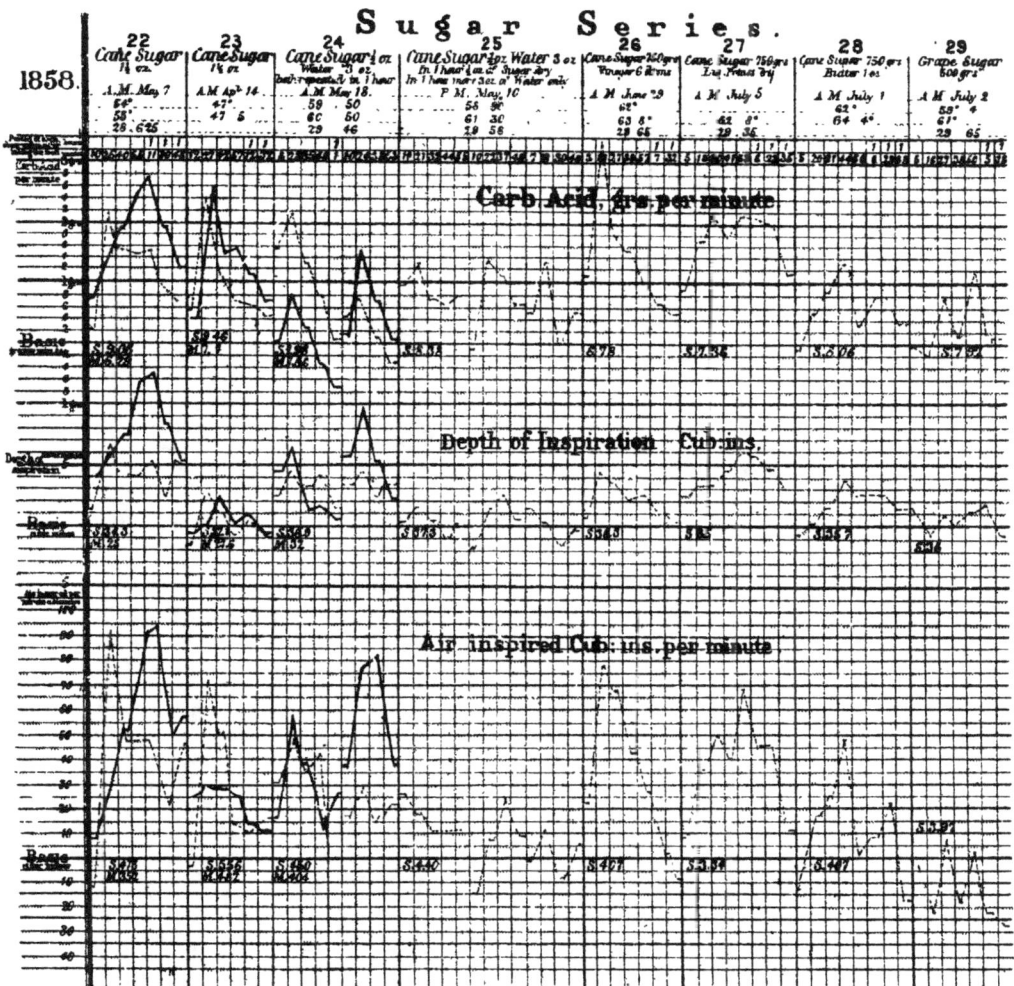

out one hour and a-half was no less than 1·24 grain per minute. The maximum increase in the air inspired was 79 cubic inches in 20 minutes (No. 100).

The addition of an alkali did not increase the effect of the sugar. Thus, 750 grains of white sugar with 40 minims of *Liquor Potassæ* and twelve ounces of water gave a maximum increase of 2·13 grains of carbonic acid per minute, but the increase in the quantity of air inspired was no less than 165 cubic inches per minute (No. 100).

The addition of 500 grains of butter lessened the effect of the sugar to a maximum increase of only 1·3 grain of carbonic acid and 48 cubic inches of air per minute (No. 100).

Milk sugar has less influence than cane sugar, and grape sugar less than either. The maximum increase of carbonic acid from the former was 1·62 grain, and with the latter 1·3 grain per minute, and of air 24 cubic inches per minute (No. 100).

There was always great ease and depth of respiration after taking sugar, so that it was so far an agreeable food, but in five minutes there was sometimes a sour taste in the mouth, and it differed from ordinary food in causing a craving for food in about a hour and a half.

Molasses, Treacle, and Golden Syrup.

Molasses is the uncrystallisable juice of the sugar-cane or other source of sugar, which is left after the crystallised sugar has been separated. It is a natural product, since some of the saccharine juice is not capable of crystallisation; but it is further an artificial product, as the result of the action of heat during the sugar-making process, which renders a part of the sugar uncrystallisable which would otherwise have crystallised.

It is not so pure as crystallised sugar, since it contains matters which are excluded in crystallisation, but it possesses remarkable sweetening properties, and except that it contains nearly 20 per cent. more water than moist sugar, is as useful as sugar when the flavour is not objectionable.

Treacle, as it is ordinarily understood, is produced when sugar is being extracted from the juice of the sugar-cane, while golden syrup is the same kind of product when the moist sugar is being refined. The latter is much lighter in colour and clearer than the former, and is sold at a higher price, but its sweetening properties are very inferior to those of treacle. The treacle which is obtained from the sugar maple is preferred to any other.

Dr. Ure found the following quantities of sugar in treacle and syrups of varied specific gravity, and established the rule that if the decimal parts of the specific gravity be multiplied by twenty-six, it will give very nearly the number of pounds of sugar in each gallon :—

No. 101.

Sp. gr.	Sugar per cent. by weight	Sp. gr.	Sugar per cent. by weight
1·0395	10	1·1110	26·316
1·0500	12·5	1·1250	29·412
1·0685	16·666	1·1340	31·25
1·0820	20	1·1440	33·333
1·0905	21·74	1·2310	50
1·1045	25	1·3260	66·666

SUGAR CANDY.

This is used rather as a luxury than a food, and requires but little notice here. It is obtained by the slow evaporation of cane sugar in large crystalline masses, which are hard and do not so readily absorb moisture as ordinary sugar. It is sweet, but by no means so sweet as sugar, and could not be used with advantage to sweeten foods. Barley sugar is prepared by boiling down rapidly a concentrated solution of sugar, and then letting it cool to a vitreous mass. The process through which the sugar has passed in the preparation of barley sugar is apt to change its chemical composition, so that it consists in great part of *caramel*, and is not a true sugar. The chemical composition of caramel differs from that of sugar by the loss of the elements of water, and is in 100 parts :—

No. 102.

C. 47·06   H. 5·88   O. 47·06

Whenever sugar is exposed to a considerable heat, as in the manufacture of fruit preserves, a part of it is converted into caramel, and by continuing the process it may be converted into glucose, and greatly lose its sweetening properties.

Honey.

Honey is a well-known product of the activity of the working or neuter bee, but not of the bee itself, for it is simply collected by that useful insect, and varies in quality according to the plants which produce it. Whilst, therefore, it is mild or bland in one locality, its flavour may be far more aromatic and stimulating to the taste in another, so that honey produced from village gardens, or from heather, or from the Swiss herbage, has very different flavours, and produces different effects.

Honey obtained from certain localities is known to be poisonous, and to produce giddiness or temporary madness, as is recorded in the report of the Retreat of the Ten Thousand; and without proceeding to that length, it must be within the experience of many that they have sometimes experienced nausea and a sense of *malaise* at the stomach after eating it. How far this affects all individuals alike is not known: but it is not unusual to hear a visitor to Switzerland say that he liked the honey, but it did not agree with him.

This substance is found usually at the base of the petal of the flower, and in the nectariferous glands. The bee sucks it by its proboscis, and transfers it to its honey bag or dilated œsophagus, whence it deposits it, without chemical change, in the hive for future use. The hive, as is well known, contains several combs, which consist of a series of hexagonal cells, adhering to each other by their sides. In a comparatively few instances cells for the reception of honey are specially made of an elongated form, but usually the honey is deposited in old cells which have become too small to hold the larvæ, or in the ordinary cells after the breeding season is over. They are open to the bees until they have been filled with honey, and afterwards they are closed with wax until the honey is required for use.

Hives vary very much in the quantity of honey which they contain, but those of our cottagers, which are robbed every year, produce twenty to forty pounds, whilst in the decayed trees of the American forests, and in Palestine, they have been known to contain more than two hundred pounds. Virgin honey is the portion which first runs off, either with or without the aid of gentle heat, whilst pressure and heat are used in the extraction of that which follows. The former is the purer and sweeter kind, whilst the latter has a deeper colour, less translucency, more impurity, and a stronger flavour. The latter, when produced from old combs, is usually of very dark colour, and often of a disagreeable sticky waxy flavour. Every kind but virgin honey is clarified by causing the impurities to rise to the surface by heat and then skimming them off.

This has always been regarded as a luscious and coveted food, so that the excellence of Palestine was represented by the spies describing it as a land that 'floweth with milk and honey' (Numb. xiii. 27). When Jacob wished to propitiate the governor of Egypt, he sent him 'a little honey' (Gen. xliii. 7). It was also well known as an article of ordinary food, so that John the Baptist's 'meat was locusts and wild honey.' It was also prepared into wafers (Ex. xvi. 31), into a kind of bread with flour and oil, and even eaten with butter, when it was the most delicious compound known. Solomon, however, was well aware that ' it is not good to eat too much honey,' and it is very curious and interesting to observe that whilst its use was so common amongst the Jews, it was forbidden to offer it as a part of their ritual, for it was said, 'Ye shall burn no leaven nor any honey' (Lev. ii. 7).

Honey is, moreover, mentioned by writers of antiquity of all nations, so that its use was very general. It was always particularly esteemed in this island, when sugar could not be obtained, and so largely was it used that the clarification of it, by boiling it with white of eggs, was a recognised branch of trade. It was used in cooking as well as to sweeten food, and was largely employed, even to a recent date, in the manufacture of fermented liquors, as mead and metheglin.

It has also been a source of profit in all time. Judah sold honey to the Tyrians (Ezek. xxvii. 17), and in the Western States of America, bee tracking and honey and wax selling are common and profitable occupations. In our own country, a large proportion of the cottagers keep hives, and the thrifty housewife dresses her children and sometimes pays the rent thereby.

The bee is found in all temperate and warm climates,

and so numerous are the varieties that no less than 250 exist in Britain alone.

The composition of honey varies within certain limits, according to the plants from which it is derived, but it is always highly saccharine, and of the consistence of very thick fluid. The sugar is found in both crystallised and uncrystallised forms, as coarse sugar, grape-sugar, and mannite; and, in addition, there is gum, wax, mucilage, extractive matters, acid, and peculiar flavours. The pollen powder of flowers is very commonly mixed with it, and is said to be the cause of certain ill effects on the system. After it has been kept for some time it ceases to be a fluid, and appears as a solid crystallised mass, in which state it has lost much of its richness and flavour.

So valuable a substance could not escape adulteration, and is often found mixed with starch and sulphate of lime. The former may be determined by the microscope (Nos. 46, 47, p. 147), and the latter by being insoluble when the honey is mixed with water.

Manna.

This substance is probably the same as that called manna in the Bible, and is clearly analogous to it, both in appearance and qualities.

'And when the dew that lay was gone up, behold, upon the face of the wilderness there lay a small round thing, as small as the hoarfrost on the ground.'—Ex. xvi. 14. 'And the manna was as coriander seed.' 'And the people went about, and gathered it, and ground it in mills, or beat it in a mortar, and baked it in pans, and made cakes of it: and the taste of it was as the taste of fresh oil. And when the dew fell upon the camp in the night, the manna fell upon it.'—Numb. xi. 7, 8, and 9.

In Kurdistan, Lebanon, and elsewhere, it is found on

the leaves of various plants, as the dwarf oak, tamarisk, and larch, and is shaken from them before sunrise into cloths which have been placed underneath. It is also sometimes found upon sand and stones, and is then whiter and purer than the tree manna; but its appearance depends upon the dews or rain, so that it is chiefly found during a period of six weeks in the end of autumn and the early spring.

Its flavour is agreeably aromatic, and like that of honey, to which substance it is made to have a further resemblance by the mode of preparing it into a paste, after it has been thoroughly cleansed from leaves and other impurities by being boiled. In the state in which it is found it resembles a coriander seed.

The honey-dew, which is found on many plants under certain atmospheric conditions, and which attracts destroying insects, to the horror of the gardener, is probably of the nature of manna. It is said to be sweetish in flavour, and sometimes is sufficiently abundant to fall from one leaf to another.

A shower of manna was recorded in July of 1872, which appeared as small seeds on the leaves of trees and on the ground.

PART II.

LIQUIDS.

CHAPTER XXXIII.

WATER.

It is needless to insist that water is a most important food, for it is found in all foods, whether solid, liquid, or gaseous, and is taken into the body to the amount of several pints daily. It, moreover, constitutes about 87 per cent. of the whole bulk of the body, and as it wastes at every moment, it must be restored by a new supply.

It is required for many purposes:—First, to soften or dissolve solid foods, so as to facilitate their mastication and digestion; second, to maintain a due bulk of blood and the structures of the body; third, to keep substances in solution or suspension whilst moving in the body; fourth, to supply elements in the chemical changes of the body; fifth, to enable the waste material to be carried away from the body; sixth, to discharge superfluous heat by transpiration through the skin, and by emission through other outlets; and seventh, to supply in a convenient form heat to, or to abstract heat from, the body. Some of these functions are performed by water in its liquid state, and others in a state of vapour.

The importance of these statements will be better appreciated by a consideration of the proportionate quantities of water which are present in various solid, liquid, and gaseous foods, and in the excretions, as shown in the following table:—

No. 103.

Water in various Foods and Excretions.

Ingesta.					
Arrowroot	. 18 per cent.	Potatoes	. 75 per cent.
Barley flour	. 15 ,,	Poultry	. 74 ,,
Beer and ale	. 91 ,,	Pure butter and fats	. 15 ,,
Butter milk	. 88 ,,	Rice	. 13 ,,
Carrots	. 83 ,,	Rye meal	. 15 ,,
Cheese	. 36 ,,	Salmon	. 77 ,,
Coffee	nearly 100 ,,	Skim cheese	. 44 ,,
Cream	. 66 ,,	Skim milk	. 88 ,,
Dried bacon	. 15 ,,	Sugar	. 5 ,,
Eels	. 75 ,,	Tea	nearly 100 ,,
Egg	. 74 ,,	Treacle	. 23 ,,
Fat beef	. 51 ,,	Tripe	. 68 ,,
Fat mutton	. 52 ,,	Turnips	. 91 ,,
Fat pork	. 39 ,,	Veal	. 63 ,,
Green bacon	. 24 ,,	Wheaten bread	. 37 ,,
Indian meal	. 14 ,,	Wheaten flour	. 15 ,,
Lean beef	. 72 ,,	White fish	. 78 ,,
Lean mutton	. 72 ,,	White of egg	. 78 ,,
New milk	. 86 ,,	Yolk of egg	. 52 ,,
Oatmeal	. 15 ,,		
Ox liver	. 74 ,,	*Estimated Egesta, daily.*	
Parsnips	. 82 ,,	Fæces	. 3 ounces
Pea meal	. 15 ,,	Transpiration	. 40 ,,
		Urine	. 40 ,,

The subject of the purity of water is one of the leading questions of the day, and vast efforts are made to show what sources are pure and what impure, and how contamination may be avoided; but we are met on the threshold of the enquiry with the question, What is pure water? and no absolute standard has, as yet, been agreed upon.

It is true that water may be made simply from two gases—oxygen and hydrogen—by mixing two volumes of the latter with one of the former, and passing an electric spark through them; but as such water does not exist in Nature, how can it be taken as the standard of purity? Water which has been distilled approaches this in composition, but it has, at least, 1·85 per cent. by volume of atmospheric air mixed with it, and not unfrequently traces of ammonia; but if it had not these additions, the same objection to its use as a standard exists, viz., that it is not found in Nature. All known natural waters possess other elements, such as salts of lime, which are found as the solid residue after the fluid has been evaporated, besides minute quantities of organic matters and ammonia; and how, then, can we avoid the conclusion that such elements are natural to water, are a part of the water, are, so far as they go, water? If so, their presence cannot render the water impure, and water containing them must be called pure water.

Further, it may readily be shown that these additional substances, and particularly the salts of lime and magnesia, are required by the body, and must be obtained from either solid or liquid food; and, if they are foods, how can they be regarded as impurities?

Viewing water practically as a food, it must be said to consist of such elements as are met with in the largest and best sources. Water is essential to life; and it is idle to suppose that man has been left to create for himself a substance which is essential to his existence, or that he could by any artificial means prepare it in sufficient quantity and of a higher degree of purity; or, in other words, with fewer elements than Nature has provided. Such water as Nature has supplied must, therefore, be regarded as normal, and its constituents as a necessary part of it.

But if this be admitted, we shall not have got rid of the question as to what is purity in water, for the same elements which exist in normal water are found in water which must be called abnormal, but in much greater quantity, and as some of these have been derived from foreign sources, and are injurious to health, they may fairly be called impure. Hence, normal and abnormal, if not pure and impure waters, are distinguished only by the amount of certain substances which are common to all.

Rain-water itself is no exception to the rule, for it is not possible to collect it on the earth without some admixture of solid matters obtained in its passage through the atmosphere.

Dr. Angus Smith, in his excellent work on Air and Rain, has delineated some of the substances which rain brings down from the air in the neighbourhood of Manchester, and Dr. Hassall and others have published drawings illustrating the same subject. The following is a sketch of substances found in the water, and of crystals produced in the evaporation of the rain-water, in the neighbourhoods of Manchester, Newcastle, and London:—

No. 104.—RAIN-WATER.

The substances found in rain-water, collected before it has touched the earth, are minute animalcules, vestiges of living creatures, portions of hair and raw or manufactured materials, phosphates, sulphates, nitrates, chlorides, and ammonia. These vary, according to the locality, so that the chlorides, being derived from common salt, are usually more abundant near the sea, whilst nearly all contaminating substances are more abundant in the air over large towns, and where there are special manufactories.

The quantities are, however, very small, and are better expressed in comparative than absolute terms, as has been done by Dr. Angus Smith:—

No. 105.

Comparative.

	Acidity	Ammonia	Albumenoid Ammonia
Ireland—Valentia	none	1·0	1·0
England—Inland country places	none	5·94	3·21
,, Sea coast	none	10·55	
Scotland—Sea coast, country places, west	1·0	2·69	3·09
,, Inland ,, ,,	2·27	2·96	1·15
,, Sea coast ,, average	9·30	4·10	3·11
,, ,, ,, east	17·61	5·51	3·1
,, Towns (not including Glasgow)	22·85	21·22	6·23
German specimens	9·22		3·59
Darmstadt	12·56		
London, 1869	27·97	19·17	6·03
St. Helen's	28·71	25·33	6·76
Manchester, 1869	60·13	35·33	6·38
,, average of 1869–70	73·44	35·94	7·38
,, 1870	86·76	36·54	8·38
Runcorn	82·40	25·72	5·59
Liverpool	83·46	29·89	4·67
Glasgow	109·16	50·55	8·22
England—Towns	61·57	28·67	6·29

As a rule, rain-water gives an acid reaction, either as it falls or directly after it has touched the earth, but sometimes it is alkaline.

The water, thus reaching the earth, dissolves animal and vegetable substances with which it comes in contact, such as the manures and other decaying substances on the fields, or the substances in the body of the earth, as it runs off quickly into the rivers or percolates through the strata, and reappears as springs or as drainage water flowing into rivers. In the latter case it will have lost much of the animal matter which it had dissolved at the surface, whilst it has gained minerals from beneath the surface.

Hence it follows that the quantity of the several substances which it has appropriated will vary according to the state of the atmosphere, the surface of the soil, the material of the subsoil; and of those which it subsequently lost by the perfection of the percolating or filtering process.

But having been collected into rivers or wells, its character may be entirely changed by subsequent additions, so that animal excreta or relicta, or decaying vegetable, or other matters known as dirt, may get into the well from surface drains, or by percolation through the subsoil, from cesspools, or such matters, with more or less of the soil itself, may be found in the rivers.

Hence, our examples of normal water should be taken from localities where the atmosphere is the most pure, animal productions rare, the volume of water large and in motion as in our lakes, or from deep artesian and other wells.

The oxygen thus found in water not only as a gas renders the water more agreeable to the palate, but by chemical action oxidises and purifies contaminating matters which may have gained access to them. The normal proportion of the oxygen to the nitrogen is as one volume to two volumes, and when it is reduced below that standard there is evidence that it has been consumed by this process. Thus, in reference to the Thames water, Professor Miller found that the proportion progressively diminished from Kingston to Woolwich, as shown in the following table:—

No. 106.

	Kingston	Hammersmith	Somerset House	Greenwich	Woolwich
	Cub. cent.	*Cub. cent.*	*Cub. cent.*	*Cub. cent.*	*Cub. cent.*
Oxygen	7·4	4·1	1·5	0·25	1·8
Nitrogen	15·0	15·1	16·2	14·5	15·5

The quantity of nitrogen may be greater than the proportion which exists in atmospheric air, as in some of the springs at Bath and on the Continent.

The power of water to absorb gases varies with the pressure of the atmosphere and the temperature. At the ordinary pressure and with a temperature of 32° it will be about $\frac{1}{50}$th of its volume of nitrogen, and a yet larger proportion of hydrogen, whilst with increased pressure the power may be increased many fold.

We will now proceed to show the actual composition of such water—of water indeed which we must regard as our normal standard.

1. *As to Atmospheric Air.*

The presence of atmospheric air gives vivacity to the flavour which is always agreeable, and in an exaggerated degree is experienced in drinking aërated waters. Water cannot be obtained without it, for even rain-water usually contains $2\frac{1}{2}$ volumes of air in 100 of water, and yet such water is sometimes called flat or insipid. Running water usually contains a larger volume of air than still water, and particularly such as ripples over stones or falls from a height by which it is mixed with air, and is far more agreeable than distilled water, and more useful as a food in certain culinary operations, as, for example, in making tea and beer.

2. *As to Mineral Matters.*

The following is the total solid residue and the degree of hardness in the waters of some English, Scotch, and Swiss lakes:—

No. 107.

	In 100,000 lbs. of water	
	Solid residue lbs.	Degree of hardness, or lbs. of mineral matter
Bassenthwaite lake	4·64	1·74
Buttermere	3·56	4·04
Crummock Water	4·06	1·01
Derwent Water	6·56	1·3
Ennerdale lake	2·16	1·45
Grassmere	4·18	2·7
Windermere	5·78	3·1
Bala lake	2·79	0·4
Banw and Eira	4·86	2·0
Colinton Water	14·1	9·17
Lake of Geneva	15·2	
Swanston Water	12·7	6·22
Zurich lake	14·3	10·61
Zug	13·2	9·03

Certain very deep wells give the following quantities:—

No. 108.

	In 100,000 lbs. of water	
	Solid residue lbs.	Degree of hardness, or lbs. of mineral matter
Deal	33·24	26·31
London—Royal Mint	83·96	17·48
„ Trafalgar Square	83·4	5·92
Northampton	57·76	10·33
Tring	28·6	26·33

Springs and other wells of less depth:—

Ben Rhydding	8·0	
Critchmere springs	6·24	2·6
Malvern—Holywell	9·37	
„ St. Ann's	5·68	
Spring at Witby	7·6	2·8

WATER. 277

Great collecting areas away from the habitations of men, where pollution seems to be almost impossible, give the following numbers in 100,000 lbs. or 10,000 gallons of water:—

No. 109.

	Total solid residue lbs.	Degree of hardness, or lbs. of mineral matter
Batley gathering ground	7·60	3·33
Halifax ,, ,,	8·16	3·20
Manchester reservoirs, in Derbyshire	6·2	3·73
Rivington Pike	8·48	3·72
Skiddaw	4·34	3·37

Rivers which supply water to important towns have the following:—

No. 110.

	In 100,000 lbs. of water	
	Total solid residue lbs.	Degree of hardness, or lbs. of mineral matter
Avon, at junction with Frome	11·1	0·35
Calder, Rochdale road	7·30	3·77
Cocker	4·62	2·15
Crawley Burn	11·28	6·08
Danube, near Vienna	14·14	
Derwent	6·00	3·37
Frome, below last mill	11·6	0·63
Garonne, at Toulouse	13·67	
Irk (one of its sources)	19·78	0
Irwell, near its source	7·80	0
Kent, above Carpet works	3·3	0
,, below Kendal and Low mills	4·2	trace
Medlock, near its source	12·80	0
Mersey, above Warrington	28·36	2·30
,, below ,,	29·04	1·80
Severn, above Newtown	8·6	3·09
,, in Wales	3·87	0·9
Thames, at Hampton	27·87	trace
,, head waters	28·25	0

The hardness of the Thames water supply to London is about 14 degrees.

Brook water varies much in the quantity of solid matter which it contains, according to the soil through which it passes, to recent rains washing soil into it, and to lowness of water in the summer season; but such matter is only held in suspension, and will subside if time be allowed. It is, no doubt, an impurity, but it is not a very deleterious one, for such waters when clear do not necessarily contain any excess of organic or mineral matters, or nitrites, or chlorides, and are often soft, of good taste, and healthful. The well water in the cottager's yard or garden is not unfrequently much worse than brook water, for independently of any special contamination from petties, the surface water is often allowed to drift in and to convey impurities with it.

But there are sources of water in which the mineral matter is in much greater quantity than those now cited, as shown in the following table:—

No. 111.

	In 100,000 lbs. of water	
	Solid residue lbs.	Degree of hardness, or lbs. of mineral matter
Bedford—Pillory pump.	140·74	54·51
„ Shallow well from upper oolite	72·46	40·77
Bristol—Spring in All Saints' lane	127·28	66·92
„ Water supply	28·66	24·46
Cornbrook	76·90	38·79
„ near junction with Irwell	142·90	64·40
Croal, near Bolton	69·20	25·67
Deal—Deep chalk well at Hill's brewery	202·14	47·25
Norwich—Artesian well, 400 ft. deep	38·14	29·64
Rochdale spring	37·04	14·38
Runcorn public fountain	60·80	25·34
Worthing—New deep well in chalk	32·44	24·69

Waters having this great quantity of salts of lime are known as hard, not only because of their flavour, but from the sensation which they give when used for washing the skin and clothing, and are less fitted than soft waters for cleansing purposes. This hardness is divided into two parts, one of which is temporary, and is due to the presence of carbonic acid gas, which keeps salts in solution, and the other permanent. The former may be readily removed by simply boiling the water and expelling the gas when the salts are deposited, as is found in tea kettles and boilers, and hence is called temporary, whilst the other cannot be removed except by distillation of the water.

This hardness has been estimated and classed in degrees, so that a specimen of water is said to be of so many degrees of hardness. This is purely technical, but it has been agreed to consider water with 1 lb. of carbonate of lime (or its equivalent of other hardening salts) in 10,000 gallons or 100,000 lbs. of water, as 1°, 2 lbs. 2°, &c.

It is well known that hard water causes soap to curd and prevents the formation of a lather, until a large quantity of soap has been added. This fact has also been used as a measure of hardness, for it has been found that 12 lbs. of best hard soap must be added to 10,000 gallons of water of 1° of hardness before a lather will remain, and, hence, that quantity used in 10,000 gallons or 1·2 lb. in 1,000 gallons, or 0·12 lb. in 100 gallons, to produce an ordinary lather, is a measure of 1 degree of hardness.

Such water may be useful as a food in certain cases, by supplying a large quantity of lime and magnesia; but, on the other hand, it is said to be a cause of disease, and particularly to induce calculi in the kidneys or bladder.

Its flavour is harsh and disagreeable to one accustomed to drink soft water, but not necessarily so to one accustomed to its use. We found that those accustomed to drink the hard lime-waters of Illinois regarded the soft sandstone waters of the Missouri as flat and tasteless, whilst the Missourian would reject the hard water, and, it is said, would be more liable to fever if he used it.

So also in a less degree is it in Scotland, and in other parts of our country; each person likes that kind of water to which he has been accustomed from his infancy.

Hard water is not, however, so useful in cooking as soft water, and for other domestic purposes it causes an enormous waste of soap, labour, and material. Hence, it has become very desirable to soften it, and this has been effected by the addition of lime, in the process known as Clark's process. One ounce of quick-lime should be added to 1,000 gallons of water for every degree of hardness. It is first to be slacked and stirred in a few gallons, and immediately poured into the whole quantity, taking care to repeat the operation, and to thoroughly mix the whole contents together. After this it should be left at rest, and it will become sufficiently clear in three hours for external use, but should not be drunk for twelve hours.

Dr. Clark patented his process in 1841, and prepared and applied his test in the following manner:—

'*Preparation of the Soap Test.*—Sixteen grains of pure Iceland spar (carbonate of lime) are dissolved (taking care to avoid loss) in pure hydrochloric acid; the solution is evaporated to dryness in an air-bath, the residue is again redissolved in water, and again evaporated; and these operations are repeated until the solution gives to test-paper neither an acid nor an alkaline reaction. The solution is made up by additional distilled water to the bulk of precisely one gallon. It is then called the " standard solution of 16 degrees of hardness."[1] Good London curd soap is dissolved in proof spirit,

[1] The 'standard solution of 16 degrees of hardness' may be obtained much more simply by dissolving in a gallon of water a quantity of selenite equivalent to 16 grains of carbonate of lime. As the formula of selenite is $Ca_2SO_4\, 2H_2O = 172$, and that of carbonate of lime $Ca_2CO_3 = 100$, the following proportion gives us the quantity of selenite required:— $100 : 172 :: 16 : x;\ x = 27\cdot 52$. The selenite should be reduced to fine

in the proportion of one ounce of avoirdupois for every gallon of spirit, and the solution is filtered into a well-stoppered phial, capable of holding 2,000 grains of distilled water; 100 test measures, each measure equal to 10 water-grain measures of the standard solution of 16 degrees of hardness, are introduced. Into the water in this phial the soap solution is gradually poured from a graduated burette; the mixture being well shaken after each addition of the solution of soap, until a lather is formed of sufficient consistence to remain for five minutes all over the surface of the water, when the phial is placed on its side. The number of measures of soap solution is noticed, and the strength of the solution is altered, if necessary, by a further addition of either soap or spirit, until exactly 32 measures of the liquid are required for 100 measures of the water of 16 degrees of hardness. The experiment is made a second and a third time, in order to leave no doubt as to the strength of the soap solution, and then a large quantity of the test may be prepared; for which purpose Dr. Clark recommends to scrape off the soap into shavings, by a straight sharp edge of glass, and to dissolve it by heat in part of the proof spirit, mixing the solution thus formed with the rest of the proof spirit.

'*Process for ascertaining the Hardness of Water.*—Previous to applying the soap test, it is necessary to expel from the water the excess of carbonic acid—that is, the excess over and above what is necessary to form alkaline or earthy bicarbonates, this excess having the property of slowly decomposing a lather once formed. For this purpose, before measuring out the water for trial, it should be shaken briskly in a stoppered glass bottle half-filled with it, sucking out the air from the bottle at intervals by means of a glass tube, so as to change the atmosphere in the bottle; 100 measures of the water are then introduced into the stoppered phial, and treated with the soap test, the carbonic acid eliminated being sucked out from time to time from the upper part of the bottle. The hardness of the water is then inferred directly from the number of mea-

powder before it is weighed. It dissolves without difficulty. I do not know who first proposed this method.'—E. T. C.

sures of soap solution employed, by reference to the subjoined table. In trials of waters above 16 degrees of hardness, 100 measures of distilled water should be added, and 60 measures of the soap test dropped into the mixture, provided a lather is not formed previously. If, at 60 test measures of soap test, or at any number of such measures between 32° and 60°, the proper lather be produced, then a final trial may be made in the following manner:—100 test measures of the water under trial are mixed with 100 measures of distilled water, well agitated, and the carbonic acid sucked out; to this mixture soap test is added until the lather is produced, the number of test measures required is divided by 2, and the double of such degree will be the hardness of the water. For example, suppose half the soap test that has been required correspond to $10\frac{5}{10}$ths degrees of hardness, then the hardness of the water under trial will be 21. Suppose, however, that 60 measures of the soap test have failed to produce a lather, then another 100 measures of distilled water are added, and the preliminary trial made, until 90 test measures of soap solution have been added. Should a lather now be produced, a final trial is made by adding to 100 test measures of the water to be tried, 200 test measures of distilled water, and the quantity of soap test required is divided by 3; and the degree of hardness corresponding with the third part being ascertained by comparison with the standard solutions, this degree multiplied by 3 will be the hardness of the water. Thus, suppose 85·5 measures of soap solution were required $\frac{85·5}{3} = 28·5$, and on referring to the table this number is found to correspond to 14°, which, multiplied by 3, gives 42° for the actual hardness of the water.

'Table of soap test measures, corresponding to 100 test measures of each standard solution:—

No. 112.

Degree of Hardness	Soap Test Measure	Differences as for the next degree of Hardness
0	1·4	1·8
1	3·2	2·2
2	5·4	2·2
3	7·6	2·0

Degree of Hardness	Soap Test Measure	Differences as for the next degree of Hardness
4	9·6	2·0
5	11·6	2·0
6	13·6	2·0
7	15·6	1·9
8	17·5	1·9
9	19·4	1·9
10	21·3	1·8
11	23·1	1·8
12	24·9	1·8
13	26·7	1·8
14	28·5	1·8
15	30·3	1·7
16	32·0	

3. *Nitrogenous Compounds and Organic Matter.*

Nitrogen is the chief element which is believed to give impurity to water, since it is derived from animal and vegetable decayed or excreted matters, and is regarded as preventible; and because some kinds are apt to excite the fermentation process both within and without the body and thus to engender certain important diseases. It is, however, necessary to distinguish between the different forms in which nitrogen appears, since the effect of each is different from the other. Nitrogen exists in water in four forms :—

1. As free gas.
2. In combination with oxygen, as nitrates and nitrites.
3. In combination with hydrogen, as ammonia.
4. In connection with carbon, oxygen, and hydrogen, in other organic forms of which albumen is taken as the type.

There is no reason to believe that free nitrogen in water is abnormal in small, or injurious in large quantities, and it forms part of the atmospheric air, which in a peculiar form is present in water. It is often very abundant in mineral springs and gives freshness to the taste of water.

Nitrates and nitrites, perhaps in every proportion, but certainly in any quantity beyond a trace, must be derived from animal and vegetable matters, and so far excite a suspicion of their injurious qualities. It is, however, a noticeable fact, that many good and healthful drinking waters contain much of these substances, and hence, however harmful may be their origin, they are themselves harmless. This results from the oxidising process through which they have passed in their course through the strata of the earth, so that they have become medicines rather than poisons, or useless rather than noxious. Their presence is not, therefore, of great moment, although their absence might be more desirable. Some remarkable examples are cited by the Rivers Pollution Commission.

Ammonia in almost every quantity is to be deprecated, since it is the result of decomposition of animal and vegetable substances, and has qualities which are not beneficial as food. It is usually found in very minute quantities in the best rain and lake waters, but not necessarily in the deep-well waters.

It was not found in the following sources at the time the examinations were made for the Rivers Pollution Commission. Worthing (old deep well), Bedford (Pillory pump), Newcastle-on-Tyne (supply from Whittle Dean), Sunderland (deep wells in dolomite), Norwich (from 'Broads' and artesian well), Colinton Water, Bassenthwaite, Ennerdale, Dublin (Custan's well), Deal (from deep well), Zug, Frome, Avon (above Longfords), and the water supply of Gloucester and part of Bath, and Windsor well, near Liverpool.

As to the English lakes, it was only 0·001 in 100,000 parts, in Grassmere and Derwent Water; 0·002 in Rydal

Lake and Windermere; 0·004 in Buttermere, and 0·007 in Crummock Water.

There are, however, other sources, where the proportion is increased a hundred or a thousandfold, and such are exclusively those which are contaminated with sewage or manufacturers' refuse; so that the Medlock, just above its junction with the Irwell, had 1·116 lb. in 100,000 lbs. of water.

Organic matter in other forms is most undesirable, since it has not been entirely decomposed into its final elements, and is in a state ready to do mischief. Its presence in a very minute quantity is tolerably uniform in the best drinking waters, and is represented in the following table under the two heads of organic carbon and organic nitrogen, two of its component parts :—

No. 113.

Pounds in 100,000 *lbs. of water.*

	Organic Carbon	Organic Nitrogen
Buttermere	0·127	0·040
Caterham	0·020	0·006
Colinton Water	0·203	0·042
Coniston Water	0·085	0·017
Crawley Burn	0·187	0·031
Crummock	0·183	0·55
Deal	0·032	0·022
Derwent Water	0·218	0·043
Grassmere lake	0·235	0·050
Katrine	0·256	0·008
Manchester	0·183	0·009
Northampton	0·168	0·024
Otter spring	0·026	0·012
Rivington Pike	0·243	0·031
Royal Mint	0·195	0·025
Rydal	0·254	0·043
Swanston	0·378	0·059
Thames at Hampton	0·260	0·024
Trafalgar Square	0·150	0·012
Tring, deep well	0·036	0·010
Welsh waters	0·289	0·004
Windermere	0·299	0·076
Zug	0·149	0·026
Zurich	0·92	0·009

From the preliminary observations of this chapter, it may be inferred that, in this proportion of organic matter no injurious influence is exerted, and indeed, it is rather to be regarded as a normal constituent of water.

But there are sources of water in which the quantity is extremely great, and from the use of which the most important diseases have been known to follow. Such are referred to in the following table :—

No. 114.

Pounds in 100,000 *lbs., or* 10,000 *gallons, of water.*

	Organic Carbon	Organic Nitrogen
Cornbrook, before junction with Irwell	4·129	0·383
Darweer, below Blackburn	2·127	0·295
Irk, ,, ,, ,,	2·452	0·352
Irwell, at Throsle-next-Weir	2·104	0·248
Roch, above Bury	4·518	0·288

On a review of this part of the subject, it appears that a knowledge of the quantity of free nitrogen and of the nitrates and nitrites is of far less consequence than that of the organic nitrogen and ammonia, and hence, that the determination of the total nitrogen is not so valuable as of that which represents ammonia and organic compounds. It is desirable that in any complete analysis, each of these sources should be enquired into; but if any part may be omitted, it must be that of the free nitrogen and nitrates and nitrites.

The importance of this part of the subject is so great, that it is desirable to refer to the methods of analysis, and so far as may be possible, to enable a non-professional person to determine approximatively the character of the water which he drinks as food.

The chemical world is now greatly divided in opinion as to the proper method of determining the purity of water, and chemists are generally ranged on two sides —one agreeing with Frankland and Armstrong in the process which they have devised, and the other preferring the process applied by Chapman and Wanklyn and Kreaus, or the ammonia process. In the recent discussions on this subject, it has appeared that many of the practical chemists of the day adopt the latter process, and also that they object to the mode adopted by Frankland, in expressing the quantity of nitrogenous elements in water by the phrase ' previous sewage contamination,' both as implying that the nitrogenous material was necessarily derived from sewage and that nitrogenous material originally derived from sewage may not have become harmless.

It is not our purpose to enter into these chemical questions further than is necessary to a work on Foods; and we shall now content ourselves with describing the methods by which the quantity of the several elements may be ascertained.

It is not necessary to determine the quantity of free nitrogen.

The quantity of nitrates and nitrites is ascertained by converting them into ammonia by means of metallic aluminium, acting upon them in the cold, and in a strongly alkaline solution, and estimating the nitrogen from the ammonia. The following is Chapman's modification of Schulze's process :—

'The process is carried out as follows :—100 c. c. of the water are introduced into a non-tubulated retort, and 50 to 70 c. c. of a solution of caustic soda added. The caustic soda must be free from nitrates, and the strength of the solution should be such that 1 litre contains 100 grm. of caustic soda. The contents of the retort are to be distilled until they do not exceed 100 c. c., and until no more ammonia comes over; that is, until the Nessler test is incapable of detecting ammonia in the distillate. The retort is now cooled, and a piece of aluminium introduced into it (foil will answer very well with dilute solutions, but we much prefer thin sheet aluminium in all cases). The neck of the retort is now inclined a little upwards, and its mouth closed with a cork, through which passes the narrow end of a small tube filled with broken-up tobacco-pipe, wet either with water, or, better, with very dilute hydrochloric acid free from ammonia. This tube need not be more than an inch and a-half long, nor larger than a goose quill. It is connected with a second tube containing pumice-stone moistened with strong sulphuric acid. This last tube serves to prevent any ammonia from the air entering the apparatus, which is allowed to stand in this way for a few hours or over night. The contents of the pipe-clay tube are now washed into the retort with a little distilled water, and the retort adapted to a condenser, the other end of which dips beneath the surface of a little distilled water free from ammonia (about 70 to 80 c. c.).[1] The contents of the retort are now distilled to about half their original volume; the distillate is made up to 150 c. c.; 50 c. c. of this are taken out, and the Nessler test added to them. If the colour so produced is not too strong, the estimation may be made at once; if it is, the remainder of the distillate must be diluted with the requisite quantity of water.'

Frankland and Armstrong adopt a different process, and destroy the nitrates and nitrites by sulphuric acid, in the following manner:—

'*Estimation of Nitrogen in the form of Nitrates and Nitrites.*— The following is the mode in which this process is applied to the estimation of nitrogen existing as nitrates and nitrites in potable waters:—The solid residue from the half litre of water used for determination, No. 1 (estimation of total solid constituents) is treated with a small quantity of distilled water; a

---

[1] Condensers are very apt to contain a trace of ammonia if they have been standing all night, and must, therefore, be washed out with the utmost care. We prefer to distil a little water through them until ammonia can be no longer detected in the distillate.

very slight excess of argentic sulphate is added to convert the chlorides present into sulphates, and the filtered liquid is then concentrated by evaporation in a small beaker, until it is reduced in bulk to two or three cubic centimètres. The liquid must now be transferred to a glass tube, furnished at its upper extremity with a cup and stopcock, previously filled with mercury at the mercurial trough, the beaker being rinsed out once or twice with a very small volume of recently-boiled distilled water, and finally with a pure and concentrated sulphuric acid, in somewhat greater volume than that of the concentrated solution and rinsings previously introduced into the tube. By a little dexterity it is easy to introduce successively the concentrated liquid, rinsings, and sulphuric acid into the tube by means of the cup and stopcock, without the admission of any trace of air. Should, however, air inadvertently gain admittance, it is easily removed by depressing the tube in the mercury trough, and then momentarily opening the stopcock. If this be done within a minute or two after the introduction of the sulphuric acid, no fear need be entertained of the loss of nitric oxide, as the evolution of this gas does not begin until a minute or so after the violent agitation of the contents of the tube.

'The acid mixture being thus introduced, the lower extremity of the tube is to be firmly closed by the thumb, and the contents violently agitated by a simultaneous vertical and lateral movement, in such a manner that there is always an unbroken column of mercury, at least an inch long, between the acid liquid and the thumb. From the description, this manipulation may appear difficult, but in practice it is extremely simple, the acid liquid never coming in contact with the thumb. In about a minute from the commencement of the agitation a strong pressure begins to be felt against the thumb of the operator, and the mercury spurts out in minute streams, as nitric oxide gas is evolved. The escape of the metal should be gently resisted, so as to maintain a considerable excess of pressure inside the tube, and thus prevent the possibility of air gaining access to the interior during the shaking. In from three to five minutes the reaction is completed, and the nitric oxide may then be transferred to a suitable measuring apparatus, where its volume is to be determined over mercury. As half a litre of water is used for the determination, and as nitric oxide occupies exactly double the volume of the nitrogen which it contains, the volume of nitric oxide read off expresses the volume of nitrogen existing as nitrates and nitrites in one litre of water. From the number so obtained the weight of nitrogen in these forms in 100,000 parts of water is easily calculated.'

Ammonia is very readily and neatly determined by the use of a reagent invented by Nessler, and hence called Nessler's test, and the process Nesslerising. The reagent causes a yellow or brown colouration or deposit when ammonia is present.

This reagent is prepared as follows:—Dissolve thirty-five grammes of iodide of potassium in a small quantity of distilled water, and add to it a strong watery solution of bichloride of mercury (corrosive sublimate), which will cause a red precipitate that disappears on shaking up the mixture. Add the solution of bichloride of mercury, carefully shaking up as that liquid is added, so as to dissolve the precipitate as fast as it is formed. After continuing the addition of the bichloride of mercury for some time, a point will ultimately be reached at which the precipitate will cease to dissolve. When the precipitate begins to be insoluble in the liquid, stop the addition of the bichloride of mercury. Filter. Add to the filtrate 120 grammes of caustic soda in strong aqueous solution (or about 160 grammes of potash).

After adding the solution of alkali as just described, dilute the liquid so as to make its volume equal one litre. Add to it about 5 c. c. of a saturated aqueous solution of bichloride of mercury, allow to subside, and decant the clear liquid.

WATER. 291

As the subject is of great importance, and the use of the test not difficult, it seems desirable to extract the directions given in Wanklyn and Chapman's practical treatise on 'Water Analysis.'—

'*Use of the Test.*—When a small quantity of the reagent is added to a solution containing a trace of ammonia, a yellow or brown *colouration* is produced. If more ammonia is present, a *precipitate* is formed; and if ammonia be added to the reagent, a precipitate is almost always obtained.

In order to use the test quantitatively, the following things are required :—

(1.) Distilled water, free from ammonia.
(2.) Standard solution of ammonia.
(3.) A burette to measure the standard ammonia.
(4.) A pipette for the Nessler reagent. It should deliver about $1\frac{1}{2}$ c. c.
(5.) Glass cylinders that will contain about 160 c. c.; they are graduated at 100 c. c. and at 150 c. c.

(1.) Distilled water of sufficient purity is generally to be obtained when a considerable quantity of water is distilled. The first portions of distilled water usually contain ammonia. After a while, on continuing the distillation, the water usually distils over in a state of tolerable purity, but towards the end the ammonia will again appear in the distillate. By collecting the middle portion of the distillate apart from the rest, it will usually be easy to obtain distilled water of sufficient purity. In order to be available, the distilled water should not contain so much as $\frac{1}{100}$th of a milligrm. of ammonia in 100 c. c. of water. If there is no opportunity of distilling a large quantity of water, and taking the middle fraction of the distillate, it may be necessary to re-distil distilled water of ordinary quality: the first part of the distillate will be ammoniacal; after that there will be water free from ammonia.

(2.) The standard ammonia should contain $\frac{1}{100}$th of a milligrm. of ammonia in one cubic centimètre of water. It is made by dissolving 0·03882 grm. of sulphate of ammonia in a litre of water. If chloride of ammonium be taken, the quantity of the chloride to be dissolved in a litre of water is 0·0315 grm. It will be found most convenient in practice to keep a solution of ten times this strength (0·3882 grm. sulphate of ammonia in a litre of water), and to dilute it when required for use.

In order to estimate ammonia, fill one of the cylinders up to 100 c. c. with the solution to be examined, and add $1\frac{1}{2}$ c. c. of Nessler reagent by means of the pipette. Observe the colour, and then run as much of the standard solution of ammonia as may be judged to correspond to it into another cylinder containing distilled water, fill up with water to 100 c. c., and add $1\frac{1}{2}$ c. c. of Nessler test. Allow the liquids to stand for ten minutes. If the colouration is equal, the amount of standard ammonia used will represent the ammonia in the fluid under examination. If not, another cylinder must be filled, employing a different amount of the standard ammonia, and this must be repeated until the colours correspond. It is very seldom necessary to make more than two such comparative experiments; and with a little practice, the operation of "Nesslerising" will become very easy and rapid. With regard to the limits of the readings, it is not difficult to recognise $\frac{1}{400}$th of a milligrm. of ammonia in 100 c. c. of water, and the difference between $\frac{19}{100}$ths and $\frac{20}{100}$ths of a milligrm. should be visible. It will be observed that $\frac{1}{100}$th of a milligrm. of ammonia will be more visible in 50 c. c. of water than in 100 c. c. of water; so that when it is desirable to detect the very minutest quantities, concentration of NH_3 in a small bulk of water is to be recommended.

With regard to the superior limit. When the ammonia becomes too concentrated, precipitation occurs. Different samples of Nessler test will sustain different quantities of NH_3 without precipitation.

The presence of a great number of substances in aqueous solution containing ammonia will interfere with the indication of the Nessler reagent, and it is always desirable to have the ammonia in pure distilled water, if that be possible. In order to do so, the solution containing the ammonia should be distilled with a little alkali, and the Nessler reagent applied to the distillate.

If a water contains much carbonic acid, it is desirable to add a little potash to the water before adding the Nessler test. Thus, in estimating the ammonia in the distillate from soda water, too small a number will be obtained, if this precaution be neglected.

When there is a necessity for the use of the Nessler test without previous distillation, a special device has to be resorted to in order to get rid of the disturbing influence on the Nessler test of the substances dissolved. Thus:—

Take 500 c. c. of water, add a few drops of solution of chloride of calcium, then a slight excess of potash. Filter. Put it into a retort and distil until the distillate comes over, free from ammonia, then make up the contents of the retort with distilled water to their original volume, viz., 500 c. c. Now take 200 c. c. of the original water, treat it with chloride of calcium and potash, as before. Then filter, care being taken to have the filter-paper well washed before commencing the filtration. In this way two samples of water are obtained, the one with the ammonia, as in the original water, and the other without the ammonia, but in every other respect the same as the former. This second portion of water is to be used in the place of distilled water to make the Nessler comparisons; as both samples contain the same impurities, they will affect the tint of the Nessler test in the same manner. Thus the error arising from the presence of salts, &c., is avoided.

The following table will be of use in converting observed amounts of ammonia into nitrogen or nitric acid :—

Table showing the Amount of Nitrogen and Nitric Acid corresponding to different Amounts of Ammonia.

No. 115.

Ammonia NH_3	Nitrogen	Nitric Acid HNO_3
1	0·82	3·71
2	1·65	7·41
3	2·47	11·12
4	3·29	14·82

Table showing the Amount of Nitrogen and Nitric Acid corresponding to different Amounts of Ammonia—(continued).

Ammonia NH_3	Nitrogen	Nitric Acid HNO_3
5	4·12	18·53
6	4·94	22·24
7	5·77	25·94
8	6·59	29·65
9	7·42	33·35
10	8·23	37·06
11	9·05	40·76
12	9·88	44·47
13	10·70	48·18
14	11·53	51·88
15	12·35	55·59
16	13·17	59·29
17	13·99	63·00
18	14·82	66·71
19	15·64	70·41
20	16·47	74·12
21	17·29	77·82
22	18·12	81·53
23	18·94	85·24
24	19·77	88·94
25	20·60	92·65

The organic matter is now generally determined by the two rival methods.

Frankland and Armstrong first destroy the nitrates by the aid of sulphate of soda, leaving ammonia and organic matter in solution. They then evaporate the water so treated to dryness in the water-bath, and afterwards burn the dry residue with oxide of copper and chromate of lead, and collect and estimate the carbonic acid and nitrogen. This is a somewhat laborious process, and it is alleged that there are two tendencies to error, one in not entirely destroying the nitrates and the other in destroying a part of the organic matter, and thus that the limit of error is often as large as the whole quantity of nitrogen in good drinking water.

The ammonia process consists in at once converting the organic matter into ammonia, and estimating the nitrogen contained in it; and it is asserted, that when the apparatus is ready the whole determination can be made in less than half an hour.

The following are the details of the method:—

'Half a litre of water is taken and placed in a tabulated retort, and 15 c. c. of a saturated solution of carbonate of soda* added. The water is then distilled until the distillate begins to come over free from ammonia (*i.e.*, until 50 c. c. of distillate contain less than $\frac{1}{100}$th of a milligramme of NH_3). A solution of potash and permanganate of potash is next added. This solution is made by dissolving 200 grammes of solid caustic potash and 8 grammes of crystallised permanganate of potash in a litre of water. The solution is boiled to expel any ammonia, and both it and the solution of carbonate of soda ought to be tested on a sample of pure water before being used in the examination of water.

50 c. c. of this solution of potash and permanganate should be used with half a litre of the water to be tested.

The distillation is continued until 50 c. c. of distillate contain less than $\frac{1}{100}$th milligramme of ammonia.

Both sets of distillate have the ammonia in them determined by means of the Nessler test, as described above.

No matter how good the water may be, it is desirable never to distil over less than 100 c. c. with carbonate of soda, and not less than 200 c. c. after the addition of the potash and permanganate of potash.

It will easily be understood that the greatest cleanliness is requisite in carrying out this process. The Liebig's condenser is especially liable to contain traces of ammonia, and should be cleaned out immediately before being used. The best way of effecting this is by distilling a little water through it. In boiling the contents of the retort it is well to use the naked flame placed *quite close* to the retort, so as not to heat one spot only. We are in the habit of using a large Bunsen burner placed close to the bottom of the retort. Persons who are not in the habit of distilling with the naked flame will probably find an argand gas-lamp with a metallic chimney to be the more convenient source of heat.

* Seldom needed except with very soft water.

With regard to the retort itself, it should be capable of holding about 1,500 c. c. when in position for distillation and filled up, so as to run over. The tubulure should be so situated as to admit of the charge being poured in when the retort is *in situ* for distillation. The charge is to be introduced by means of a funnel, so as to avoid dirtying or cracking the retort.

Very bad specimens of water may be conveniently examined as follows:—

About 600 c. c. of recently-distilled water are put into the retort, and distilled with 15 c. c. of saturated solution of carbonate of soda, until the distillate comes over ammonia free. Then 100 c. c. of the very bad specimen of water are added, and the operation proceeded with, as has been described. Of course the values must be multiplied by 10, in order to obtain the quantities of ammonia yielded by a litre of the specimen.

As has been already mentioned, the first portion of ammoniacal distillate contains both the free ammonia of the water and that obtained from the decomposition of any urea that may exist in the water. Usually it is quite unnecessary to make any separation of the free ammonia from the ammonia present as urea. In the case, however, of very foul water, as, for instance, in the Thames water taken at London Bridge, it is sometimes worth while to make this distinction. When this is desired, a determination of the free ammonia actually present in the water must be made. The difference between the amount of ammonia evolved by carbonate of soda and the ammonia, present as such, is equal to the ammonia obtained from the urea.'

It was formerly the practice to determine the organic matter by two other methods, which may be named. One was by taking the solid residue after evaporating the water to dryness, and after having weighed it, sub- ed it to great heat, and the loss of weight indicated quantity of organic matter. The other was the use solution of permanganate of potash, which was lorised in proportion to the quantity of organic matter to which it was exposed. Both have been largely employed; but they are not so accurate as the methods above described.

A careful consideration of this subject shows that the real difference in the methods rests upon the value which should be attached to the nitrates and nitrites, and thence of the whole nitrogen in drinking water, and if it will suffice for sanitary purposes and as a food to regard only the free ammonia and the combined nitrogen in organic or albuminoid compounds, which may also be readily converted into ammonia, the analysis is simplified and the range of the sanitary question lessened.

There can be no doubt of the fact, that nitrogenous compounds may be rendered comparatively or entirely harmless, when oxidised and converted into nitrates, whether in the soil of the earth, in the tissues of plants, or in the laboratory, and that so far 'previous sewage contamination' need not be a cause of alarm, but they are valuable as indicating a source of contamination. Should they not be altogether oxidised, some portion will appear as organic albuminous compounds, which may be determined as already indicated, and to them should be attached nearly all the importance of the subject.

Hence, regarding a given specimen of water, as fit or not fit for food, it will suffice to determine the presence or absence of ammonia and organic nitrogen; but if a complete description of the water be desired, it will be needful to carry the analysis much further.

4. *Chlorine.*

Chlorine is derived almost entirely from common salt or chloride of sodium, and therefore it exists in sea water and in all springs containing common salt. Hence, its presence is not necessarily abnormal; but if it be not due to either of those causes, it is probably derived from a contaminating source, as the urine and other excreta of animals, which contain common salt.

Common salt (and therefore chlorine) is required by the body and taken largely as food, and hence the presence of a proportion in drinking water is not necessarily injurious, although it may have been derived from an impure source. Its presence is, in fact, important rather as indicating the probable presence of other animal matters already referred to, than any injury which it can cause, and if no such animal matters are found, the water may be drunk with impunity.

The following shows the quantity of chlorine which is present in the more important sources of drinking water:—

No. 116.

In 100,000 lbs.	lbs.
Lake of Zurich	0·17
„ Zug	0·27
Bala lake	0·70
The Rhine, above Schaffhausen	0·20
Grassmere lake	0·79
Rydal lake and Olleswater	0·69
Windermere	0·99
Buttermere, Colinton Water and Crummock Water	0·89
Ennerdale and Lancaster gathering grounds	0·99
Derwent Water and Bassenthwaite	1·29
Cocker, Derwent and Skiddaw	1·09
Coniston Water	1·89
Thames, at Kew	0·84
„ „ London Bridge	1·83
Rhine, at Basle	0·10
„ „ Bonn	1·01
Elbe, at Hamburg	2·75
Rivington Pike	1·53
Tring Deep well	1·39

The next table indicates a few sources which must be regarded with great suspicion on several grounds.

No 117.

Pounds in 100,000 *lbs. of water.*

	Nitrogen as Nitrates and Nitrites	Total Nitrogen	Chlorine
Liverpool—Bevington Bush well, 1868	8·678	8·721	12·61
,, Soho well, 1868	2·195	2·220	7·51
,, Water Street well	1·975	1·989	7·94
Congleton—Town pump	1·076	1·122	3·18
Rochdale—Spring near Churchyard	1·813	1·860	2·98
Leyland	2·466	2·524	
Kidderminster—Shallow well, 1870	3·069	3·222	8·38
,, Another ,,	5 322	5·378	8·20
Leamington—Mr. Jones' well, 1870	6·086	6·111	14·20
Durham—Private well	6·268	6·313	9·75
Darlington—Blackwell pump	6·724	6·757	8·45
Kendal—Shallow well	2·465	3·090	17·00
Witney—Well in Wiggin's yard	4·432	4·880	22·90

The spring in All Saints' Yard, Bristol, already referred to, is a most notable instance of pollution from sewage contamination, and the following number of pounds in 10,000 gallons cannot fail to arrest attention when contrasted with those of the Buttermere water :—

No. 118.

Bristol	Total solids 127·28	Nitrogen 4·745	Chlorine 7·10
Buttermere	,, 3·56	,, 0·043	,, 0·89

Having thus considered the subject in as much detail as is necessary for our purpose, it will be convenient to append a complete analysis of our best drinking waters, with a view to regard them as standards of composition.

No. 119.

Pounds in 100,000 *gallons.*

	Total Solids	Organic Carbon	Organic Nitrogen	Ammonia	Nitrogen in Nitrates and Nitrites	Total combined Nitrogen	Chlorine	Total Hardness	
Bassenthwaite	4·64	·154	·037	·0	·0	·037	1·29	2·83	
Buttermere	3·56	·127	·040	·004	·0	·043	·89	1·01	
Cocker	4·62	·069	·022	·001	·0	·023	1·09	2·15	
Crawley Burn	11·28	·187	·031	·001	·0	·032	1·04	6·08	
Crummock Water	4·06	·183	·055	·007	·0	·061	·89	1·30	
Derwentwater	6·56	·218	·043	·001	·0	·044	1·29	1·74	
Derwent	6·0	·219	·041	·004	·0	·044	1·09	3·37	
Ennerdale	2·16	·042	·017	·0	·0	·017	·99	1·45	
Grassmere	4·18	·235	·05	·001	·0	·051	·79	2·70	
Kent	6·48	·149	·020	·0	·044	·064	·90	3·90	
Rhine	15·80	·108	·012	·003	·0	·015	·20	10·76	hard water
Rivington Pike	8·48	·243	·031	·004	·0	·034	1·53	3·72	
Rydal	4·44	·254	·043	·002	·0	·045	·69	3·10	
Severn, above Newtown	6·60	·123	·016	·003	·010	·028	1·35	3·09	
Skiddaw	4·34	·132	·024	·001	·0	·025	1·09	3·37	
Swanston Water	12·70	·378	·059	·001	·0	·060	1·39	6·22	
Windermere	5·78	·299	·076	·002	·018	·096	·99	4·04	
Zug	13·20	·149	·026	·0	trace	·026	·27	9·03	hard water
Zurich	14·30	·092	·009	·002	·0	·011	·17	10·61	hard water

There are other very extensive sources of contamination of drinking waters to which it is necessary to refer for a moment, viz., that resulting from throwing the refuse of manufactures into rivers, including such important metals as arsenic, and colouring matters to the extent of rendering the water such as might be used for writing purposes. It is not requisite that we should enter into details; but it will be interesting to cite one, not an extreme, instance, as showing the difference of composition in water above and below the source of impurity, and thereby the extent of the contamination :—

'The river Irwell, at its source, held in 10,000 gallons 7·80 lbs. of solid matter, 0·187 lb. and 0·025 lb. of organic carbon and nitrogen, 0·049 lb. total nitrogen, 1·15 lb. of chlorine, and 3·72 degrees of hardness; whilst below Manchester these numbers were increased to, solid matter, 55·80 lbs.; organic carbon, 1·173 lb.; organic nitrogen, 0·332 lb.; total nitrogen, 1·648 lb.; chlorine, 9·63 lbs.; and total hardness, 22·92 lbs.'

The examination by the Rivers Pollution Commission of fifteen samples of waters contaminated by the cotton and woollen mills in Yorkshire showed the following quantities of material in 100,000 lbs. of water, which were thrown into the rivers :—

No. 120.

Total solid matters . . 337	Total combined Nitrogen . 20·015
Organic Carbon . . 64·783	Metallic Arsenic . . 0·011
Organic Nitrogen . . 10·384	Chlorine . . . . 21·94
Ammonia . . . . 11·647	Mineral matters . . 474·84
Nitrogen as Nitrates and Nitrites . . . 0·041	

They also give a page in their Report of 1871, showing in *fac-simile* a letter written with the water of the river Calder at Wakefield, which equals in depth of colour that from a watered ink, and similar examples might have been made from the river at Bradford.

It is, perhaps, scarcely necessary to add that a proportion of the solid contents of such waters, as well as of waters in their natural state, is not in solution but suspension; and that, with perfect quietude of the mass of water, the latter portion will in due time subside, and may be removed.

It is, however, possible to aid the process of clarification of waters by the addition of a substance which will cause the suspended particles to attach themselves to it. Thus the deliciously soft waters of the Missouri as they enter the Mississippi contain much sand in suspension, and by the addition of raw eggs, well stirred in a hogshead of water, the whole will be deposited in the course of twelve hours, and the water become clear and bright. The addition of alum, chips of wood, bitter almonds, nuts of *Strychnos potatorum*, and various other substances, exert the same influence over certain kinds of water.

Before taking leave of this part of our subject, it is desirable to quote the instructions which have been given for the examination of drinking-waters, by two very competent authorities.

The Rivers Pollution Commission have laid down the following particulars for enquiry into the quality of potable waters, and have framed their Tables upon them:—

'1. Organic carbon } both due to organic substance.
 2.   „   nitrogen
 3. Ammonia.
 4. Nitrogen in combination with oxygen, as in nitric and nitrous acids.
 5. The total combined nitrogen.
 6. Chlorine.
 7. Hardening constituents.'

The same Commission, in a review of the subject in 1870, arrived at the following conclusions, and consider any liquid as unfit to enter a stream which has the following characteristics in 100,000 parts, or pounds in 10,000 gallons:—

'1. Containing in *suspension* more than 3 parts by weight of dry mineral water, or 1 part by weight of dry organic matter.

WATER. 303

2. Containing in *solution* more than 2 parts by weight of organic carbon, or 0·3 part by weight of organic nitrogen.
3. Which shall exhibit by daylight a distinct colour, when a stratum of it one inch deep is placed in a white porcelain or earthenware vessel.
4. Containing in *solution* more than 2 parts by weight of any metal, except calcium, magnesium, potassium, and sodium.
5. Containing, whether in *solution or suspension*, in chemical combination or otherwise, more than 0·05 part of metallic arsenic.
6. Containing, after acidification with sulphuric acid, more than one part by weight of pure chlorine.
7. Containing more than one part by weight of sulphur in the condition either of sulphuretted hydrogen or of a soluble sulphuret.
8. Possessing an acidity greater than that which is produced by adding two parts by weight of real muriatic acid to 1,000 parts of distilled water.
9. Possessing an alkalinity greater than that produced by adding one part by weight of dry caustic soda to 1,000 parts by weight of distilled water.'

Messrs. Wanklyn and Chapman in their work, already referred to, devote a chapter to this part of the subject; and, as the directions are short and clear, we shall, with their permission, quote it entire:—

'Examine the water as to clearness. This is best done by filling a good-sized flask, 1,200–1,500 c. c. capacity, with the water. The flask is now to be held in front of a dark-coloured or black wall, a strong light falling on the flask from one side or from above. Any small particles floating in the water will now become readily visible. Care must be taken not to confound minute bubbles of air with suspended matter.

The colour of the water should also be noted. It is best seen by placing the flask containing the water on a sheet of white paper, and placing by its side a similar flask filled with pure distilled water. The two flasks should stand in good diffused daylight. Very minute shades of colour can be seen in this way, and as the glass of which flasks are made is very thin, and but very slightly coloured, we are not liable to mistake the colour of the vessel for that of the water. Dr. Letheby recommends the use of a long cylinder for the purpose of ascertaining the colour, and if such a cylinder of clear thin colourless glass can be obtained, it is a very good plan. Unfortunately, however, most cylinders are made of thick glass, with a decided purplish or green colour. Such vessels are very liable to mislead.

Should the water contain much suspended matter, or be very dark in colour, it may, we think, be said to be unfit for drinking purposes in its then state, though filtration may render it quite good.

Observe the smell of the water. This is best done by shaking up some of the water in a flask with a short and wide neck about one-third full, and then inhaling the air in the upper part of the flask. Should it smell disagreeable in any high degree, the water may be said to be unfit to drink.

Now warm the water slightly and smell again. Warming will often bring out the smell of a water when none could be detected in the cold.

Now add a little caustic potash to the warm water; should this cause any unpleasant smell, we may be pretty sure that the water contains organic matter in some quantity. Notice if a precipitate occurs on the addition of the potash; if so, whether much or little, whether coloured or white. The occurrence of a precipitate indicates hardness; the colour may either be caused by organic colouring matter in the water, or by iron.

Add Nessler test to about 100 c. c. of this water, either in a cylinder or small flask. Should this produce a yellow or brown colour, or a brown precipitate, the water contains ammoniacal salts. This is a most suspicious circumstance, and is almost enough in itself to condemn the water for drinking purposes.

Add iodide of potassium, acetic acid, and starch paste to

100 c. c. of the water. A blue colour indicates nitrites; this also is a most suspicious circumstance, and should the colour be at all deep, the water can hardly be fit to drink. It is to be noted that inasmuch as iodide of potassium often contains iodate, the acetic acid, starch and iodide should be mixed before adding them to the water, so as to make sure that the colour is really produced by the water, and not by any iodate that the reagent may contain.

Boil about 100 c. c. of this water in a flask with a few drops of sulphuric acid, remove from the source of heat, and add sulphuretted hydrogen water. Should a brown or black colouration be produced, the presence of lead or copper may be inferred, and the water condemned (bismuth, mercury, and silver, would of course give the same reaction, but are hardly likely to be present). Should no colour be detected, add a little ammonia or potash. Should this produce a blackish precipitate, iron is almost sure to be present.

Boil a little of the water for a few moments with red litmus. Should the litmus not turn blue, repeat the operation with blue litmus. We learn from this whether the water has an alkaline or an acid reaction. This observation is seldom of importance, except when the water comes from a manufacturing district; it is then often of the greatest value.

The preliminary examination described above takes up a very short time, and gives us much information. It does not require more than 500 c. c. of water, and may be conducted with less. The water used in the examination for clearness, colour, &c., is not reckoned, because it can be employed afterwards in other parts of the analysis.

We may here remark, that if a water contains suspended matter, it should, in our opinion, be analysed with that suspended matter in it. If the nature and quantity of the suspended matter be required, the water should be examined both before and after filtration. The difference between the two results is the value for the suspended matter. This double examination extends only to the total solid residue and the organic matter. The nitric acid and chlorine will not be affected by the suspended matter. A slight difference will sometimes be found between the hardness before and after filtration, but it is not of sufficient moment to render a second determination requisite.'

In selecting water, where selection is practicable, it is desirable that it should be clear and bright, without smell or disagreeable taste, cool and soft, and of smooth and soft flavour, but the latter quality will necessarily vary with the nature of the soil or rock from which the water is obtained.

As a rule, there is an unpleasant smell and not unfrequently an unpleasant taste from water contaminated with animal matter, either when first drawn from the well or after having been set aside for a time, and such should never be drunk. It is, moreover, not unfrequently turbid, or leaves a deposit more or less slight, after having been left at rest; but sometimes water sufficiently impure to induce disease may have none of these characteristics.

Turbid water, if from a brook may be harmless, since the turbidity may be due only to soil or sand with which it is mixed, and which may entirely subside; but all turbid water should be regarded as suspicious, either in reference to healthfulness or hardness. Turbid water from wells almost always implies contamination.

Unfiltered water may also contain animalcules or the lower forms of vegetable life, and particularly, if it have been derived from a watershed or allowed to remain without much motion in uncovered tanks. Such additions are extremely rare in deep-well water, and very frequent in pools. It must not be inferred that such is necessarily injurious, since it is a question of degree, and their importance is rather in indicating the other conditions of the water in which they were generated, or to which they were simply added. A state of water which could engender such organisms must either be comparatively stagnant or fed by animal impurities, and may therefore be hurtful, whilst water into which they have accidentally gained access may be otherwise pure.

It is, of course, desirable that all such impurities should be extracted by filtration, or rendered harmless by boiling, and it may be laid down as a general rule, that whilst no kind of water is injured by filtration through a clean filter, otherwise than by the absoption of the enclosed air, nearly all may possess substances which might advantageously be removed by that process.

Wherever there is reasonable ground to believe that the water is impure from animal matters, and where the water has a disagreeable smell and taste, it is desirable that it should be boiled, and if possible filtered, before being used as drinking water.

There are two objects to be attained in filtering water, viz., to remove any gases upon which a disagreeable smell may depend, and to arrest any particles of matter which may be suspended in the water, viz., to deodorise and clarify the water. A third object is attained by some filters, as for example that of Medlock and the Spongy Iron Company, in which it is sought to decompose the animal matter by the presence of iron.

Filters of sand were formerly in common use, and are still employed when the filtering works are on a great scale, and such may aërate or clarify water and remove all organisms. Sand is not, however, a deodoriser or a bleacher, since it does not absorb gases, and therefore the use of charcoal has been preferred on good grounds. Animal charcoal is to be preferred to vegetable charcoal, since it will absorb a very much larger volume of gas and destroy animal matter.

The varieties of filters are very great, and many are nearly equal in value. It is, however, necessary to restore the purifying power of filters from time to time, and this may be effected in the following manner:— Take two wine-glasses full of Condy's crimson fluid undiluted, with ten drops of sulphuric acid and a tablespoonful of pure muriatic acid, and add them to from two to four gallons of water. Then place the whole in the filter for a few hours, after which pass three gallons of pure soft water through. Charcoal should be renewed.

Foulness of tanks is a very common source of foulness of water both from materials thrown into them, particles of solid matter deposited from the atmosphere, and deposits from the water which may ultimately ferment. An instance was recorded by us many years ago, in which a very violent type of scarlet fever was associated with drinking water from a tank in which herrings in a state of decomposition were found. A disagreeable taste and smell in tank water is daily traced to the ordinary deposits in the tank.

It is essential that the tank be covered, and cleared out at intervals varying from three to twelve months, according to the character of the water supplied; but a better plan is to have direct continuous service without storage in the house at all.

The practice of taking drinking water from a tank which supplies a watercloset is very reprehensible, since gases pass through the trap between the closet and the tank. Such water will soon give a disagreeable flavour or odour, and may induce typhoid fever or other serious disease. Steps should be taken by sanitary authorities to supply special tanks to the waterclosets.

Mineral Waters.

The subject of mineral waters scarcely falls within the scope of this work, since they act rather as medicines than foods; but certain of them are refreshing and agreeable beverages. Among those which are most commonly associated with food are soda water, seltzer water, and Vichy water, all of which may be factitious, but the two latter may be natural. Of late years many other agreeable waters have been introduced, such as Apollinaris, Taunus, Belthal, Rosbach, &c. They are all highly charged with carbonic acid gas, and emit it more or less freely when the cork is removed from the bottle.

The following table, extracted from Watts's 'Dictionary of Chemistry,' shows the chemical composition of the best-known springs, and may be useful both to the healthy and the invalid.

LIQUID FOODS.

	Vichy[1]	Chaudesaigues[2]	Bilin[3]	Ems[4]	Fachingen[5]	Teplitz in Bohemia[6]	Spa[7]	Wellbach[8]	Karlsbrunnen[9]
Name of Spring	Pints Carré	Par	Joseph's quelle	Kessel-brunnen	..	Stein-bad	Pon-toon	..	..
Elevation above sea level.	..	..	..	..	..	..	1000	1100	2353
Temp. cent.	43·75°	81°	..	46°	..	44°	..	14°	8°
Sp. Gr.	..	..	..	1·0034	..	1·00065	..	..	..
Solid contents, parts in one million parts	5160	943	5081	2781	3298	626	563	3977	915
Calcium	117·1	18·4	159·7	59·4	130	25·9	51·3	289·7	241·8
Barium	..	..	..	·3	..	..	..	..	..
Strontium	1·7	..	·5	·6	..	..	..	6·4	..
Magnesium	64·2	4·	44·5	29·3	64·4	11·	41·7	239·8	7·4
Sodium	1814	319·2	1686·7	1122·9	1156·6	195·2	66·	954·5	..
K.	162·7	trace	101·4	34·7	..	·4	4·6	..	..
Li.	..	..	..	..	..	..	..	..	..
NH^4	..	..	..	..	..	..	..	..	..
Al.	..	..	..	trace	..	..	1·1	..	..
Iron	2·	..	·7	1·6	5·5	2·1	23·2	..	31·9
Mn	trace	..	..	·2	..	..	3·2	..	..
CO^3 SO^4	2415	368·4	2030	952	1573	262	265	1667	378·8
Sulphurous Acid	196·8	22·	769·7	38·7	14·7	48·5	9·	75·2	39·6
Fluorine	..	..	..	·1	..	..	..	..	..
Chlorine	324·8	85·2	227·9	487	..	33·5	35·5	621·6	14·1
Bromine	..	..	..	..	..	..	..	..	..
Iodine	..	..	..	..	..	..	..	..	..
Sulphur	..	..	..	..	..	..	..	..	..
NO^3	..	..	..	..	..	..	..	..	..
PO^4	16·7	..	3·8	Al ·2	Na ·9	Na 1·9	1·8	..	..
BO^3	trace	..	..	..	..	..	..	..	..
SiO^3	68	..	46·2	53·9	11·3	41·9	60·5	122	72·1
Arsenious Acid	1	..	..	..	..	..	..	..	..
Organic	..	..	..	..	..	..	..	..	..
Carbonic Acid	445	..	305	98	1348	..	744	..	406
Nitrogen	..	..	..	·4	..	..	..	..	..
Oxygen	..	..	..	..	..	..	..	..	..
Sulphuretted Hydrogen	..	..	..	..	..	..	..	..	..

(Cub. cent. per litre)

The above was analysed by:—[1] Bouquet. [2] Chevalier. [3] Struve. [4] Struve. [5] Bischof. [6] Berzelius. [7] Struve. [8] Jung. [9] Meissner.

WATER.

122.

	Pfäffers[10]	Töplitz[11] Illyria	Carlsbad[12]	Pullna[13]	Saidschütz[14]	Seidlitz[15]	Pyrmont[16]	Bath[17]	Cheltenham[18]	Wiesbaden[19]	Selzers[20]	Aix[21]	Harrogate[22]
	..	..	Sprudel	..	..	..	..	King's bath	Royal well	Kesselbrunnen	..	Kaiserquelle	Old sulphur well
	2116	..	1170	..	..	..	..	..	..	..	800	..	..
	45°	36°	74°	..	..	..	16°	..	14°	68°	15°	..	9°
	1·0044	..	1·00497	..	..	..	..	..	1·0064	..	..	..	1·01113
	252	295	5455	32771	23285	16406	2685	2062	8174	8262	3658	4101	15478
	48·4	57·3	125	139·6	385·8	722·9	588·3	386·7	179·5	363·2	113·9	63·4	493·6
	..	..	..	..	..	..	..	..	..	..	·1	..	..
	..	..	·5	..	..	..	1·2	..	..	..	1·4	..	..
	10·3	16·8	50·3	3319	2813	2918	119·1	53·9	8·1	54·8	51·2	14·7	198·7
	24·	10·5	1793	5222	1974	..	90·4	160	2701	2687	1233	1419	4940
	·3	..	..	344·	239·6	..	1·9	29·8	..	79·4	47·8	69·5	479·7
	..	..	..	..	..	..	·2	..	..	..	trace	..	..
	..	..	..	..	..	..	..	..	..	5·6	..	..	..
	..	20·9	trace	..	..	..	trace	..	..	..	trace	..	..
	·3	..	1·7	..	Mn Fe Co} 24·9	..	30·2	7·4	4·1	2·6	trace	4·5	..
	..	..	·4	..	..	..	2·9	..	..	·3	trace	..	..
	85·9	96·9	1028	656·2	463·7	904	534·7	86·9	639·5	261·5	753·8	503·2	104·8
	24·1	32·4	1749	21154	14273	11568	1138	1029	2259	63·7	28·4	276·1	18·2
	..	..	1·5	..	..	..	..	..	..	..	1·5	..	..
	40·5	29·4	630	1913	211·1	292	108·4	265·3	2066	4687	1388	1691	9187
	..	..	..	..	trace	..	..	..	..	23·2	3·	..	2·7	..
	..	..	..	..	4·3	..	..	..	..	..	..	·4	..
	..	..	..	..	..	..	..	..	..	..	..	3·9	87·7
	..	..	..	..	2746	..	..	..	..	Ca ·4	..	..	..
	..	..	Al,Ca ·4	Ca ·3	..	..	Al 1·8	..	Fe 2·6	..	Ca, Al ·4	..	..
	18·2	13·	75·1	22·9	4·7	..	65·5	42·6	14·5	60·4	39·2	66·	3·4
	..	..	9·1	..	..	..	..	..	240·7	..	..	75·1	..
	38	..	1100	69	200	..	1680	91·6	125	197	1087	75·2	80·3
	..	35·	..	..	..	..	..	..	..	..	..	168	10·6
	..	13·	..	..	..	..	..	..	..	..	..	..	C²H⁴ 21·3
	trace	..	..	..	..	..	..	..	..	..	..	1·2	19·6

[10] Pagenstecher. [11] Graf. [12] Berzelius. [13] Struve. [14] Berzelius. [15] Nauman. [16] Struve. [17] Merck and Galloway. [18] Abel and Rowney. [19] Fresenius. [20] Struve. [21] Liebig. [22] Hofmann.

CHAPTER XXXIV.

MILK, CREAM, BUTTER-MILK, AND WHEY.

MILK.

THIS is one of the most important foods which nature has supplied for the use of man, since it contains all the elements of nutrition within itself, and in the most digestible form, and it is so widely provided that wherever the mammalia are propagated it may be obtained by their young if not by man.

Whilst there are certain variations in the composition and character of milk, according to the animal and its food, they are limited in degree, and it is scarcely incorrect when speaking in a general manner to say that milk has always the same qualities. Thus, whatever may be the source, milk contains in its natural state a nitrogenous element, as casein, carbonaceous elements as fat and sugar, acids and salts.

Cow's milk, whether the ordinary cow of Europe or the buffalo of India, is that best known to the world, inasmuch as the cow can be kept and propagated more readily and economically than any other animal in relation to its meat and milk-producing qualities; and hence it is taken as a standard of comparison, but it differs from human, goat's and asses' milk, as well as from the milk of other animals, in the proportion of its casein, fat, sugar, salts and water.

The milk of the cow is not, however, of uniform quality. Certain kinds, as the Alderney, give a very large proportion of butter and but little milk, whilst some yield much milk, others a large proportion of casein or cheese, and others still, an unusual proportion of water. Moreover, soon after calving the milk is unusually rich in colostrum, whilst when the cow is again in calf the milk becomes poorer in casein as in all its solid constituents. The kind and quantity of food also greatly influence the quality of milk, so that with good dry food the milk is relatively richer in solids, whilst with good grass it abounds in fat. Artificial feeding is also important, for oil cake will cause a yield of far more solids then when grains from wort are supplied, and a penalty is inflicted in Boston for the sale of milk of cows fed on the latter.

The test of quality is commonly the proportion of cream, since butter is the product most appreciated by consumers. As cream is lighter than water the more cream the less is the specific gravity of the milk, other constituents being the same, and it is estimated by a specific gravity bulb. But in truth the other constituents are equally valuable as food, and the advantage of the test is not in reference to milk in its natural state, but to show whether a given specimen of milk has been robbed of a part of its cream. It is rather a test as to a particular fraud, than of natural quality.

The average proportion of cream is 10 to 12 per cent., and it is the practice with very large consumers to agree with the milk-dealers as to the standard to be adopted. At the Liverpool Workhouse they adopt a standard of 10 per cent. and pay $\frac{1}{2}d.$ per gallon for each degree in excess, and deduct the like amount for each degree in defect.

As the composition of cow's milk cannot be always the same, the recorded analyses differ greatly, and it is not possible to do more here than give the analysis of an average specimen, as follows, in 100 parts (Letheby) :—

No. 123.

New milk	.	. Water	86	Nitrogenous	4·1	Sugar 5·2	Fat 3·9	Salt	0·80
Skimmed milk		,,	88	,,	4·0	,, 5·4	,, 1·8	,,	0·8

The real difference between skimmed milk and new milk is in the removal of nearly all the cream from the former, and it is needless refinement to give any other variation in the analysis than such as results from that condition. But a further change takes place with time, for chemical actions proceed rapidly in milk, and a portion of the sugar becoming transformed into lactic and other acids so far lessens its nutritive value. There is, however, a practical error in the usually-received opinion, as to the value of skimmed milk, for many deem it of little value even to the poor, much less for use in their own houses. When, however, it is considered that the addition of $\frac{1}{4}$ oz. or $\frac{1}{2}$ oz. of fat, as suet, to a pint of skimmed milk renders it equal in nutritive value to new milk, such an opinion cannot be sustained. It is a mistake on the part of the housewife to decline to use it with such an addition in her puddings, whilst for the poor, to whom economy is important, and the remaining constituents of the milk of the utmost value to their children, every encouragement to its use should be offered, and particularly when new milk cannot be obtained.

There are in good new milk 640 grains of carbon and 57 grains of nitrogen in the imperial pint, whilst in good skimmed milk the quantities are 516 and 55 grains.

The effect in my experiments (No. 124) of eating onè pint of good cow's new milk was to give a maximum increase of 2·26 grains of carbonic acid per minute in the exposed air, with a maximum increase of 96 cubic inches in the volume of air inspired. The rate of pulsation was also increased.

The maximum effect of the same quantity of skimmed milk was an increase of ·84 grain of carbonic acid in the expired air, and 21 cubic inches in the inspired air.

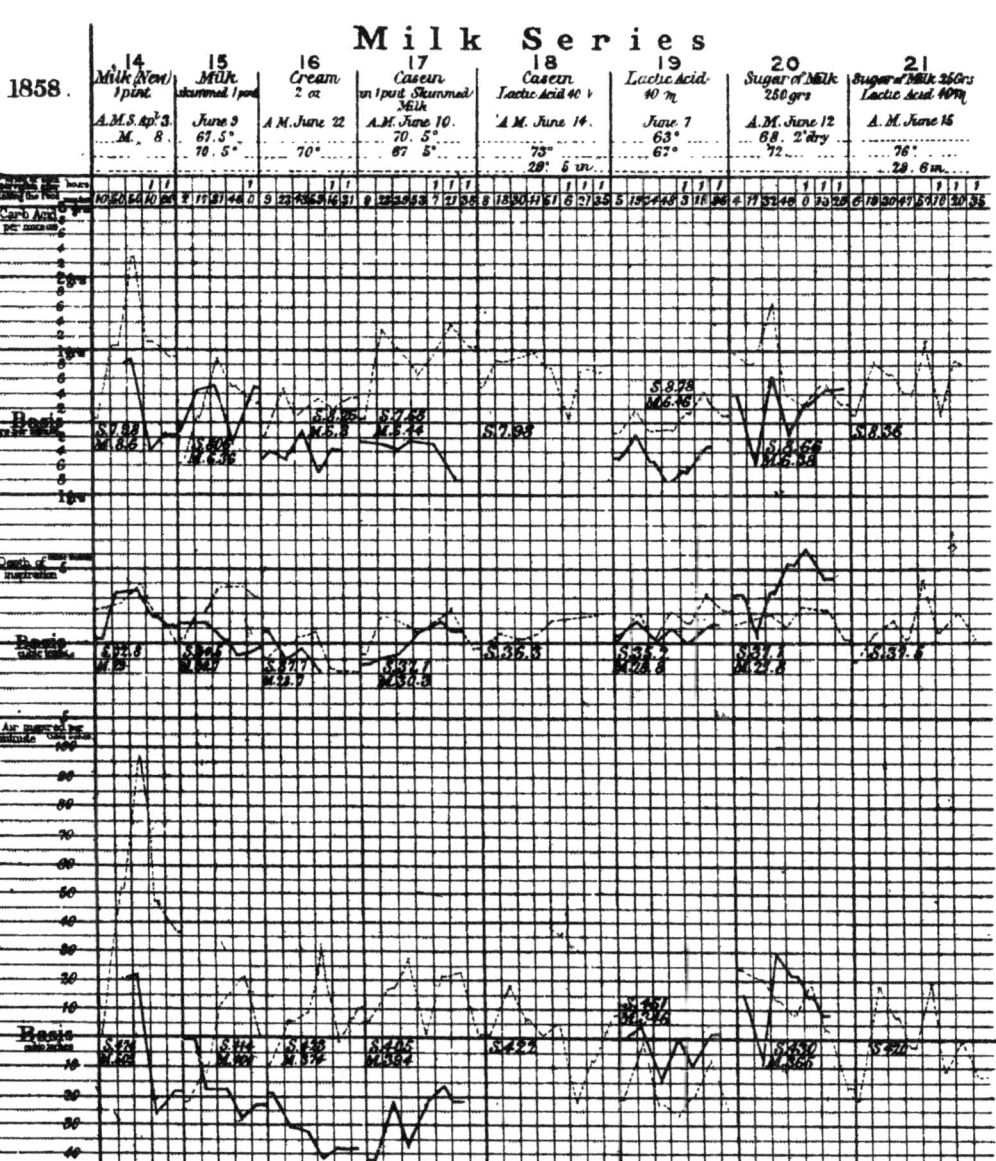

Hence the effect within three hours of new milk is much greater, and of skim milk equal to, that of flesh.

The lactic acid in milk is not by any means without importance, for it is readily converted into carbonic acid, and the effect upon me of taking the proportion which is found in one pint of skim milk, was to increase the carbonic acid evolved by ·42, and on another person to ·8 grain per minute. (No. 124.)

The effect of 250 grains of milk-sugar dissolved in 10 oz. of hot water was to give a maximum increase of 1·62 grain of carbonic acid per minute, whilst the volume of the air inspired was increased by 24 cubic inches per minute. (No. 124.)

There is no milk which is so agreeable and so little disagreeable to the taste as cow's milk, for it has a fuller flavour than human milk, or that of the mare or ass, whilst it lacks the strong flavour of the milk of the buffalo or goat, and is not so surfeiting as that of the sheep. Some of these peculiarities in other kinds of milk depend upon the quantity of a nutritive material which may be readily determined; but others, as the flavour of goat's and buffalo's milk, depend upon an acid which is not so easily measured and is not nutritious.

The following is the chemical composition of several kinds of milk (according to Vernois and Becquerel):—

No. 125.

	Sp. Gr. +1,000	Water	Solids	Casein and Nitrogenous compounds	Sugar	Fat	Salts
Goat	33·53	84·49	15·51	3·51	3·69	5·68	0·61
Sheep	40·98	83·23	16·77	6·97	3·94	5·13	0·71
Buffalo							
Mare	33·74	90·43	9·57	3·33	3·27	2·43	0·52
Ass	34·57	89	10·99	3·56	5·05	1·85	0·54
Woman	32·67	88·9	10·92	3·92	4·36	2·66	0·13
Cow	33·38	86·4	13·59	5·52	3·3	3·61	0·66

The salts in milk are as follows, per cent. :—

No. 126.

Phosphate of lime	.	. 0·3	Chloride of potassium.	.	0·17
,,	,, magnesium	. 0·06	,, sodium	.	. 0·03
,,	,, iron .	. 0·007	Free soda . .	.	. 0·04

As milk is so essential a food for infants, and particularly when the mother's milk cannot be obtained, it is desirable to prepare a kind which may resemble the latter in composition.

Cow's milk differs from human milk chiefly in having a larger proportion of fat and casein and a less proportion of sugar. If therefore a mixture be made of two-thirds of cow's milk and one-third of warm water, to which half an ounce of sugar of milk be added to the pint, we shall obtain a composition very similar to that of the mother's milk. If sugar of milk be not obtainable, it may be substituted by somewhat more than half the quantity of refined cane sugar.

Asses' milk differs from human milk chiefly in having more sugar and less fat, so that whilst it is not equal to human milk as a nutrient it is the best natural substitute for it, but its use is recommended rather in cases of disease than of health, when it is desirable to modify the composition of the mother's milk. Equal parts of asses' milk and cow's milk approach closely in composition to human milk.

The effect of insufficient food on the quantity and constituents of milk is shown in a very interesting enquiry which Dr. De Caisne, made on 43 nursing women during the late siege of Paris, and communicated to the Académie des Sciences. He divided the women into three classes, according to the quantity and quality of milk, and its effect upon the children is detailed in the following table :—

No. 127.

		Butter	Casein	Albumen	Salts	Sugar
1. Much milk, Children thrive	Fasting	3·10	0·24	2·20	0·20	6·24
Women get thin	Well fed	4·16	1·05	1·15	0·30	7·12
2. Little and bad milk	Fasting	2·90	0·18	1·95	0·16	7·05
Children had diarrhœa	Well fed	5·42	1·15	0·95	0·25	7·05
3. Scarcely any milk	Fasting	2·95	0·31	2·35	0·31	5·90
Children died	Well fed	4·10	1·90	1·75	0·31	5·95

Cow's milk is especially rich in salts, as compared with human milk, so that it may be interesting to state the composition of these salts.

Thus in 100 parts of the salts from cow's milk, there are:—

No. 128.

Potash	.	. 23·46	Chloride of potassium.	.	. 14·18
Soda .	.	. 6·96	,, ,, sodium	.	. 4·74
Lime .	.	. 17·34	Phosphoric acid	.	. 28·40
Magnesia .	.	2·20			

The absence of ordinary cow's milk is a great deprivation to Englishmen and Englishwomen (particularly to those who are not in good health) when travelling in Egypt, and in many parts of India, and Africa, for the milk of the buffalo is much stronger, and even repulsive until the palate has become accustomed to its use. In like manner, there are many persons who cannot take goat's milk, which can alone be obtained in some parts of Switzerland; but as a taste for it may be acquired, and the use of it would be very economical and beneficial to the poor in this country, the keeping of goats should be recommended.

10 grains of new milk when consumed in the body produces sufficient heat to raise 1·64 lb. of water 1° F., which is equal to lifting 1,266 lbs. one foot high.

The milk of all animals is more easily digested when eaten hot, and particularly by those who have the impression, not to say conviction, that it does not agree with them, and by invalids. This is not, however, due to any marked chemical change effected in the milk by heat, for the only effect is to coagulate the albumen and to raise it as a scum upon the surface, but to the stimulating effect of heat, both on the palate and stomach. It may, however, be added that the peculiar flavours of goat's and buffalo's milk are lessened by boiling the milk, and may be readily disguised by the addition of other foods or condiments.

Although the quantity of milk consumed in this country is considerable in the whole, it is small when computed on the basis of the population, and even milk-eating people, as those of Cheshire, take little, as compared with the inhabitants of many other countries. The peasantry of Sweden and Norway, of Switzerland and the Tyrol, the Bedouin of Arabia, the people of Kurdistan, and other mountainous countries in the East, live in great part on milk, and are said to take from four to seven pints a day per man, whilst the same is true of the wandering people found on pasture-lands bordering the Great Sahara. The inhabitants of the Western States of America, as of all newly-peopled countries take it largely, whilst, in my enquiries into the dietary of the labouring classes in this country, I found that the weekly consumption amounted to only 32 oz. in England, 85 oz. in Wales, 125 oz. in Scotland, and 135 oz. in Ireland, or from one-quarter to about one pint per head per day.

Camel's milk is in common use amongst the Arabs, and is often mixed with flour into a paste, and boiled, when it is called *ayesh*. It is also drunk largely by those living on the Thull, a sandy unproductive district at Lera, in India, and is stated to have a brackish flavour and but little fat.

Mare's milk is used more commonly than cow's milk

in Tartary, but is inferior as a nutrient in nearly all its elements. It is, however, very generally used in the preparation of a nutritious fermented and intoxicating beverage as thick as pea-soup, called *koumiss*. The curds and whey are separated by adding a little old and sour *koumiss* to the milk, but they are again mixed by vigorous shaking.

Adulteration of Milk.

Fraudulent changes in milk are no doubt frequent, but have been exaggerated, and are now probably less than they ever were. They consist in the addition of water, the subtraction of cream, and the addition of mineral substances to deepen its opacity, and of animal substances to impart a richness of flavour.

The addition of water, and the subtraction of a part of the cream, are to be reprehended, since they lower the nutritive value of the food without lowering the cost. Injurious substances are added; but it is desirable that tests of specific gravity and the amount of cream should be regularly applied by a purchaser. The lactometer shows the specific gravity, and a glass vessel graduated to 100 parts shows the amount of cream. This should not be less than $\frac{8}{100}$ths to $\frac{12}{100}$ths by measure. Specific gravity alone is valueless, because removal of cream increases the weight, and addition of water diminishes it; so that a skilful manipulator can easily cheat the consumer in two directions, viz. in loss of cream, and by dilution of the milk.

It is, however, certain that even in London it is possible to obtain milk as good as could be procured in the country; for in an experience of many years at the West-end we have been supplied with milk of excellent quality. This is, however, the case chiefly when the milkman is himself a cow-keeper in the country, or obtains milk in a large quantity from the country, whilst the smaller dealers who supply the poor are less to be relied upon. Hence, it is desirable to aid the establishment of milk companies, who may obtain large supplies, and be managed by respectable and honourable persons.

The addition of foul water to milk, and the use of such water in the cleansing of the vessels which contain milk, has been known to communicate infectious disease.

It is worthy of remark, that the milk of diseased animals does not appear to contain any known germs of disease, and that it differs from that of healthy animals only in the lessened proportion of all the nutritive elements.

Preserved Milk.

The preservation of milk so that it may be transported from the producing ground to distant localities has recently become an important branch of trade, in America, Switzerland, and our own country. It is not pretended that this is effected on the ground of economy, and it can be approved only on that of necessity or convenience, for not only is the cost greater, but the product is neither so agreeable nor so nutritious as fresh milk. In many districts of England, it is not easy to obtain fresh milk and in some the difficulty is practically insuperable, so that any reasonably good substitute for it must be welcome; but on the lines of railway and in towns as well as in by far the greater part of England, such a substitute if useful at all is simply as a convenience.

It is prepared by the addition of refined sugar and an alkali, and by gentle evaporation in vacuum pans, until it has a very thick semi-fluid consistence, when it is placed in tins, which are sealed by solder, and may be transported and kept for many months. When such a tin is allowed to remain open for some weeks, the substance becomes drier and nearly solid on its surface, but its character does not change.

Mr. A. Willard, lecturer in the Maine State Agricultural College, has published a very instructive paper on this subject in the 'Journal of the Royal Agricultural Society of England,' Vol. 15, which merits careful perusal.

It appears that this process has long been carried on in America in two forms: the one producing plain condensed milk, which will keep good for from one to four weeks; and the other the ordinary preserved milk, which is said to be good for any period. The former is simply milk evaporated, so that four pints become one pint, without the addition of any preserving substance; whilst the other is evaporated in the same degree, and receives a preparation of fine refined sugar for every gallon of milk.

The following is the process at Dr. Crane's factory:—

'The milk as it comes to the factory is carefully examined, and if all right it is received and weighed. The cans are then placed upon the car, which runs on rails to the cooling vat. Here the milk is drawn into large tin pails, eight inches in diameter and eighteen inches long, holding twenty quarts each. About eighteen quarts are put in each pail, which is then placed in the vat containing cold spring water. After the milk has been cooled to 60°, the pails are immediately plunged into the water of the heating vat, which has a temperature of from 185° to 190° F.

'The best refined sugar is then added at the rate of four pounds for each pail. The pails are kept in the vat of heated water about thirty minutes, when the milk is drawn into the large condensing pan. This pan has fifty corrugations, and is set over water and upon a furnace in the adjoining room. Directly over the pan are arranged two large fans, which are kept in motion by machinery. The temperature of the milk while evaporation is going on is uniform at 160° F.

'The fans carry off the water, forcing it through ventilators out of the building as fast as it is formed into vapour. Under this process it takes about seven hours to condense the milk, seventy-five per cent. of its original bulk in water being driven off.

'The faucets at each end of the pan are opened, and the condensed fluid passes through fine wire-strainers, or sieves, into large cans. These cans, when filled, are rolled away to the tables at the back of the room, where their contents are drawn off into small tin cans holding one pound each, and these are immediately sealed up to exclude the air.'

The process at the Elgin factory, which is noted for its product, is very similar.

'After the milk is received it passes through a strainer to the receiving vat; from this it is conducted off, going through another strainer into the heating-cans, each holding about twenty gallons. These cans are set in hot water, and the milk is held in them till it reaches a temperature of 150° to 175° F. It then goes through another strainer into a large vat, at the bottom of which is a coil of copper pipe, through which steam is conducted, and here the milk is heated up to the boiling point. Then the best quality of granulated sugar is added, in the proportion of one and a quarter pound to the gallon of milk, when it is drawn into the vacuum-pan, having a capacity of condensing 3,000 quarts or more at a time. The milk remains in the vacuum-pan, subjected to steam for about three hours, during which time about 75 per cent. of the bulk of water is removed, when it is drawn off into cans holding 40 quarts each. The cans are only partially filled, and are then set in a large vat containing cold water, the water being of a height equal to that of the milk in the cans. Here it is stirred until the temperature of the condensed fluid is reduced to a little below 70°; it is then turned into large drawing cans with faucets, in order to facilitate the filling of the small cans. The milk is drawn from the faucets into the small cans, holding a pound each, when they go to the tables, and are immediately soldered to exclude the air.'

The business is becoming very extensive in America, and the greatest care is taken, by extreme caution and cleanliness, to prevent the development of fungi or vegetable organisms of any kind, so that the milk may keep perfectly good.

Preserved milk thus prepared, whether in America, England, or on the Continent of Europe, contains about one-third of its weight of sugar.

This preparation has been recommended as a food for infants, and it is much liked by them; but it is an error to assume that a given quantity when dissolved in water will yield new milk or be as useful as new milk in feeding infants and young children, and it should never be used as a substitute in such cases whenever new milk can be obtained.

Dr. Daly has raised a very important question in the columns of the 'Lancet,' of November 2nd, 1872, and asserts that such milk causes an undue development of fat, leads infants to refuse food of a more simple flavour, and renders them less able to resist disease. His remarks are sufficiently important to merit quotation.

'Children like the condensed milk far better than cow's milk; I suppose from its sweetness. Its general effect is that infants thrive remarkably, and get much fatter as a rule on it than even on the breast. To look at, children brought up on condensed milk are extremely healthy.

Now arises the question, which has induced me to trespass upon you with this letter, viz., are all the children really strong and well nourished, or are they only strong-looking? That they are well fattened is manifest, but is the fatness only the result of a sort of converse Bantingism? The continued use of the so-called Banting regimen will, no doubt, soon reduce a man's adipose tissue and weight, but it will do so at the expense of deteriorating the whole system, of dangerously lowering the vitality, and therefore the power of resisting the disease. So likewise I have observed in a number of cases, carefully watched during the past eighteen months, that while condensed milk fattens, and while children apparently thrive upon it, the vitality of the child is below par to a very dangerous degree.

I found that during the summers of 1871 and 1872 children taking condensed milk sank rapidly of diarrhœa, which was not at all severe or such as to cause alarm, and that the prostration was out of all proportion to the severity of the disease, and came on almost immediately. Indeed, as far as my experience goes, it has been invariably the case that children fed on condensed milk and attacked with diarrhœa at all severely, almost immediately get into a semi-collapsed state, and if brandy be not at once given, they die. I have observed the same with regard to other diseases, as for instance measles, whooping-cough, and bronchitis. The resisting power of the child has been bad, and those children brought up on the impure London fed cows' milk will resist an attack of acute disease better than children fed on condensed milk. I can give no explanation of this. Whether my observations be correct, or only the result of a strange series of coincidences, it is a matter well worthy of more extended investigation.

I have now before me notes of a number of cases confirming the above views. In one child that I saw fed entirely upon condensed milk from a month old, collapse set in from a few hours' diarrhœa, which was not at all severe, and the child only recovered by the free use of brandy. This child was in perfect health, to look at, before the attack. Another child, taking condensed milk, and who was on the brink of the grave from one day's slight diarrhœa, by my advice was afterwards fed on goat's milk, and while taking goat's milk it had another attack of diarrhœa, more severe than the first, but the alarming symptoms of exhaustion did not come on, and the child was soon well. I could relate many such cases. I have seen several children sink of slight diarrhœa in a few hours who were being brought up on the condensed milk, the children being healthy to look at, and the diarrhœa such as generally gets well. I do not wish to draw any positive conclusions from these cases; I only state what I have observed. Again, I have invariably found that children brought up on this milk are most backward in walking, no doubt due to the inferior muscular nutrition, and also that the anterior fontanelle is very late in closing, arising from the osseous system not being properly nourished. The children generally, too, have the abdomen rather large. I may mention also that it is most difficult to get children using condensed milk to take other kinds of food, and as they grow old enough to eat farinaceous puddings they will not touch them unless they are saturated with sugar.'

Such is a medical practitioner's view of this matter, and it deserves the attention of mothers and of all who have the care of infants and young children. Without explaining the medical aspect of the question (which would be out of place here), I remark that as a food the addition of nearly two ounces of sugar to the pint of cow's milk greatly lessens its nutritive value, and induces a tendency to starvation of muscle-forming element. Then, whilst in natural cow's milk the proportion of nitrogen (flesh forming) to carbon (fat forming) is 1 to 12, in the preserved milk it is not much more than one half or about 1 to 20. If the object were to feed an animal for the market it would be attained by this method, but if to make infants into strong muscular men and women, the proportion which Nature has provided must be supplied.

For office use, or when travelling it is not convenient to obtain milk at the required moment, or for family use when milk cannot be obtained at all, it is a valuable food.

Artificial Milk.

Many attempts have been made to prepare milk artificially, and thus to supply a want which is often experienced, but they have not been successful so far as to please the appetite of those who consume it.

As the composition of cow's milk is well known, there is no difficulty in combining the same elements artificially, so that the compound may be as nutritious as natural milk. Thus M. Dubrunfant has suggested the following recipe:—

Half a pint of water, $1\frac{1}{4}$ ounce of cane or grape sugar, $\frac{1}{2}$ an ounce of dry white of egg, and 15 grains of crystals of carbonate of soda, to be made into an emulsion whilst warm with $1\frac{3}{4}$ to 2 ounces of the finest olive oil or some other pure fat. This compound will be as thick as cream, and another half pint of water must be added to make it of the consistency of milk. The addition of a little gelatin, say 30 grains to the pint, will increase the resemblance of the compound to cream, and will allow more water to be added when artificial milk is to be produced.

It is, however, needless to say that such a compound could not for a moment compete with natural milk if both could be obtained; and it may be of interest to add, that when natural milk is divided into its component parts, the effect of all is less than that of the combination in the milk.

CREAM.

Cream consists of minute globules of oil which are distributed throughout the mass of milk and give opacity to it, but after the milk has been kept at rest for a time the globules, by reason of their less specific gravity, rise to the surface and accumulate as a layer of cream. A considerable time is requisite to cause all the globules to rise, and thus leave the milk free from them; but it is shortened, and the operation rendered more complete by the application of a gentle heat, as in the preparation of clotted cream in Devonshire, and in that of *shor* or *malai* in Bengal.

The cream thus separated may be removed by skimming, but a portion of milk is also taken away in the operation. The quantity and purity of the cream depend upon all the causes already mentioned, as affecting the quality of the milk, as also on the duration and completeness of the repose of the milk, and the care exercised in skimming it, and practically they vary much.

When thus separated, it consists of fatty matter with a proportion of milk, and as the cream is often allowed to accumulate for some days before it is converted into butter, the milk becomes sour. When the cream has been churned, the butter collects in masses, and the watery-looking residue is butter-milk; but a proportion of butter-milk is retained by the butter in the inverse ratio of the washing and pressing of the butter. Hence, as butter is scarcely ever entirely free from butter-milk, it cannot be absolutely free from a trace of nitrogenous matter.

The usual chemical composition of cream in 100 parts is water 66, nitrogenous 2·7, sugar 2·8, fat 26·7, salts 1·8. Thus, about one quarter of the weight is butter, but in that respect the composition varies from 22 to 33 per cent.

The effect upon myself of taking 2 ounces of good fresh cream was to cause a maximum increase of ·48 grain and an average increase of ·24 grain of carbonic acid per minute by the respiration, and a maximum increase in the quantity of air inspired of 33 cubic inches per minute. (No. 124.)

Butter-milk.

Butter-milk is the residue of cream after the butter has been made, and is about two-thirds of the whole weight of the cream. In general chemical characters it is analogous to skim milk, for it retains all the elements of milk except the fat, but it may have acquired free acid in addition.

The following is the chemical composition of butter-milk in 100 parts:—

No. 129.

| Water 88 | Nitrogenous 4·1 | Sugar 6·4 | Fat 0·7 | Salts 0·8 |

It is evident that the causes of variation in the composition of milk apply to butter-milk, and the composition cannot be reduced to a fixed standard. Moreover, it frequently happens, and particularly in those countries where milk is abundant and labour scarce, that small lumps of butter are allowed to remain in butter-milk, by which the proportionate quantity of fat will be increased and the total nutritive value exceed that of well-skimmed milk. This, therefore, is a very valuable article of food, and as it is usual to regard it 'as even inferior to skim milk,' it is clear that there is not a just appreciation of it in the dietary of the poor. It enters largely into the dietary of the labourers in Ireland, Scotland, and South Wales, whilst it is commonly used in Cheshire, and is of the greatest value where there is a deficiency of other nitrogenous food. The Irish labourer obtains about four pints a day, which alone offers a large quantity of nutriment, whilst its flavour is a most agreeable addition to that of boiled potato or Indian meal stirabout. Its use is universal in India; and as a food it is regarded as so important by the pastoral tribes of Meerut that they say 'a man may live without bread; but without butter-milk he dies.' It should be everywhere distributed to the poor where milk cannot be obtained by them in sufficient quantity.

There are 490 grains of carbon and 55 grains of nitrogen in each pint of butter-milk.

When butter-milk is added to boiling whey and the two are well mixed, a soft curd is thrown down. This mixture is called *fleetings* in Wales, and is eaten when either hot or cold with the addition of bread.

Whey.

Whey is the residuum of milk after the manufacture of cheese, and is therefore nearly destitute of casein and fat, whilst it contains the sugar, acids, and salts of milk. If the cheese be made from new milk there will be a small proportion of fat left in the whey, and probably the whole of the curd will not have been removed whether the milk be new or skimmed.

Hence whey as an article of food is valuable, chiefly from its salts and free acids, and is often an agreeable addition to other foods in the dietary of the sick.

According to Professor Voelcker, it yields 193·2 grains of carbon and 14·5 grains of nitrogen per pint.

The use of curd and whey was well known to our Saxon and Norman ancestors, and the fritters made with milk was by no means to be despised. The following is the recipe :—' Take of crud'd and p̃sse out the wheyse. do' thto sū whyte of ayreū (eggs) fry hē, do' thto and lay on su (sugar) and messe forth.'

CHAPTER XXXV.

TEA, COFFEE, CHICORY, COCOA, AND CHOCOLATE.

TEA.

WE have on several occasions, in the course of this work, referred to the marked changes in the food of so fixed and stable a population as that of this country, all of which have been associated with a gradual refinement of taste, increased pecuniary means, and diminution in the price of products, as the results of improved modes of manufacture. The most striking change hitherto mentioned has been the substitution of fine wheaten flour for coarse wheaten flour, barley-meal, rye-meal, and even for oatmeal, and the cause is due to, or at least essentially connected with, the reduced cost of wheat. But it is probable that, of all changes of diet, none can compare with the extension of the use of tea. From the sixteenth century, when it was sold at ten guineas a pound, to our day, when we have it of excellent quality at 2s. a pound, its use has been extended from a hundred people to the millions of Great Britain and the English-speaking race, wherever they can obtain it. Other races have also adopted it, but much more tardily, because of their lack of maritime commerce, the difficulty and expense of obtaining it, or their more limited pecuniary means; and, even at this day, the difference in the use of this article by the Latin races and the Anglo-Saxon and allied races is extremely marked. If to be an Englishman is to eat beef, to be an Englishwoman is to drink tea; and, although the excessive drinking of beer has been regarded as a characteristic opprobrium of our race, the badge seems now to be transferred, or to be in course of transfer, to the great Teutonic race—a race, allied to us in many mental and bodily qualities, as well as in social habits.

No one who has lived for half a century can have failed to note the wonderful extension of the tea-drinking habits in England, from the time when tea was a coveted and almost unattainable luxury to the labourer's wife, to its use morning, noon, and night by all classes. The caricature of Hogarth, in which a lady and gentleman approach in a very dainty manner, each holding an Oriental tea-cup of infantine size, implies more than a satire upon the porcelain-purchasing habits of the day, and shows that the use of tea was not only the fashion of a select few, but the quantity of the beverage consumed was as small as the tea-cups.

So also as to the strength of the infusion which is drank, whilst it is very weak in the tea-shops of China and Japan, it is so weak in the cup of the professional tea-taster, when pursuing his calling, that only the weight of a sixpence is used at a time; and whilst it is still necessarily weak in the cup of the poverty-stricken wife, it is the fashion among the wealthy classes to make it so strong that a tea-cup of dry tea is thrown into the tea-pot, which is to produce, perhaps, half-a-dozen cups of the beverage. This change can be without effect, only on the presumption that the action of tea is slight and unimportant, within the limits of toleration, inherent to our nature.

Such changes—so wide and so great—must have left their mark upon the physical, mental, and social habits of our day, whether they are regarded positively or negatively. Whether, in spite of them, or in consequence of them, the changes of the day have been towards health of body, activity of mind, cultivation of taste, and refinement of personal habits—that is to say, towards the subjugation of the body to the mind and conscience.

The cultivation of the tea plant and the preparation of tea have been surrounded by some mystery, from the distance at which they occur, and the few who have transmitted descriptions to us. There is still an air of romance about the Flowery Land, which is the fruit of our ignorance as much as of our veneration for the antiquity of that race and government, and we are apt to extend it to all its productions and associations.

The affair is, however, a very simple one in all that does not relate to minute mechanical details, many of which may not be really essential to the process. There is but one kind of plant from which tea is made wherever it is found, although by cultivation it may have produced varieties; and the whole principle involved in the process of manufacture is not that of any important chemical change, but simply the drying of the leaf for preservation and for future use with the least possible injury.

The tea plant, *Thea sinensis*, is closely allied to the camellias, with which we are familiar in this country. The leaf is, however, more pointed, and is lance-shaped, and not so thick and hard as that of the camellia. The same plant produces in one place green and in another black tea, and is called *Thea viridis* or

Thea Bohea, as though green tea were made from the one and black from the other, but in certain districts the *Thea viridis* is grown for the production of black tea. The Indian tea is produced from the *Thea Assamensis*, but it is the same plant as that obtained from China.

Efforts are made to introduce the tea plant into various other countries, as, for example, into Ceylon, Australia, Brazil, the Carolinas, and California. It is usually raised from seed, and the following account of the process, extracted from the 'South Australian Register,' may be as useful as interesting:—

'The plants are usually placed at five feet distance from each other, which takes about six pounds of seed to the acre. If the seed be fresh the general practice is to plant it out at once in the spots where it is intended to permanently remain. This is done by digging a hole with a pickaxe or mattock one foot deep, filling it with the loosened earth, and sticking the seed about three inches below the surface. In cases of experiment with inferior seed or upon untried land, it is most advisable to plant in nursery beds about six feet in width, into which the seeds are dibbled or drilled at two or three inches apart from each other. The most suitable season for this operation is the commencement of the winter rains, as the moisture materially assists their germination. Some will germinate in four or six weeks, but many will lie dormant for four months, and in most cases none will grow until the advent of the rains.

Young small seedling plants stand extreme cold badly; if, therefore, they germinate at such a time they ought in a new climate to have a top dressing of manure, and to be protected by branches or grass, which, when the sun shines, should be removed to give them the advantage of his beams. In this manner small plants may become fit for removal in two months.

As soon as the young plants are three inches high they ought to be weeded, for if weeds are allowed to remain they draw them up, and then make the plants thin and weakly. It matters not though the weeder in removing weeds disturbs seed germinating, as they can easily be put in the ground again. Moreover, the advantage given to the young plant by opening up the soil around them is great, and more than compensates for the injury done.

Plants are sometimes raised from layers and cuttings. The most favourable season for these operations is when the sap is beginning to rise and in full action, as during the rains. Seedling plants ought not to be transplanted until eight inches high, and in planting them on the slopes of steep hills care should be taken to place them horizontal to the dip of the land, as the earth is not so liable to be washed away from the roots by the heavy rains.

Tea leaves should not be gathered until the third year, and then during the rains, in order to allow the bushes to attain a considerable size. If the plants send out long leading shoots in the second year, these ought to be nipped off in order to induce the plant to throw out lateral shoots and become of a thick and bushy form, yielding an abundance of leaves.

The leaves are gathered during the rains, the number of pluckings of each plant being four, and in every wet season five, with an interval of from four to six weeks between each. Each plant will yield in the third season about half a pound of raw leaves or two ounces of manufactured tea of a superior quality, giving an average of about 80 lbs. to the acre. Two years more will increase the yield tenfold, being $1\frac{3}{4}$ lb. of manufactured tea to the plant.

The process of gathering the leaves is extremely simple, and is chiefly performed by women and children. All old and fibrous leaves are left upon the tree. The young leaves are stripped off with the hand, an inch or so of the soft and succulent stalk being taken with them. The finest kind of black tea is prepared from the tender buds at the extremity of the twigs, and is kept from the rest. A woman accustomed to the work will gather in a day from 16 lbs. to 20 lbs. of raw leaves.

The proper time for sowing the seed in South Australia is June to August.'

The tea plant is cultivated in China on the hill sides, at an elevation extending to 4,000 feet where the soil is rich and deep, the drainage good, and sunlight abundant, through about 11 degrees of latitude. It will grow in almost any temperate climate, and therefore further north and south than the 24° or 35°, but it is not then so valuable for the preparation of tea. It requires also good manuring and cultivation; and when the leaves become hard and tough the old wood must be cut out and new shoots produced, and thus the tree will remain useful for the period of a generation. The plant would grow to the height of thirty or forty feet and have a stem more than a foot in diameter, but by proper pruning it is kept down to a height of from three to five feet. The leaves when full grown measure from five to nine inches in length. About 320 lbs. of dried tea are produced on an acre, and 1 lb. of dried tea is prepared from 4 lbs. of green leaves. In some respects the cultivation reminds us of the grape on the sides of the Rhine; and in another particular the two products are similar, namely, that as the produce of adjoining vineyards may materially differ in the flavour and quality of the wine produced, although the soil, climate and cultivation may seem to be the same, each tea plantation obtains its own character for produce.

Other things being equal, the qualities of teas vary with the time of picking, and therefore with the development of the leaf. The earliest pickings of the buds or youngest leaves in April yield fine young Hyson, with its thin leaf, which may be closely rolled together, its greenish colour and delicate flavour—that is to say, with a large proportion of juices in relation to the solid substance of the leaf, and with the substance in the most pliable form. This is, however, a kind of tea which is prepared with difficulty, that it may remain without change of colour, and be preserved perfectly; and as it is very apt to set up the process of fermentation, it is rarely conveyed in large masses, as in our great tea ships, but in small quantities on horses' backs, or by caravans to Russia; and no inconsiderable part of the crop is reserved for the use of the wealthy Chinese. At this early stage the petiole of the leaf and even the delicate stalk is sometimes broken off and prepared with the leaf.

The later pickings, extending through April to May, are not more valuable in nutrition because the leaf is more matured, for the value consists not in the structure of the leaf (which is much the same as that of other leaves), but in the juices, and they are as perfect after the first rise of the sap as at the middle of the season. Hence the finer and more delicate the texture of the leaf, or, in other words, the less it is developed and aged, the better is the quality of the tea. This value is connected not only with the positive advantage of the juices referred to, but with the amount of tannin which is present in the leaf, and which gives an astringent and bitter taste to the infusion. There is a proportion of tannin in all tea leaves of whatever age, and it is probable that the quality of the tea is improved by thus gaining in what is called body; but when it is in large quantity, as in the older leaves, its flavour masks that of the tea in the infusion, and with greater depth of colour and fulness there is much less delicacy of flavour.

Having thus obtained leaves either at various times or of various sizes (both of which circumstances may indicate difference in quality) the next step is that of drying and preserving them. The details of the process are very minute, so that the accounts vary, or they themselves may vary, without affecting the product, but the principles involved are, first, to dry them; second, to render them supple for the further process of rolling; third, to preserve the colour if intended for green tea, or to allow the colour to change if intended for black tea.

The leaves are dried in pans, which are heated to the proper degree with straw or charcoal without the least smoke, and the quantity in the pan is such as that the heat may be equally applied to all with only such movement of the leaves as may be effected by the hands or by shaking the vessel. The exposure to heat is regulated not only by the heat of the pan, but by the removal of the leaves and the re-application of the heat at different stages of the process.

The physical character of the leaf as to suppleness is modified by manipulation and the regulation of the heat.

The colour of green tea is retained by the rapidity of the drying process, whilst to produce black tea the leaf is exposed to atmospheric changes for a longer time, so as to set up the process of fermentation. A face is put upon such green tea as may be desired, by heating and manipulating it with a very small quantity of Prussian blue mixed with gypsum and indigo, but it is less frequently performed than formerly on the higher class of green teas.

During this process of preparation the teas are sorted according to the size of the leaf: a classification which also indicates quality, so that the term *Su-chong*, with which we are now familiar, indicates one of such classes. The smallest, however, are called *Pha-ho*, the second size *Pow-chong*, and the fourth or largest *Toy-chong*.

The leaves thus prepared for the market have acquired a new appearance and flavour, and have experienced a change quite as great as that between roasted meat and raw meat, or old dried hay and grass, but the principle involved is much the same in all the cases.

It is evident from the foregoing that the product of different farms, and even of the same farm in different seasons, may vary with more or less care of cultivation and age of the plants. The whole produce is called a *chop* of tea, and consists of some or all the varieties just indicated; but green tea and black tea are not usually produced on the same farm or even in the same locality.

It is also clear that, as great skill is required in the manufacture of the tea, the quality will vary much as the care and skill varies; and it is in this respect that the production of tea in other countries, as in Japan, Java, India, the Brazils, North America, and Australia, is difficult, and excellence is attained only after great delay. Just as there are good and bad farmers in every country, so will it be in China and India; but the long continuance of the tea cultivation in China by the same families tends to maintain a degree of uniformity which would not be found in our changing country.

The Government of India have for nearly forty years made great efforts to extend the growth of the indigenous Tea plant and the production of tea in various parts of that great empire, and have succeeded in Upper and Lower Assam, Kumaon, Cachar, Arracan, Dehra Doon, Darjeeling, Sunderbund, Chittagong, and other districts, and great expectations are formed as to the fitness of the Himalayas. The tea produced has a more penetrating flavour, and contains a larger proportion of chemical elements than much of the China tea, so that when used alone it is not agreeable, but as a mixing tea it has a high value. It is therefore probable that, so far as the plant and its cultivation is concerned, they have the requisite knowledge and art, and the comparative failure lies rather in the picking and manufacture of the leaves.

The flavouring of tea is a well-known process, and is carried on with the middle and inferior qualities of teas exclusively. It is easily effected by placing the tea leaves in process of manufacture with the aromatic flowers of plants; as, for example, the *Olea fragrans*, in preparing the varieties of scented Pekoe. Such odours are very evanescent, but delicate and agreeable, and give a pleasing variety to the flavour of the inferior tea with which they are usually associated. They do not, however, add to the chemical or dietetic value of the tea.

Hence tea is divided into three great classes, green, black, and scented, and each is subdivided according to the size of the leaf; but the names vary somewhat from year to year.

Of green teas there are:

Gunpowder, Hyson, young Hyson, Imperial, Twankay, Japan and Java, coloured and uncoloured.

Of black teas there are:

Congou, Moning and Kaisou; Souchong; Oolong; Orange Pekoe, Canton and Foo Choo; and Caper, plain and scented.

With regard to Indian teas, it has been recommended to classify them under eight heads, viz. :—

1. Fine Pekoe, or all flowery leaf.
2. Pekoe, little flowery, with small leaf.
3. Pekoe Souchong, large leaf, few ends.
4. Souchong, larger leaf, without ends.
5. Congou, all coarse, dark, leafy sorts.
6. Broken Pekoe, siftings of fine Pekoe Pekoe.
7. Broken black siftings of Pekoe Souchong Souchong.
8. Fannings, siftings of Congou, old leaf.

There are also inferior kinds of tea in which the quality is infinitely deteriorated. Thus, the dust of the leaf is mixed with clay and manipulated into the form of the ordinary leaf and sold as lie tea. Tea leaves which have been already used are again manipulated and rolled into shape and sold as genuine tea. The leaves of other plants are added to those of the tea plant, and thus the quality of the whole is deteriorated, or an undue proportion of stalk is added to the leaf, and the weight thereby increased disproportionally to the chemical value.

It is stated in the Planter's *Price Current* that one sample of tea proved to be wholly composed of the following substances:—

'Iron, plumbago, chalk, China clay, sand, Prussian blue, turmeric, indigo, starch, gypsum, catechu, gum, the leaves of the camellia, sarangua, *Chloranthus officinalis*, elm, oak, willow, poplar, elder, beech, hawthorn, and sloe.'

The following list of trees from which leaves are obtained for this purpose in North America, Abyssinia, Tasmania, the Mauritius, the Isle of Bourbon, and other countries, has been compiled by Prof. Johnston, and is worthy of perusal, since it must not be assumed that the Chinese tea plant alone possesses agreeable, stimulating, and yet not intoxicating properties.

No. 130.
LIST OF SUBSTITUTES FOR CHINESE TEA AND MATÉ.

Name of the Plant	Natural order	Where collected and used	Name given to it
Hydrangea thunbergii	Hydrangeaceæ.	Japan.	Ama tsja or Tea of Heaven.
Sageretia theezans	Rhamnaceæ.	China.	?
Ocymum album	Labiatæ.	India.	Toolsie tea.
Catha edulis	Celastraceæ.	Abyssinia.	Khat or chaat.
Glaphyria nitida	Myrtaceæ.	Bencoolen. (flowers used).	Tea plant and Tree of Long Life.
Correa alba	Rutaceæ.	New Holland.	Tea plants, and Tasmanian tea.
Acæna sanguisorba	Sanguisorbiaceæ.	Do.	
Leptospermum scoparium, and L. thea	Myrtaceæ.	Do.	
Melaleuca scoparia, and M. genistifolia	Myrtaceæ.	Do.	
Myrtus Ugni	Myrtaceæ.	Chili.	Substitutes for Paraguay tea.
Psoralea glandulosa	Leguminosæ.	Do.	
Alstonia theaformis	Styraceæ.	New Granada.	Santa Fé tea.
Capraria bifolia	Scrophulariaceæ.	Central America.	?
Lantana pseudothea	Verbenaceæ.	Brazil.	Capitaô da matto.
Chenopodium ambrosioides	Chenopodiaceæ.	Mexico and Columbia.	Mexican tea.
Viburnum cassinoides	Caprifoliaceæ.	North America.	Appalachian tea.
Prinos glaber	Aquifoliaceæ.	Do.	
Ceanothus Americanus	Rhamnaceæ.	Do.	New Jersey tea (medicinal).
Gaultheria procumbens	Ericaceæ.	Do.	Mountain tea.
Ledum palustre, Ledum latifolium	Ericaceæ.	Do.	Labrador tea, or James' tea.
Monarda didyma M. purpurea	Labiatæ.	Do.	Oswego tea.
Angræcum fragrans	Orchidiaceæ.	Mauritius.	Bourbon or Faham tea.
Micromeria theasinensis	Labiatæ.	France.	?
Stachytarpheta jamaicensis	Verbenaceæ.	Austria.	Brazilian tea.
Prunus spinosa, ¼ mixed with ¾ Fragaria collina, or F. vesca	Drupaceæ. Rosaceæ.	Northern Europe.	Sloe and Strawberry tea, one of our best substitutes for Chinese tea.
Salvia officinalis	Labiatæ.	Do.	Sage tea.

Many of these plants are believed to possess the narcotic properties already referred to in maté and coffee leaves, and have more the character of medicine than food; whilst aromatic leaves, as mint and heather, have been used in this and other countries, to prepare beverages with an agreeable flavour or odour without other properties. The general use of China tea, and the low price at which it may be obtained, have almost entirely displaced such substances, except with a medicinal object, and no further reference to them is necessary.

Tea, of whatever quality, being thus prepared, it is desirable to put it into the market at the earliest moment and in the best possible condition, for although it is said that the Chinese do not drink it until it is a year old, the value of new tea is superior to that of old; and the longer the duration of a voyage in which a great mass of tea is packed up in a closed hold, the greater the probability that the process of fermentation will be set up. Hence has arisen the great strife to bring the first cargo of the season to England, and the fastest and most skilfully commanded ships are engaged in the trade both for the profit and honour of success. Such clippers have made the trip in ninety days, and the whole cargo has been dispersed and in the retailers' hands within a week of their arrival. Hence, those who desire it may now obtain tea of the highest quality and in the best condition in which it can be obtained in England. Many years ago, in Papers at the Society of Arts, we showed that the value of tea is determined in the market by its flavour and body—that is to say, by the aromatic qualities of its essential oil and the chemical elements of the leaf, rather than by the chemical composition of its juices, and but little estimate is made of the quantity of the peculiar chemical principle which the juices possess. Delicacy and fulness of flavour, with a certain body, are the required characteristics of the market, and the former is due chiefly to the volatile oil which is present to the extent of 1 per cent., but the quantity of theine—the active principle of tea—is the character required by the chemist and physiologist. Hence the two methods of estimation are not necessarily identical, but there is a general agreement between fulness of flavour and the dietetic quality of a genuine tea.

The tea-taster prepares his samples from a uniform and very small quantity, viz., the weight of a new sixpence, and infuses it for five minutes with about four ounces of water in a covered pottery vessel, and in order to prevent injury to his health by repeated tasting, does not swallow the fluid. He must have naturally a sensitive and refined taste, and should be always in good health, and able to estimate flavour with the same minuteness at all times. He regulates the value of the sample simply by the rules already mentioned.

In selecting tea it must be borne in mind that nearly all genuine teas possess approximately the same amount of theine, and for dietetic purposes, all, whatever their price, are practically equal. This is the hard utilitarian view of the matter, and one which would lead the thrifty housewife to select the cheaper teas, provided they are genuine. But nearly all persons go further than mere utility, and seek for luxury in the flavour of the tea, and in that direction may proceed from 2s. a pound for a genuine Congou to two guineas a pound for Russian Caravan tea, but practically they are limited to prices which should not be wider apart than 2s. and 6s. per pound. Fortunately, the lowest priced genuine tea is of good flavour—good enough to satisfy the desires of ordinary consumers; whilst more refined tastes and larger means may enjoy a wider range of selection.

It is desirable to determine first whether to purchase a green or a black tea, and to bear in mind that the finest green tea is the purest tea obtainable in this country. Of green teas it matters not which variety is selected, provided the quality of Hyson be of the highest. Ordinary *faced* green teas are but little met with now in the market. Of black teas Congou should be preferred for economy, and also as a foundation for a mixed tea. A higher class of tea for ordinary use may be composed of three parts of Congou and one part of Assam or Oolong; whilst for the best kinds a mixture may be made of one part of Kaisow and three of fine Souchong, or of two parts of Kaisow, three of Souchong, and one of Oolong orange-flavoured Pekoe, or fine Assam; or equal parts of Souchong, Kaisow, and flowery Pekoe may be taken. It is possible that one kind of tea, as Souchong, may be so fine as to give fulness of both flavour and body, but these qualities are more certainly obtained by an admixture of several kinds of tea. The leading idea in this selection is that ordinary Congou is prepared from the lower and older leaves, and has therefore a rougher and more earthy although a stronger flavour than the Souchong or Kaisow, whilst the Souchong does not always possess the depth of colour and strength of body which many deem to be desirable.

Oolong is a black tea, but yields a very light-coloured infusion, with a penetrating flavour, so that it is rather a green tea in quality, and is more adapted to use as a mixing tea than to be used alone. In many respects it is practically similar to some of the Indian teas, as, for example, those of Darjeeling.

Pekoe is a fine tea as respects its flavour, but is not of full body, and is therefore rarely used alone. Flowery Pekoe contains a proportion of the flower of the plant, and is said to have the most delicate flavour, and is well fitted for use as a mixing tea.

Scented teas are never used alone, but are mixed with stronger teas. A tea drinker who appreciates the natural flavour of tea would not select them, but if he did drink them he would not mix them with a good common tea; whilst others may prefer to hide the quality of a fine tea by mixing it with a flavour which does not belong to tea. If mixed with fine teas, they should be used very sparingly.

But whatever may be the kind of tea selected it is desirable that the leaf should be closely rolled, and that there should be little or no stalk with it. There is no disadvantage in using broken leaves provided they are of fine quality and can be obtained economically.

The use of tea should be regulated by weight, for it is well known that bulk and weight are not convertible terms. Thus a finely rolled Gunpowder will give a far greater weight to bulk than a loosely rolled Oolong or Congou; and equal bulks, even if the qualities were equal, would give very different infusions. In 1861, when discussing the subject at length at the Society of Arts, I prepared the following table to show the relation of bulk to weight:—

No. 131.

Black Teas.

Kind of Tea	Weight of a moderate sized caddy spoonful (Grains)	Number of such spoonsful to the pound
Oolong	39	179
Congou, inferior	52	138
Flowery Pekoe	62	113
Souchong	70	100
Congou, fine	87	80

Green Teas.

Hyson skin	53	120
Twankay	70	100
Hyson	66	106
Fine Imperial	90	77
Scented Caper	103	68
Fine Gunpowder	123	57

Hence a given bulk of Gunpowder will be more than three times heavier than the same of Oolong, and twice as heavy as of Flowery Pekoe. It is not probable that for ordinary use tea will be taken by weight, yet even then it is desirable to remember the simple rule that closely rolled are much heavier than loosely rolled teas; but when tea is made in very large quantities it is more convenient to use it by weight.

The chemical composition of tea consists, in 100 parts, of a crystallisable salt, called theine, 2 to 3, casein 15, gum 18, sugar 3, tannin 26·25, starch, aromatic oil 0·75, fat 4, vegetable fibre 20, mineral substances 5, and water 5. The quantity of theine is not usually so great as 3 per cent., yet it is stated that it has been found so high as even 6 per cent. in some green teas. It has the following formula :—

No. 132.

C. 8.   H. 10.   N. 4.   O. $2 + H^2 O$.

The nutritive value of tea is therefore small if we use it simply as an infusion, for if we take a fairly good tea, and carry out the rule of a spoonful for each person and one for the pot, we shall have only, say, 100 grains of tea; and as the active principle or theine is only 3 per cent., and is composed of 28·83 per cent. of nitrogen, it follows that we should obtain less than one grain of nitrogen, which is a quantity quite inappreciable as nutritive matter. If the leaves be eaten, as in Tartary, they would still yield less than a quarter of an ounce of matter, of which the vegetable fibre to the amount of one-fifth would be indigestible.

We must not therefore regard tea as a nutrient in the sense of supplying material to maintain structure or generate heat by its own decomposition; but tea is nevertheless a very valuable article of diet, as all infer from experience, and as has been proved by direct experiments on the vital functions.

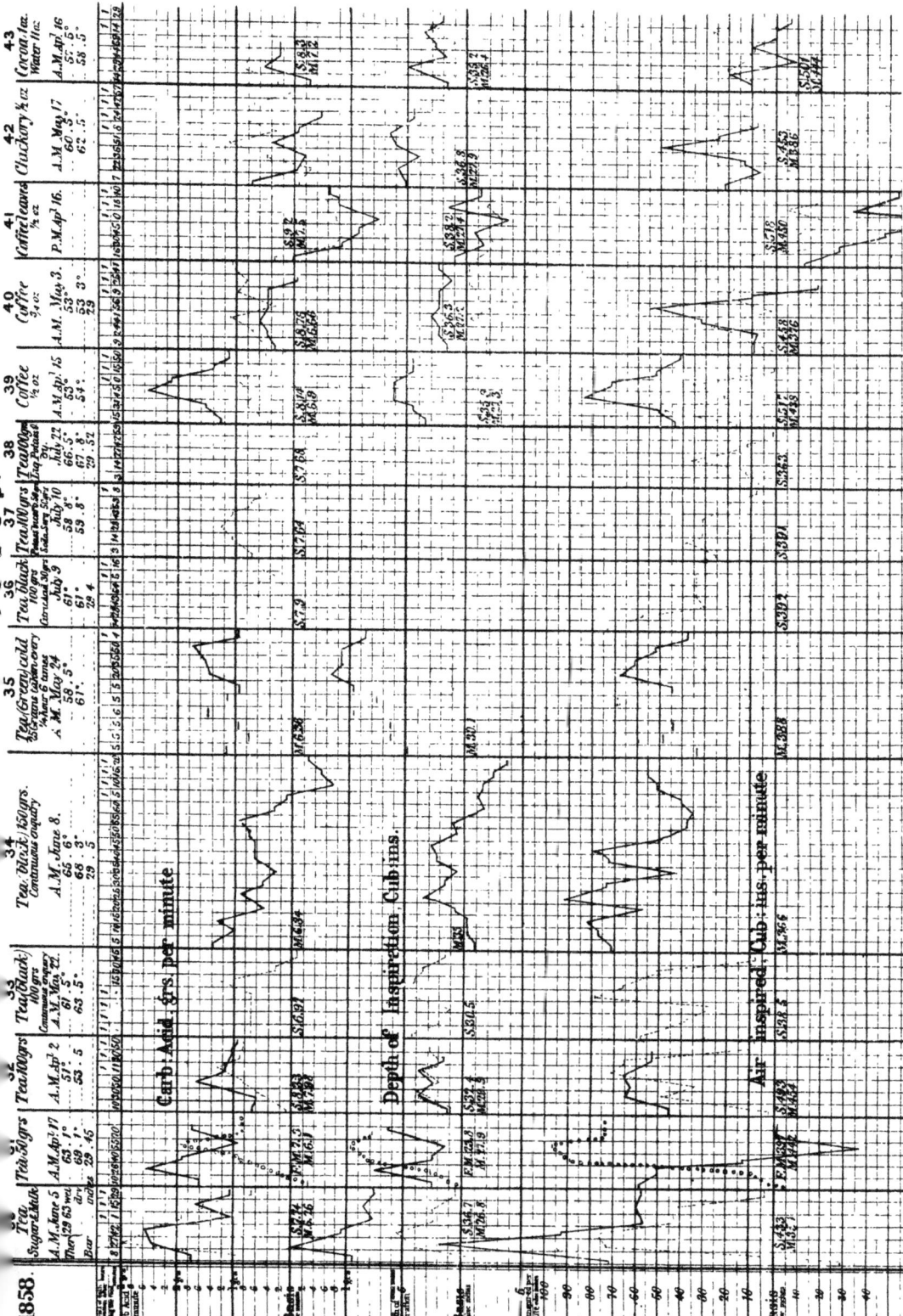

I performed a very extended series of experiments on myself and others, which proved that tea excites vital actions, and is particularly a respiratory stimulant. These were published in the 'Philosophical Transactions' of 1859; and as the subject is an important one, I propose to cite the chief results with the illustrations, in the form of Diagram, No. 133.

1. *As to the Carbonic Acid evolved in respiration.*

1. One hundred grains of the finest black tea gave a maximum increase of 0·87 grain and 1·72 grain per minute in 50 and 71 minutes on two persons.

2. Fifty grains gave to four persons maxima of increase of 1·08, 1·38, 2·58, 1·6, 2·0, and 0·69 grains per minute, under different experiments.

3. One hundred grains of the finest green tea, drunk when cold, gave maxima of increase of 0·9, 2·58, and 0·64 grains per minute on three persons.

4. Twenty-five grains of green tea, drunk when cold, and after having been infused several hours, and repeated every quarter of an hour, for five doses, gave an average increase of 1·2 and a maximum increase of 1·8 grain per minute. The total increase, as shown by ten observations, was no less than 193 grains; and at the close of the experiment the increase continued at the rate of 54 grains per hour.

5. When 150 grains of black tea, infused in one pint of water, were taken, and the whole carbonic acid was collected for 65 minutes, it was shown that there had been an excess of 51·35 grains evolved, which was not more than one-fourth of the total increase when the tea had been divided into repeated doses.

6. When we took 100 grains of black tea, and the whole carbonic acid was collected during 1 hour and 50 minutes, the total increase was 70·40 grains.

2. *As to the Volume of Air inspired.*

There was an average increase in the quantity of air inspired in every experiment but one. Thus pursuing the order of the above-mentioned experiments,

In No. 1, the quantity was increased by 71 and 68 cubic inches per minute.
",, 2, the increase was 34, 39, 50, 72, 95, and 26 cubic inches per minute.
",, 3, the average increase was 120 and 50 cubic inches.
",, 4, the average increase was 66 cubic inches.
",, 5, the maximum increase was 92 cubic inches.
",, 6, the average increase was 47·5 cubic inches.

3. *As to the Depth of Inspiration.*

The rate of respiration either did not increase or was lessened; and as the volume of air inspired was increased, the depth of inspiration was greater so that the increased volume of air inspired at each inspiration varied from 3 to 10·6 cubic inches. With this increased depth, there was also a sense of greater freedom of respiration.

4. *As to the Rate of Pulsation.*

The rate of pulsation followed that of respiration, but in a less degree, and was either not increased or was slightly decreased.

Hence, it was proved beyond all doubt that tea is a most powerful respiratory excitant. As it causes an evolution of carbon greatly beyond that which it supplies, it follows that it must powerfully promote those vital changes in food which ultimately produce the carbonic acid to be evolved. Instead, therefore, of supplying nutritive matter, it causes the assimilation and transformation of other foods.

The effect was not proportionate to the quantity of tea taken, and in this respect it is analogous to the action of a ferment. The effect of repeated small doses was much greater than that of one large one, and yet even in that class of experiment the greatest increase in both the quantity of air inspired and carbonic acid evolved occurred in the earlier half of the enquiry.

The sense of ease in respiration and increase of general comfort after taking tea is well known, as is also the fact that tea tends to induce perspiration, and thereby to cool the body.

Hence, in reference to nutrition, we may say that tea increases waste, since it promotes the transformation of food without supplying nutriment, and increases the loss of heat without supplying fuel, and it is therefore especially adapted to the wants of those who usually eat too much, and after a full meal, when the process of assimilation should be quickened, but is less adapted to the poor and ill fed, and during fasting.

To take tea before a meal is as absurd as not to take it after a meal, unless the system be at all times replete with nutritive material; and the fashion of the day of taking tea at about five o'clock can only be defended when there has been a hearty lunch at one or two o'clock, and an anticipated dinner or supper at seven or eight o'clock. For those to take tea before dinner who eat little or no lunch, must be so far injurious and tend to promote irritability of the stomach.

As a matter of comfort, however, it is to be observed that a cup of tea is in health always refreshing, and, to those accustomed to its use, always welcome.

It may, also, be added that, whilst tea promotes assimilation, there is no ground for believing that it promotes the digestion of food in healthy persons, and therefore it is not usual to take it with, but after, a principal meal. Indeed, but few persons could tolerate a tea-dinner as a daily habit, however agreeable it may be as a change of diet; and by the universal consent of mankind, tea is less fitted to accompany meat than starch and fat.

I have not referred to the effect of tea on the mind, because it is not capable of proof by weight and measure; but it is an action which is universally admitted, and is quite in keeping with the action of tea upon the respiratory tract as a respiratory excitant. There can be no doubt that under certain circumstances it quickens the intellect, both in thought and imagination, and takes away the tendency to sleep, so that in experiments which we made hourly through three days and nights, tea taken twice during the night prevented any desire for sleep. This is not always the same on any person, neither is it uniform on different persons, nor does it accurately correspond with the quantity of tea taken. Moreover, it appears to be measured by the mode of cooking the tea, so that a strong infusion which has been poured off the leaves and kept hot for a considerable period has greater effect.

The Chinese say that it tends to 'clear away impurities, to drive off drowsiness, and to remove or prevent headache;' and it is familiarly known all over the world as 'the cup which cheers and not inebriates.'

It also appears as if muscular activity were increased at the same time, for there is a greater readiness for and ease in making exertion after taking tea; but if it be indulged, a greater sense of exhaustion follows.

It has been affirmed that tea causes waste of muscular tissue, because in certain experiments its use appeared to be followed by a diminished excretion of urea, but such is not a correct inference.

Another action of tea, which is of great importance, is that upon the skin and mucous membranes, which we pointed out in 1858. To have a dry state of the mouth or a dry skin as the result of taking tea would be a great anomaly, and quite opposed to experience; whilst a moist skin and a moist state of the mouth are exceedingly common. The production of perspiration by a cup of hot tea is a familiar fact, and the relief to the sense of heat in hot weather is well known. This may be attributed to the hot water with which the tea is taken; but however true that may be in particular instances, it is equally true when cold tea is taken, and the effect is attributable to the tea as well as to the fluid. The increased action of the skin, by causing an increase in the sensible or insensible perspiration, renders a large quantity of heat near the surface latent, by converting fluid into vapour, and thus powerfully cools the skin.

This action varies with the state and requirements of the system in relation to atmospheric conditions, and will be beneficial or otherwise as it is desirable to lessen the heat of the body and to increase the sensibility of the skin; but it will clearly be the least useful to those who have thin, active and sensitive skins, to the ill fed and feeble, and in cold weather. The problem is not, however, a simple one, for it involves the questions of the amount and kind of other foods taken, in addition to those of constitution, clothing, occupation, and temperature and moisture of the atmosphere.

There is a growing belief that strong infusions of tea may act as a narcotico-irritant poison, and Dr. Arlidge has very recently affirmed that the quantity of tea which is consumed by women of the working-classes has an influence which is injurious in that direction. This lends importance to the subject, and demands some consideration.

The perceptible effects of full doses of tea which are generally, if not universally, admitted are :—.
1. A sense of wakefulness.
2. Clearness of mind and activity of thought and imagination.
3. Increased disposition to make muscular exertion.
4. Reaction, with a sense of exhaustion in the morning following the preceding efforts and in proportion to them.

The first may be due to direct action upon the cerebral system, which can be measured only by the inability to sleep; but it is associated with, if not due to increased respiratory action, and it should be borne in mind that subsidence of respiration is necessary to healthful sleep. The fourth naturally results from the three preceding, for if there have been increased nervous action and loss of sleep, a sense of exhaustion must follow. But it has not been shown that these are evidences, or that there are any other evidences, of a poison in the action of tea, and we should therefore pause before we admit the truth of the assertion.

There never was a period when strong tea was used so generally by the middle and higher classes as at present, and therefore we should find abundant illustrations of the evil tendency in those classes if it exists.

It is assumed that the poison, so called, is the *theine*, or the peculiar alkaloid of tea; but experiments have not shown that it acts as a poison in even larger proportions than can be obtained from the ordinary use of the infusion. If we assume that there is three per cent. of theine in tea, and that half an ounce of tea is used for the preparation of a single cup, as is now sometimes the case, there will be two and a quarter grains of theine taken by the rich man with comparative impunity; but the labourer's wife uses one small tea-spoonful, or, say, sixty grains, for a whole meal, and if she drank that quantity thrice a day, she would take less than that which is consumed by her wealthier neighbour in one strong cup after dinner.

It is now, moreover, the practice to allow the infusion to draw for a longer time than formerly, so that there is time to extract both the theine and the essential oil, and it does not appear that in this respect there is now much difference in the practice of the different classes of society. At the same time, we have often noticed that taking tea at certain Societies, where it was prepared of good strength and kept hot for half an hour or longer, the effect in maintaining wakefulness seemed to be greater than when home-made tea was drunk, and it might thence be inferred that with longer brewing there was more theine extracted, and with more theine there was more wakefulness. This is, however, mere theory; but the matter merits attention, with a view to a complete exhibition of the effect of tea, yet for practical purposes it suffices to adopt the usual procedure.

The question as it is associated with poverty is a very complex one, involving, first, the quantity of fluid taken at a meal and per diem; second, the heat of the fluid; third, the insufficiency of good solid food in relation both to the wants of the body and the quantity of fluid drank, besides the anxieties and the thousand causes tending towards ill health in the wives of our poorest classes.

Our experiments have sufficed to show how tea may be injurious if taken with deficient food, and thereby exaggerate the evils of the poor; but they have not shown that tea is a poison even to the rich, much less to the poor. That tea is frequently associated with indigestion is well known; but this is not attributable to the theine, and the indigestion more commonly originates in other causes, and may be maintained by the use of tea. It is not evidence of a poison.

The conclusions at which we arrived after our researches in 1858 were that tea should not be taken without food, unless after a full meal; or with insufficient food; or by the young or very feeble, and that its essential action is to waste the system or consume food, by promoting vital action which it does not support, and they have not been disproved by any subsequent scientific researches.

The foregoing experiments refer to tea when taken alone, as it is in China and some other tea-producing countries; but it is customary in all other countries to add something to the infusion. Thus in England we add sugar and milk or cream, all of which tend to increase the respiratory action of the tea, whilst they offer nourishment to the body. Hence whilst tea alone may not offer any appreciable quantity of nutritive material, the addition of sugar and milk may make a cup of tea into a true food.

The effect of taking 50 grains of black tea with milk and sugar was to give us a maximum increase of 2·96 and 2·56 grains of carbonic acid in the exposed air, and 78 cubic inches of air per minute, and thus to produce a greater effect than tea alone.

It is the practice in Russia to use lemon juice instead of milk or cream, and to do so does not injure the tea. Thus in a series of twenty-five experiments with 100 grains of black tea and 30 grains of citric acid the maximum increase of carbonic acid varied from 0·88 to 1·86 grain per minute; but there was greater variation in the quantity of air inspired. Opposed as sugar and vegetable acids are in taste, their effect is the same, for they are both ultimately changed into carbonic acid.

The addition of a caustic alkali destroyed the respiratory action of tea and made it worthless, so that 40 minims of liquor potassæ added to 100 grains of tea infused in 7 ounces of water gave no increase in the carbonic acid evolved. It is, moreover, a familiar fact, that the addition of a large quantity of soda 'spoils the tea,' both as to flavour and effect.

The mode of preparation of tea for the table has always elicited discussion, and particularly as the drinker prefers body to flavour, or *vice versá*. The aim should be to extract all the aroma and the dried juices containing theine with only so much of the substance of the leaf as may give fulness, or, as it is called, *body* to the infusion. If the former be defective, the respiratory action of the tea and the agreeableness of the flavour will be lessened, whilst if the latter be in excess there will be a degree of bitterness which will mask the aromatic flavour. As the theine is without flavour, its presence or absence cannot be determined by the taste of the tea.

All agree, therefore, that the tea should be cooked in water, and that the water should be at the boiling point when used; but there is not an agreement as to the duration of the infusing process. If the tea be scented or artificially flavoured, the aroma may be extracted in two minutes, but the proper aromatic oil of the tea requires at least five minutes for its removal. Up to this point all agree that the temperature of the water should be kept near to the boiling point, and that if the water be allowed to cool materially, the infusion will be weak in all its constituents, but it has not been held desirable to boil the tea even for five minutes lest the aromas should be dissipated.

A further prolongation of the process of infusion with water at a high temperature will extract a larger proportion of the theine, on which its respiratory action depends, and of the tannin which gives bitterness, and hence the effect will be different according to the class of tea. If very fine young leaves be infused, the prolonged action will have the good effect of extracting a large quantity of theine and not much tannin, and is therefore very desirable, and indeed essential, in obtaining the best infusion; but if older leaves be taken, the tannin will be increased, and spoil the flavour of the tea. If flavour is to be considered, it is clear that an inferior tea should not be infused so long as a fine tea.

The rule should be to procure the fine thin leaf, whether green or black, and infuse it from ten to fifteen minutes; but if common tea be selected the infusion should not stand more than five to ten minutes. In all cases the pot should be kept quite warm and covered with a cosy.

In some experiments which we have made it appears that tea of the best quality, in both aromatic oil and theine, is obtained by putting the leaves into ordinary cold water, in a covered vessel, and allowing it to remain until the water boils, when it is ready for use. The effect will vary with the duration of the heating process before the boiling point is gained, and the period should not be less than ten minutes. This is a very convenient method when the water is boiled by gas or a spirit lamp on the table. When the tea is ready it should be passed through a strainer, such as is usually suspended from the spout of the tea-pot in Germany, if it is desired to separate it entirely from the leaves.

The kind of water is believed to have great influence over the process, and soft water is preferred. The term soft is not intended to imply the presence of an alkali, as soda softens water, but the absence, or moderation in quantity, of that which hardens water, viz., salts of lime. So far the best would be distilled water, but it is not found in practice that it makes agreeable tea, because of the absence of air and minerals. The Chinese direction is 'take it from a running stream, that from hill springs is the best, river water is the next, and well water is the worst'—that is to say, take water well mixed with air. Hence avoid hard water, but prefer tap water, or running water, to well water. It is the practice of a good housewife in the country to send to the brook for water to make tea, whilst she will use the well water for drinking.

Tea-tasters use only water which has been newly boiled, and this agrees with the practice of the Chinese, for they say 'the fire must be lively and clear, but the water must not be boiled too hastily. At first it begins to sparkle like crabs' eyes, then somewhat like fishes' eyes, and lastly it boils up like pearls innumerable, springing and waving about.'

Do not therefore keep the kettle boiling long before the tea is made, but have fresh water put into the kettle, and use it directly it is boiled.

The addition of a little carbonate of soda to the water is a very general practice, and is doubtless useful with hard water, since the hardness retards the extraction of tannin and colouring matter, and the infusion appears weak. If the water be sufficiently soft and boiling, and the tea be kept hot for ten to fifteen minutes, this addition is not necessary, unless the drinker wishes to have a bitter and black infusion—or, in other words, to spoil the tea—and when used, it should never exceed five grains in quantity.

OTHER KINDS OF TEA.

Numerous beverages are used under the name of tea, besides that which we have now discussed, and although they are not in use in England, or in any English-speaking country, it is desirable to briefly refer to them.

That which is the most generally used is known as Maté, or Paraguay tea, and is prepared from the Brazilian holly (*Ilex Paraguayensis*). It grows wild, and is not cultivated for the production of this article of food, but is regarded as valuable, and the right to take the leaves is sold by the proprietors.

It is found in three qualities according to the age and size of the leaf. The first is the choicest, and consists of the unexpanded buds; the second of the full-grown leaf, with the exception of the woody fibre in the vascular system; and the third of the whole leaf roughly roasted, by placing the branches over a fire until the leaves become brittle, and may be readily beaten off the branches. It has not the delicacy of flavour, whilst it has more bitterness and astringency than the China tea, but it possesses an aromatic oil, gluten, tannin, and a proportion of theine.

It is prepared for the table as an infusion, and is flavoured by lemon juice and burnt sugar, and, as it is cheap and is drunk at every meal, about as much is consumed by a working man or woman in a day as the same class of persons would drink of China tea in a week. As with many other fluids in hot climates, it is commonly sucked through a reed, straw, or tube.

The true action of this substance has not been experimentally determined, but it appears to be more narcotic than Chinese tea, and to resemble coffee leaves.

Coffee leaves are used for the preparation of a beverage in Sumatra and other countries, and have been imported into this country for the same purpose. They are rapidly dried, and being then brittle, easily break. They contain a proportion of caffein, which is a substance identical with theine.

By the favour of Mr. Hanbury, I was enabled to determine the action of the leaves upon the respiratory function, but the results were opposed to those of tea and coffee, and showed that coffee leaves are not respiratory excitants.

I infused half an ounce of the dried leaves in 10 ounces of boiling water, and found that there was a maximum decrease of 0·84 and 0·89 grain of carbonic acid in the expired air, and 25 and 51 cubic inches of air inspired in myself and friend. The rate of pulsation and respiration fell. Hence there is no probability of their supplanting either coffee or tea in the duty which these substances perform in the animal economy, but they may have valuable medicinal qualities in lowering the respiratory action, and be of service in climates hotter than our own (No. 133).

COFFEE.

Having entered at length into the uses of tea, I do not deem it needful to occupy a similar amount of space on the subject of coffee, for in their chief actions they are analogous foods.

The coffee plant (*Coffœa Arabica*) was originally a native of Arabia and Abyssinia, but has been naturalised over a large part of the tropics, including Bourbon, Berbice, Demerara, Sumatra, Java, the West Indies, Martinique, Ceylon, Batavia, the Brazils, Saint Domingo, and Natal, and grows well in the hothouses of this country.

It requires a deep good soil, with plenty of moisture, and a temperature not lower than 65°, and is usually grown on the hill sides, very much after the manner of the tea plant. It belongs to the Cinchonaceæ, or the order of plants which produce the Peruvian bark, and although it has several congeners, it is the only species from which good coffee beans can be obtained.

The plant is very prolific, for it remains in flower during eight months of the year, and produces a succession of crops of fruit, so that there are usually three harvests annually, but at the same time the fruit is in all stages of development, and the picking of it requires care. It is pruned so as to remain about six feet high, by which process it throws out a large quantity of branches at its lower part, and produces more fruit.

The fruit is called a bean or berry, but the former is the more correct expression, and is a corruption of the Arabic word, 'bunn.' The beans are in pairs, which are placed face to face, and enclosed in a hard coriaceous membrane, and surrounded by a pulpy pericarp. The seed itself is hard and tough, so that it requires the use of machinery in breaking the pericarp and freeing it from the coriaceous covering, and is at length cleaned by the process of winnowing.

The following graphic description of coffee planting in Ceylon is extracted from the 'Illustrated London News' of June 22, 1872:—

'Coffee planting is a tedious and expensive branch of agriculture to invest in at first. From the day the planter claims his first bit of jungle for the reception of the coffee seeds which are to form the plants for his plantations, until the first account of sales from England reach him, a period of five years will have elapsed. The first operation is to cut down the heavy forest. About 100 acres is usually the extent undertaken in the first year. After the fallen trees have lain for three months, and have been thoroughly dried by the scorching sun, a match is applied, and terrific is the conflagration that follows. When the ground is well cleared, the operation of lining is begun. A long rope, with tags of white cloth at every sixth foot, is stretched as close to the ground as the blackened trunks of the old forest trees will permit. A peg is placed at each tag, and when the first line is pegged off a coolie at each end of the rope moves off to the right or left, and measures a distance of

6 feet with a wand. Pegs are again laid so as to have a peg in every 6 feet square, and the whole field is thus lined off. Men are now set to dig a small pit, about 18 inches diameter, and the same in depth, at each peg. When the rains set in, these pits are filled with the fine surface soil, and the young coffee plants are then dibbled in and firmly trodden down by the coolies' bare feet. The operation of planting is now complete, and the planter has to wait patiently many months without seeing much symptom of growth in his plants. The work of tracing roads and erecting permanent buildings may occupy his time and help to break the monotony of a jungle life, far away from home and friends.

In the third year the plants (or trees, as they are now called) are fit for topping—that is to say, the plant is cut down to a height of from 3 to $4\frac{1}{2}$ feet, according to the elevation, whether exposed to wind, or sheltered. This is a convenient height for the coolies to have full command of the plants in gathering the fruit, and in pruning the bushes after the crop. In the height of the crop the fruit is taken to the pulping house at midday and again in the evening. The task given to the coolie is to bring a bushel of berries at each collection. Off good-bearing coffee trees some quick hands will gather as much as four bushels a day, for which they of course get extra pay. The cherries are very much like our own cherries in England, and it would puzzle most people to distinguish a heap of coffee cherries from the edible fruit. Instead of the stone, however, as in our cherry, the coffee fruit contains two seeds. These coffee beans are enveloped in a thick leathery skin, which gets the name of parchment. After the thick pulp has been removed, the seeds are left in a cistern till such time as fermentation sets in; the mucilage is easily washed off, and the coffee is then in a fit state to be carried to the drying-ground.

The drying of the coffee is a most important process. A shower of rain will discolour the bean, and depreciate its value much. A constant watch must therefore be kept for the signs of rain clouds, and dreadful is the noise and hurry when such appear and threaten in a few minutes to break over the precious parchment coffee on the barbacues. When thoroughly dried the "parchment" is put into the bushel bags and despatched to Colombo. It there undergoes another drying preparatory to being relieved of the husk, which is done by being placed in circular troughs, where heavy rollers touch the coffee sufficiently to break the skin without injuring the bean. The coffee is then sized—that is, the large beans, medium-sized beans, and small are separated. This is done for the sake of having an equable roasting. A small bean would be burnt into charcoal by the time a large one was sufficiently roasted. This is a very important point, and much care is given to it by the Colombo merchant, who undertakes this part of the preparation for market.

The quality of the coffee depends very much on the district and the elevation at which it has been grown. The greater the elevation the finer the quality. Matura has long been famous for the superior quality of its coffee, and the plantations are all upwards of 4,000 feet above sea level. The climate is delightful, and most of our home flowers and vegetables grow remarkably well.'

Ceylon sends us annually more than a hundred million pounds of coffee, but of this quantity one-half finds its way to the Continent. The consumption in this country is about 1 lb. per head, whilst in Denmark, Belgium, Holland, and Germany it is from 7 to 14 lbs.

The finest coffee is that obtained from Mysore in India and from Arabia Felix, known as Mocha. The bean is smaller and yellower, and when properly prepared has a finer and fuller aroma than other kinds. A good quality is also produced in Java and Bourbon, whilst the Ceylon and Saint Domingo product is less esteemed.

It is common in certain places to make an infusion of the raw coffee, but it is almost universally roasted, and, unlike tea, has its aromatic qualities generated in the process. The object of roasting is, therefore, not only to render it friable, so that it may be readily ground, but to create or develope this aromatic volatile oil, and care is required to limit the operation so that the good effect of the latter may not be destroyed by burning the substance of the bean. It is effected by placing a quantity of it in an iron cylinder, which is slowly turned round over a fire so that all the beans may in turn be exposed to the same heat. The natural colour is a dull pale green, but it acquires three colours in roasting according to the degree, namely, yellowish-brown, chestnut-brown, and black. The first is not considered sufficient, and induces a loss of $12\frac{1}{2}$ per cent. in their weight, but the loss is increased with the chestnut-brown to 20 per cent., and when black to about 23 per cent. 112 lbs. of raw when fairly roasted yield 92 lbs. of roasted coffee.

Hence the proportion of the chemical elements vary as the drying, but that of raw coffee is stable and as follows, per cent.:—

No. 134.

Caffeine	0·8	Caffeo-tannic and Caffeic acids	5·0
Casein or Legumin	13·0		
Gum and Sugar	15·5	Woody fibre	34·0
Fat and Volatile oil	13·0	Water	12·0
Mineral matter	6·7		

Roasted coffee contains 1 per cent. of caffeine and a small proportion of volatile oil and tannin, and if it be roasted with butter as is common on the European continent, some portion of fat will remain. Schroeder found in highly roasted coffee $12\frac{1}{2}$ per cent. of extract possessing much of the properties of raw coffee, 5·7 per cent. of an oxygenated extract, 2 of a fatty matter and resin, 10·4 of blackish brown gum, and 69 of burnt vegetable fibre. When roasted coffee is distilled in water the aromatic principle is obtained.

The proper degree of roasting is that of a chestnut-brown, and when the colour approaches to black it gives a burnt dry flavour to the infusion. The colour varies, however, a little with the original colour of the bean, which itself varies from a whitish or pale yellow in the Bourbon coffee to the yellow colour of East Indian, and the greenish colour of Mocha and Jamaica.

The largest beans are those from Surinam, and the smallest are obtained from Yemen in Arabia Felix.

In preparing it for the table it is requisite to have it freshly roasted and finely ground, and to pour boiling water over it. There is some difference of opinion as to the advantage of boiling it, and there can be no doubt that when mixed with chicory, as on the Continent, boiling for a short time improves its flavour, and produces a better infusion. Such, however, is not the practice here, except so far as to boil the infusion immediately before it is drawn, but it is well understood that the water should be kept as near the boiling-point as possible.

The proper period of maceration is not fixed with absolute precision, but varies from five to fifteen minutes in England, and is longer on the Continent.

The modes of preparation are almost infinite, but all combine two principles, namely, to extract the greatest amount of aroma and body, and to render the fluid quite clear and separate from the grounds.

I have seen exceedingly good coffee prepared simply in a jug which is covered and placed near the fire for ten or fifteen minutes, but care is required to prevent grounds from passing over with the last portions of the infusion. This was the rude, but not altogether unsatisfactory, method adopted by our grandmothers, and is very common in many new countries.

A modification of this is the common tin coffee-pot without a strainer, which is placed upon the fire and out of which the coffee is poured, but I do not think it equal to the older procedure.

The French cafétière has its special advantage almost exclusively in its power to produce a very clear infusion, whilst Ash's Kaffee-kanne by the use of a jacket, which is first filled with boiling water, keeps the water hot and produces a strong infusion, and at the same time it can be placed upon the fire to boil the coffee for a minute before it is served. The infusion produced by it is strong and clear.

Various hydraulic coffee-pots have been invented with a view to draw the hot water through the ground coffee, and thus to expose the whole of the latter to its influences, whilst a clean infusion is obtained. Such as that patented and largely sold by Mr. Beard.

Others, as the Vesuvian and the Venetian, add to this the application of heat, so that the water is drawn through the coffee at a boiling temperature and excellent coffee is produced, but the method is not economical.

I have regularly used Ash's Kaffee-kanne for some years, and have found it a most effectual and economical apparatus.

Coffee, like tea, is a powerful respiratory excitant, and has a crystallised nitrogenous element called caffeine, quite identical with theine, upon which this action chiefly depends.

A large series of experiments on the respiratory functions were made by me on this substance, as on tea. Of twenty-three experiments on myself and others there was from half an ounce of coffee an increase in the quantity of carbonic acid evolved of 0·98, 1·02, 0·9, 0·4, 1·16 and 2·54 grains per minute at different times, whilst the quantity of air inspired was increased 40, 34, 35 and 84 cubic inches per minute with the same experiments. Three-quarters of an ounce of coffee did not give a greater increase, but the actual increase was 0·68 and 1·68 grain of carbonic acid and 28 and 54 cubic inches of air per minute. (No. 133.)

There was, however, this difference in the effect of coffee and tea, that the former caused an increased rate of respiration, so that the depth of inspiration was but slightly increased, and there was an increase in the rate of pulsation.

There is also another difference in the action of these allied substances, viz., that coffee does not increase the vapourising action of the skin, but decreases it, and therefore dries that organ. I pointed out long ago that whilst both tea and coffee agree in increasing the respiratory changes, tea by increasing the action of the skin lessens the force of the circulation, cools the body, and does not cause congestion of any of the mucous membranes, and particularly that of the bowels; whilst coffee, by diminishing the action of the skin, lessens also the loss of heat of the body, but increases the *vis a tergo*, and therefore the heart's action and the fulness of the pulse, and excites the mucous membranes. In one of our experiments, after having taken an infusion made from two ounces of coffee, we fell to the floor and remained unconscious for some minutes, the result, as was subsequently shown, of a very large quantity of fluid having been thrown into the intestines by which the volume of the blood in circulation was suddenly reduced.

The conditions, therefore, under which coffee may be taken are very different from those suited to tea. It is more fitted than tea for the poor and feeble. It is also more fitted for breakfast, inasmuch as the skin is then active and the heart's action feeble; whilst in good health and with sufficient food it is not needful after dinner, but if then drank should be taken soon after the meal. Hence in certain respects tea and coffee are antidotes of each other, and we know that they are not taken indiscriminately, although in a chief action they are interchangeable.

Coffee has also been said to lessen metamorphosis of animal tissues, because the emission of urea was supposed (doubtfully) to be lessened, but this implies that urea is a measure of tissue change, which cannot now be supported by facts. This introduces one of the most interesting discoveries of the day, in which my own researches have led the way by showing that the emission of carbonic acid by the lungs is the true measure of muscular exertion. I proved in my paper in the 'Phil. Trans.' for 1859, that even the movement of the hand could be measured by that standard; whilst in my paper in the 'Phil. Trans.' for 1862, I showed that the violent exertion of the treadwheel caused scarcely an appreciable increase in the emission of urea, and supported the results simultaneously obtained by Bischoff and Voit in experiments on dogs.

The discussion of this subject will be more fittingly pursued in the work on Dietaries; and I shall here be content with simply indicating the change of basis on which the estimation of muscular waste must now be made.

Coffee is an excitant of the nervous system, but not in the same degree as tea. It produces sleeplessness in many persons when it is taken at night, probably by exciting the heart's action, and preventing that fall which is natural at night, and requisite to permit sound sleep. I do not think that there is the same degree of reaction after taking strong coffee as follows strong tea. It is needless to add, that none of these effects may be marked if the infusion be very weak, as is common among the poor, and in this respect it resembles very weak tea.

Strong coffee is a valuable antidote in poisoning by

opium; and may be used as a corrective of the action of tea in persons whose skin is very active.

The addition of milk, which is so universal, forms a more perfect food with coffee than with tea; for both the former have the same kind of action on the skin and respiration, and therefore aid each other, whilst milk counteracts, in a degree, the action of tea upon the skin.

The addition of chicory to coffee does not increase its action upon any part of the system, but rather lessens it, and can be approved only as modifying the flavour of the coffee.

The adulterations of unground coffee are not numerous.

CHICORY.

Chicory (*Cichorium intybus*), with which coffee is, and has long been, mixed, both in this country and on the Continent, is the product of a composite plant growing in all European countries, and now cultivated largely for this purpose. It is said to have little property in common with coffee, and to be useful only by giving colour and a certain body to the infusion of coffee; but there can be no doubt that it possesses an aromatic oil, starch, sugar, nitrogenous substances, and salts; and, however inferior to coffee, the direction of the action of both is the same (No. 133).

We found in our experiments that half an ounce of chicory with 8 oz. of boiling water gave a maximum increase of 1·17 and 0·66 grain of carbonic acid, and 27 and 42 cubic inches of air inspired per minute, in ourself and friend, whilst the depth of inspiration was increased almost as much as from tea. Hence, it cannot be regarded as valueless, and indeed the price at which it is sold shows that it is appreciated as a food.

The root from which the powder is prepared is long and tapering, and, after having been cut into pieces, is roasted with fat, precisely as coffee beans are roasted, until it is of a brown colour and sufficiently dry to be ground into powder.

It is said to be much adulterated with roasted rye, and with substances which sometimes yield a disagreeable smell and flavour.

Cocoa and Chocolate.

These well-known substances are valuable foods, since they are not only allied to tea and coffee as respiratory excitants, but possess a large quantity of fat and other food materials.

Chocolate is produced from the seed of the cocoa tree (*Cacao Theobroma*). A spurious substance is also made from the pods of the ground-nut (*Arachis hypogœa*), and from other sources.

The seeds of the cacao are enclosed in a fruit somewhat like a cucumber in size, and are extracted by burying the fruit in the earth until the pulpy matter becomes rotten, or by first fermenting the fruit and then extracting the seeds by hand and drying them in the sun. They are about the size of an almond, and when broken into small pieces are subjected to great pressure until they are reduced to a rough powder, after which they are mixed with sugar and rolled into a very thick paste, or into a very fine powder, called Chocolate.

Cocoa nibs are the nuts roughly broken, and may be boiled in that state, but the mass is not so soluble as that which results from a more perfect system of preparation.

The peculiar active principle of cocoa and chocolate is the same, viz., theobromine, which resembles theine and caffeine, and has for its formula $C^7 H^8 N^4 O^2$. There is also a very large proportion of oil or fat, which is the chief nutritive element.

The following is the analysis of the cacao bean from various localities by Tuchen (per cent.) :—

No. 135.

	Surinam	Caraccas	Para	Trinidad
Theobromine	0·56	0·55	0·66	0·48
Cacao, red	6·61	6·18	6·18	6·22
Cacao, butter	36·97	35·08	34·48	36·42
Gluten	3·20	3·21	2·99	3·15
Starch	0·55	0·62	0·28	0·51
Gum	0·69	1·19	0·78	0·61
Extractive matter	4·18	6·22	6·02	5·48
Humic acid	7·25	9·28	8·63	9·25
Cellulose	30·00	28·66	30·21	29·86
Salts	3·00	2·91	3·00	2·98
Water	6·01	5·58	5·55	4·88

The fleshy part of the fruit is not used for the preparation of cocoa or chocolate, but is fermented and a vinous liquor made from it.

This substance in its action is less exciting to the nervous system than tea or coffee, and at the same time it contains a much larger proportion of nutritive material. Moreover, its flavour is not lessened by the addition of milk, so that it may be boiled in milk only, and thus produce a most agreeable and nutritious food. There are therefore many persons, states of system and circumstances, in which its use is to be preferred to either tea or coffee.

It is essential when using cocoa nibs to boil them for many hours in water, but the prepared cocoa or chocolate is soluble in boiling water.

So valuable a substance is liable to adulteration, and one of the most harmless is the admixture of starch or flour, but this may be readily detected by the form and figure of the starch (page 147).

We have already referred to the admixture of sugar with prepared cocoas, and with sugar at $3\frac{1}{2}d.$ per lb., and cocoa at more than twice that sum, the admixture far more than repays the cost of manufacture.

There are 3,934 grains of carbon and 140 grains of nitrogen in 1 lb. of unsweetened cocoa or chocolate.

CHAPTER XXXVI.

ALCOHOLS.

WE now approach the consideration of a class of substances usually regarded as foods, of perhaps greater importance in their effects upon a community than any other, and whilst they afford pleasure and health to some, give pleasure and disease to many, and both on the question of their right to be called foods and upon moral grounds, are driving civilised nations into two hostile camps.

It is impossible on any occasion, when these agents are considered, to omit all reference to moral effects, and scarcely possible to doubt that the abuse of them by so many does not more than overbalance the good produced on many more; but our space does not allow us to enter upon this branch of the subject, and, whilst entertaining strong views that a further limitation in their use would be a great advantage, we shall restrict our observations solely to the scientific and dietetic aspects of the subject.

It is first necessary to insist upon the facts, that alcohol does not represent alcohols, and that alcohols cannot be regarded as a homogeneous class of fluids because they have one element in common.

It seems strange that it should be necessary to insist upon the first statement, for with one element among many it is irrational to assume that it should give identical, nay even similar, characters to all, and more particularly when the experience of mankind in the use of that one element alone is almost absolutely *nil*.

Alcohol alone is perhaps altogether out of the reach of the consumers of alcohols, since in the distillation of spirits of wine other products besides alcohol pass over; but admitting that spirits of wine is sufficiently pure for our argument, it is not used by alcohol drinkers.

The forms in which alcohol is prepared as food are almost infinite, but they differ in flavour, strength, and composition, and each preparation might be reasonably expected to have its own special properties. Yet whilst all this is reasonable, and is admitted when asserted, it was not adverted to by scientific men until my experiments on the action of alcohols over the respiratory functions, in 1856-9, and is not now so generally allowed as it ought to be. It is true that a few acute observers, as Hogarth, had drawn a distinction between the inhabitants of Gin Lane and Beer Street, by which they had made it appear that the state of health and the appearance of the two classes were different. It had also been known that a brandy drinker in rum-producing countries found an earlier grave than the rum drinker. Rum had been selected as the spirit to be given to sailors, but the reason for it was perhaps not known. Was it cheaper or more readily attainable than gin? No; for gin costs scarcely more than half the price of rum, and is of home manufacture, whilst rum must be imported. Was it that the flavour was preferred to that of gin? If so the tastes of sailors differed from those of the spirit drinkers of large towns. Was it that rum had been proved to produce effects different from gin, or to be less injurious than gin? None of these had been proved, and yet the selection was a right one, and based upon sound principles. The effects of gin differ from those of rum, and rum is far the less injurious agent of the two.

In pursuing the subject, we shall adopt the usual classification of these substances, and consider in their order spirits, wine, ale, and other fermented drinks.

Ardent Spirits.

The chemical composition of anhydrous alcohol is: carbon 52·2, hydrogen 13·0, and oxygen 34·8 per cent., and is represented by the formula $C_2 H_6 O$.

The proportion in which it exists in various spirits differs extremely. In its strongest distilled form it contains 10 per cent. of water; but there is a legal standard of strength of spirits which is called proof, and consists of more water than alcohol, viz., 50·76 parts by weight of water to 49·24 parts by weight of alcohol, in 100 parts. The specific gravity of spirits of proof strength is 0·919, and as the spirit increases, the specific gravity declines. Thus tables have been constructed by which the quantity of alcohol in any spirit may be ascertained from its specific gravity. A spirit having a specific gravity of 0·838 is said to be 54 to 58 over proof, because it would require that number of volumes of water to reduce it to proof strength and specific gravity, 0·919. By suspending the strongest alcohol in a bladder in a warm place, the water will exude whilst the spirit is retained, and thus make the alcohol stronger, as was adopted in the good old smuggling times; or by heating alcohol with salts which have the strongest affinity for water, as fused chloride of calcium, or carbonate of potash, or quicklime, the water may be abstracted and 'absolute alcohol' procured. Such alcohol has a specific gravity of 0·793 at 60°, and cannot be frozen by any known degree of cold. Even at a temperature so low as 130° below zero it appears as an oily liquid, and at 146° as if it were melted wax, and at 166° it does not congeal.

By the Food and Drugs Adulteration Amendment Act of 1879, it is directed that whiskey, brandy, and rum shall be sold not weaker than 25° under proof, and gin not weaker than 35° under proof; the former, therefore, must contain a minimum of 39·4 per cent. of alcohol by weight, and the last 36·5.

Alcohol does not exist ready formed in nature, but is always the product of art. Two sets of processes are required, by one of which the component parts of vegetables are converted into alcohol, and by the other the alcohol is separated from other compounds. The former is known as fermentation, and the latter as distillation. The essential compound out of which alcohol is made is sugar, and all saccharine substances may yield it. Sugar is a natural product, and is found in nearly all vegetables, but it is also readily produced by chemical action from starch, which is much more abundant in nature than sugar, by the process of fermentation.

When sugar is fermented it breaks up, and is re-formed into two compounds, viz., carbonic acid and alcohol in nearly equal weights, so that 1 lb. of sugar yields somewhat more than a $\frac{1}{2}$ lb. of 'proof' spirit, or more than a $\frac{1}{4}$ lb. of absolute alcohol. The carbonic acid escapes as a gas, whilst the alcohol remains in the Tater. This change is set in motion by adding a ferment as yeast (*Torula cerevisiæ*), to a solution of sugar, or of substances containing sugar, in water, and maintaining a temperature of about 80° F.

The substances which contain sugar with the least

admixture of starch are, besides sugar and treacle, the juices of plants, as for example the sugar-cane, the sugar-grass, beet root and parsnip, the sugar maple, the palm, and fruits of all kinds, including the grape, from which alcohol was very early obtained, and in these the fermentation of sugar and the production of alcohol are the most readily effected.

But large as these sources are, they are quite insufficient to meet the wants of the market, and we are therefore thrown a step backward, and must select those substances which contain starch largely and afford it cheaply, such as grain of all kinds, including barley, wheat, oats, rye, rice, millet, and various fresh vegetables as potato and beet.

The grains are ground and steeped in water, or mashed, at a temperature of 160°, in order to convert the starch into sugar, but a ferment must be added (if it does not already exist) to set it in motion. This is known as diastase or glycogen, which is itself a compound containing nitrogen, but of undetermined composition; it exists in malt, so that when the operation is conducted on other grains, it is usual to add a proportion of barley in order to supply it. The operation is so very rapid, that in a very few hours the starch will have been converted into sugar. The rapidity varies, however, with the proportion of the diastase and the temperature employed, and at a temperature of 140° to 150° it is most remarkable. Thus 100 parts of starch made into a paste with 39 times its weight of water, and mixed with 6·13 parts of diastase in 40 parts of water, produced 86·9 parts of sugar in one hour. The infusion is called *wort*, and when ready is drawn off from the grain and cooled. It is then prepared for the introduction of yeast, to set on foot the process of fermentation of the sugar, and the production of alcohol.

Tables showing the proportion of starch in various vegetables, have been given in this volume (page 147), and it has been found by experiment that 100 lbs. of corn yield 3·47, or very nearly 3½ gallons of proof spirit. As a rule one bushel of malt produces two gallons of spirit, but as the composition of barley as well as other grain varies with the season or climate, the production of alcohol from it cannot be uniform, but it is not known that more than 2½ gallons are ever produced from one bushel of malt.

Undried barley contains 58·4 of starch, and 4·9 per cent. of sugar; but the proportions are greatly changed in malted barley.

One hundred lbs. of starch are equal to about 70 lbs. of sugar, and yield nearly 8 gallons of proof spirit.

It must not be inferred that the spirit thus distilled is pure and fit for every purpose. The first portions which pass over are the purest, whilst those succeeding contain an increased quantity of fusel-oil, but a proportion of various essential oils is formed during the process and gives flavour to even the best qualities of spirits. The finest spirit is obtained by redistillation, and is employed in the production of aromatic essences, as Eau de Cologne.

Fusel-oil is regarded as deleterious, and the lowest kinds of spirits which contain it largely are used in the preparation of varnish, and should not be drunk; but the use of them in adulteration, whether in this country or in America, is by no means unknown.

Spirits of wine has a neutral taste, and is not sold at taverns as an ardent spirit. It is unknown to spirit drinkers; but it may be obtained from chemists, and it is said that the sale of it to ladies for drinking, whilst somewhat secret, is on the increase. As it forms the

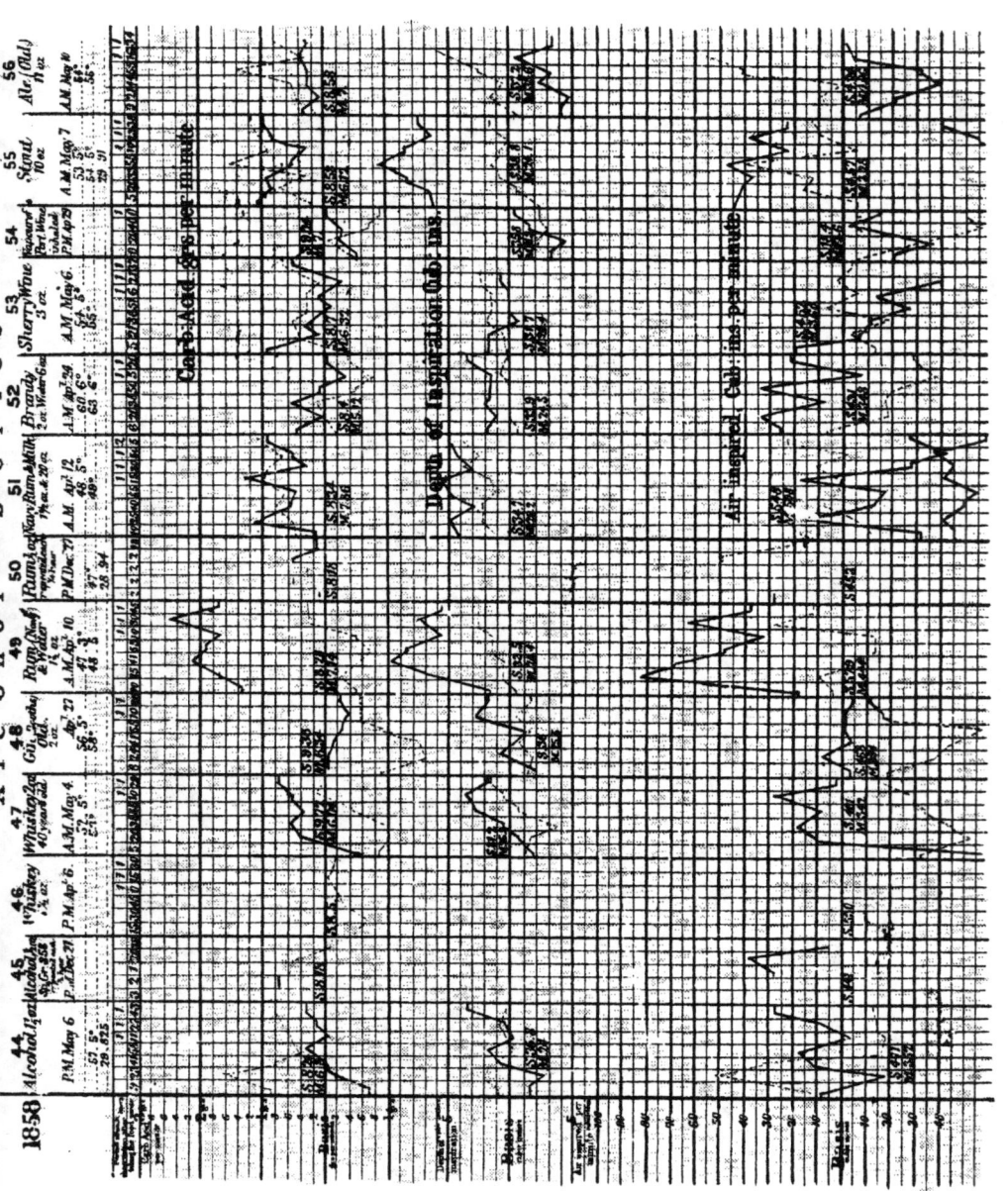

basis of all ardent spirits, it is desirable to show its effect upon the system.

The effect in my experiments on its action upon the respiratory process was to show that alcohol differed from alcohols, that one member of the class differed from another in its effect, and that each member must be considered apart from any other. This is a fundamental question, and I propose to place it as the ground-work of the following observations. The original observations are recorded in the *Phil. Trans.* 1859, and the diagram No. 136 is extracted from those well-known records.

Nearly 1½ oz. of spirits of wine, containing 76 per cent. of pure alcohol, with 6 oz. of cold water, taken at one dose, in the absence of food, caused an average increase in the carbonic acid exhaled by the lungs of 0·18 and 0·8 grain, and a maximum increase of 0·46 and 1·64 grain per minute. The quantity of air inspired was increased by 47, 53, and 26 cubic inches per minute in different persons. The rate of respiration varied (No. 136)

Half an ounce of the same spirit, with 2 oz. of cold water, taken every quarter of an hour for three hours, caused an increase of 0·74 grain of carbonic acid and 37 cubic inches of air per minute. The rate of pulsation rose, but the whole direct effect was exhausted in 70 minutes. The quantity of vapour exhaled by the lungs was increased from 3·12 to 3·76 grains per minute; perhaps in proportion to the increase in the amount of air expired (No. 136).

Hence, pure alcohol in this moderate dose, acted as a general stimulant, and small doses, frequently repeated, produced greater effect than a large one; but the effect was not so uniform as that of an ordinary food. The dose taken, although small for a spirit drinker, was a fair one for myself, and as it was taken on an empty stomach, it caused considerable oppression of the senses.

It was desirable to ascertain if the inhalation of the alcohol and of the volatile properties of other alcohols had the same effect as the drinking of them. The effect of inhaling spirits of wine for 15 minutes was to lessen the expiration of carbonic acid by 0·34 grain, and the air inspired by 11 cubic inches per minute, whilst the vapour exhaled was increased from 0 755 grain to 0·973 grain in one, and from 0·78 grain to 0·91 grain in another experiment, per 100 cubic inches. Hence, the effect on the emission of vapour by the lungs was the same from both methods, whilst the elimination of carbonic acid was opposed in the two sets of experiments. As the inhalation of so strong a vapour as that of alcohol is not usual, the results are open to the explanation that the effect was local, and may not indicate a general action.

Whiskey is so named from the Gaelic word *uisge*, or water. It is prepared in a manner very similar to that of spirits of wine, and is indeed imported largely for re-distillation, and the preparation of alcohol, gin, and brandy. It is made from any kind of grain and starch-bearing vegetables, as potatoes and turnips, which may grow in the neighbourhood, and for ages has been a chief source of the smuggler's profit. It is divided in Scotland and Ireland into small-still and large-still whiskey; the former being chiefly produced in an illicit manner by the peasantry of the country; and the latter in the distilleries under the supervision of the Government. The former has always been regarded as the better kind, in reference to flavour, and as having less of the acrid essential oils which cause so much irritation in the throat and stomach when the spirit is drunk without or with very little water.

The flavour of whiskey differs from that of spirits of

wine by reason of the volatile oils which fall over in distillation, and not unfrequently by the admixture of substances of a special character, as pine-apple flavour. The peaty flavour which is found in certain kinds is said to be due to the peaty water which is obtained in the mountains for the use of the still, or to the smoke of the fuel which is used. Especial attention has been given of late years by the large manufacturers to produce whiskey of an approved flavour and colour, and to render it more mellow and less acrid, and the quality of the large-still has been brought much nearer to that of the best small-still whiskey of a quarter of a century ago; whilst it is far superior to the kinds which were then ordinarily sold.

It is, however, most desirable that whiskey should be kept some years before use, so that it may generate volatile oils, and obtain mellowness of flavour. The most approved course is to fill a sherry cask with it and to leave it quiescent, by which it will acquire colour and mellowness of flavour. The greatest quantity is made in the United States of America, where the peach whiskey, the Monongahela whiskey of Pennsylvania and that from Bourbon County, Kentucky, are accounted the best. It is made in every State, but particularly in Ohio, Illinois, Indiana, Kentucky, Pennsylvania, and New York, and at so cheap a rate that when I travelled on the Mississippi more than twenty years ago, it could be obtained at eightpence to tenpence a gallon. Such spirit is rough and fiery, and tastes much like turpentine, and not only quickly intoxicates, but produces disease of the mucous membrane of the stomach, as well as of the liver, spleen and kidneys. It caused furious drunkenness among the Red Indians, when it was first given to them, and has been the chief cause of their degradation and extermination.

The quality of the spirit which is generally obtained in the British Islands is certainly far superior to that generally drank in the Western States of America, and the injurious effects are in the same degree less. No one can have travelled in the hills of Scotland or Ireland without being sensible that he could drink whiskey with an impunity which would be perfectly impossible in the lowlands or in England; and even ladies, who turn from it with disgust at home, relish the whiskey-and-water which is handed to them several times a day after a long walk in Scotland, or after exposure to the drenching rains of Ireland.

It is, however, very difficult to fix upon a standard of quality which shall represent the kinds of whiskey in general use, for the varieties are infinite, both in the quantity and kind of essential oils, and in the quantity of alcohol; and hence, uniformity of action from different specimens of the fluid, cannot be expected.

The effect of whiskey in my experiments was less uniform than that of spirits of wine, and varied somewhat with the age of the liquor. One and a half ounce of whiskey, containing only 45 per cent. of alcohol in 6 oz. of cold water, caused an average decrease in the carbonic acid evolved of 0·33, and a maximum decrease of 0·7 grain per minute; whilst the rate of respiration was stationary, and that of pulsation fell. Two ounces of the same whiskey caused an average decrease of 0·57, and a maximum decrease of 1 grain per minute on myself; whilst there was an average increase of 0·29, and a maximum increase of 0·66 grain per minute in my friend. The rate of respiration generally declined (No. 136).

Two ounces of the finest whiskey, containing 69 per cent. of pure alcohol, and bottled more than twenty years, when taken with water, gave me an average increase of 0·29 grain of carbonic acid, and my friend of 0·22 grain per minute; but there was considerable variation in the effect during the course of the experiment. The rate of respiration fell, and that of pulsation rose (No. 136).

Hence, the result was rather that of disturbance of the vital functions than of a steady influence in either direction.

Brandy is or should be the choicest and most agreeable member of the class of ardent spirits. It should be prepared by distillation from wine; and 1,000 gallons of wine yield from 100 to 150 gallons of brandy, consisting of about 50 to 54 per cent. of absolute alcohol. It is said that a stronger spirit is obtained from red than from white wines, but the flavour depends essentially on the quality of the wine, so that the finest is obtained from the best white wines in the Cognac and Armagnac districts of France. It would, however, be readily assumed, that strong inferior wines would be substituted in this process for delicate and high flavoured wines; and in Spain and Portugal the substitution descends yet lower, and consists of the refuse wine lees, grape skins, and remains of wine in bottles.

A very large proportion of the brandy consumed all over the world is, however, made with little or no wine, and is simply alcohol distilled as in the preparation of whiskey, and coloured and flavoured with oil of Cognac. A very large quantity of common fiery potato spirit is sent to France from Germany to be redistilled and forwarded to us as French brandy, whilst various qualities of spirits are made in this country into British brandy.

Brandy is as colourless on distillation from wine as alcohol distilled from malt, and can readily be coloured by adding burnt sugar. Its flavour is due to volatile oils and œnanthic, acetic, and numerous other ethers, some of which are derived from the distillate, whilst others are produced in the process of distillation and by age. These vary very widely with the wine from which the brandy is made, and it is said that a dealer should be able to determine by the taste the district of its manufacture.

It is therefore evident that brandy differs from whiskey, and other moderately good spirits, only in the flavours which it possesses, and the inferior kinds are in no degree superior to them; but when prepared from fine flavoured wines and kept for several years it has a delicacy and richness of bouquet, which are both agreeable and peculiar to it. It is usually distilled at a lower degree of proof with a specific gravity of about 0·930, containing half its weight of water, and when redistilled loses much of its fine aromas from œnanthic and acetic ethers.

The effects of brandy will, therefore, be very much those of the spirit of which it is made, and the purer the spirit the nearer will the effect approach that of alcohol. The ethers are, however, regarded as of much value in medicine, and their action does not appear to be in the same direction as that of the spirit, but the quantity of them may vary greatly and descend to almost *nil* in the common kinds. Perhaps there is no preparation used as food or medicine in which there is so great a variation in quality, whilst at the same time the changes are due to subtile substances, but little known to the wisest, and inappreciable by the unrefined and untutored taste.

In my experiments, 1½ oz. of excellent brandy diluted with 6 oz. of water caused an average decrease of 0·2 grain, and a maximum decrease of 0·38 grain in the carbonic acid expired per minute. In another series the average decrease was 0·38, and the maximum decrease 0·71 grain per minute in myself and of 0·02 and 0·62 grain respectively in my friend. The quantity of air inspired fell 42, 37 and 34 cubic inches per minute, in different experiments. My rate of pulsation and respiration fell in all the experiments (No. 136).

Gin differs from all the other ardent spirits in being entirely an artificial compound, and prepared from recipes adopted by each distiller with the unknown additions of the retailer.

It consists of any spirit distilled in the ordinary manner (but usually of an inferior spirit and containing much fusel-oil), of which, about 80 gallons of proof strength is distilled with 10 gallons of water, $3\frac{1}{2}$ lbs. of common salt, and 5 fluid ounces of turpentine, with or without essence of juniper berries and creosote. There is reason to believe that the properties of gin do not depend upon ethers, or any volatile compounds created in the manufacture, but simply upon the substances used in its composition.

For ordinary use it is desirable that the compound should have a soft flavour; but for the gin drinker of the lowest classes a strong acrid fiery spirit is prepared, and hence the manufacturer may make an agreeable and perhaps not injurious mixture, but the retailer in the lower class of gin shops increases its apparent strength by adding various aromatic essences, as coriander, carraway, capsicum, cardamoms and lemon, or creosote, sulphuric acid and salts of tartar. It is certainly very desirable to obtain this spirit as it leaves the producer, and to select that distillery which produces gin of approved flavour.

'Hollands' is not necessarily a purer spirit than home-made gin, but it is made from unmalted rye, and malted bigg, and is more highly flavoured with juniper berries and less mixed with turpentine. The production of it is very large, but the proportionate consumption in this country is diminishing in favour of ordinary gin.

The effect of gin in my experiments was to produce the greatest decrease in vital action of any spirit under enquiry. With 2 oz. of fine old gin in 6 oz. of water, the fall was so great as 1·52 grain of carbonic acid per minute as a maximum, and 0·65 grain per minute on the average of 86 minutes. On another occasion with a newer gin the maximum decrease was 0·46 grain of carbonic acid per minute. The diminution in the air respired was no less than 56 and 43 cubic inches per minute (No. 136).

There was, moreover, a diminution in the quantity of vapour exhaled by the lungs of from 3·12 to 2·7 grains per minute, but the proportion was very nearly that of the diminution in the quantity of air inspired since the proportion in 100 cubic inches was 0·667 against 0·699 grain. The rate of respiration fell in myself, but varied in my friend (No. 136).

We have now arrived at the consideration of a spirit whose effect is in direct opposition to that of gin, one which is very generally popular, is regarded as healthful, and is prepared without intermixture.

Rum is almost exclusively a West Indian product, and is made from fresh cane-juice and the scum which rises in the manufacture of sugar, and contains volatile and essential oils, which are produced both by the sugar-cane and the process of manufacture. It is also made from a mixture of the scummings and the uncrystallisable residue of saccharine juice, or molasses, and even from molasses alone; and although there is this unity in its several sources, the two latter are by no means equal to the first in volatile oils and ethers, and therefore in the flavour of the rum. The wort is prepared from old and new lees with new sweets added, and has a strength of about 12 per cent. of sugar. After 10 to 15 days' fermentation and distillation it yields about 10 per cent. of rum, or 1 gallon from 10 gallons of wort, and care is taken to neutralise the acid due to the acetous fermentation. Every plantation produces sugar, molasses, and rum, the proportions of which vary according to the market price of each, but usually it is 1 gallon of rum to 6 cwt. of sugar. The rum thus distilled is colourless, and is sometimes sold as white rum, in which state it is a very agreeable and delicately flavoured cordial. It is, however, generally coloured for the market by the addition of caramel or slightly burnt sugar, which has lost much of its sweetening properties, but is not in any respect improved by the addition. By the addition of pine-apple in the process of fermentation a new flavour is obtained, and the product is sold as pine-apple rum.

This spirit improves by age as much or more than any other spirit or spirituous liquor both in what it loses and in what it gains. New rum is generally strong, and readily produces intoxication, and is said to be injurious to health by inducing a tendency to fever in hot and to disease of the liver in cold climates; whilst old rum has lost spirit and gained œnanthic ether, and has been greatly improved in softness and flavour. Rum of moderate age may properly be esteemed the purest, and most healthful, member of this class, and is, I believe, the most perfect cordial with which we are acquainted.

In our experiments 1½ oz. of Navy rum containing 58 per cent. of alcohol when taken with water in the afternoon gave a maximum increase of 0·78 and 0·7 grain of carbonic acid per minute in two persons, and an average increase of 0·26 grain per minute in myself.

When the experiment was repeated in the morning the maximum increase was 1·24 and 2·14 grains per minute in two persons. Two oz. of very fine old rum containing 69 per cent. of alcohol gave a maximum increase of 0·1 and 1·5 grain of carbonic acid per minute in two persons (No. 136).

Half an ounce of moderately good rum containing only 37½ per cent. of alcohol with 2 ounces of cold water taken every quarter of an hour gave an average increase on the whole of each quarter of an hour of 0·14, 0·43, 0·2, and 0·66 grain of carbonic acid per minute with an increase of 49 cubic inches of air per minute after the last dose. There was a progressive decrease of vapour in each 100 cubic inches of air expired in each quarter of an hour of 0·7, 0·67, 0·65, and 0·67 grain (No. 136).

Thus, on a review of the experiments with moderate or small doses of these strong alcohols properly diluted and taken on an empty stomach, it is shown that the vital actions are generally increased with pure spirits of wine and rum, whilst they are lessened with brandy, and greatly lessened with gin. Whiskey varied more than the other alcohols, but generally its tendency was to lessen vital actions. But it is needful to repeat the observation, that there is much greater disturbing influence excited by these agents than by ordinary foods, and there was not that regular progression of increase or decrease usually observed with other agents. This extended even to the rate of pulsation and respiration.

When rum was added to milk as in a well-known compound, the effect on the respiratory functions was that of a true food both in degree and persistency, but it was not greater than that of milk alone. Thus 1½ ounce of Navy rum added to 1 pint of good new milk produced an average increase of 0·73 and 0·65 grain, and maxima of 0·9 and 1·38 grain of carbonic acid per minute in myself and friend. The quantity of air respired was increased by 25 and 18 cubic inches per minute and the depth of respiration was increased. The rate of respiration declined whilst that of pulsation increased (No. 136).

It has been frequently remarked in this chapter that our knowledge of the essential oils which give flavour to spirits is very limited, but I have proved by my experiments that they are powerful agents, and I may add another illustration in the fact that only a few drops of a disagreeable essential oil which is distilled from the *murc* of the grape would suffice to destroy the flavour of a whole pipe of choice brandy if not carefully excluded.

The inhalation of the volatile vapours of such spirits as when taken by the stomach caused an increase in the quantity of carbonic acid evolved by expiration did not cause an increase but a decrease.

The proportion of the vapour exhaled by the expired air was not materially changed.

There are many other preparations of ardent spirits in use in those parts of the world where our kinds of spirits are not used, to which reference may be made.

Marwa is a raw spirit prepared from rice in Assam.

Hurreah is a spirit used at Chyebana, in the district of Barrackpore.

Arrack is produced from rice, but it is also a generic name, which is sometimes used in India to describe all kinds of ardent spirits.

Mead and metheglin are spirituous liquors made from honey in Saxon and even to our own times, but their use is now nearly limited to Sussex.

WINE.

Wines have been known in all ages and by all civilised nations, but they have been made from a great variety of fruit according to the productions of each country and the means of importation. They were well known to our forefathers at a very distant period, and were either prepared in this country from elderberry, cowslip, coltsfoot, honey or sugar, or were imported from France and Southern Europe. The use of wine in the preparation of food for the wealthy classes in the 14th century was very frequent, since it enters into a very large proportion of the domestic recipes of that period, but it was commonly Greek or Cyprus, and not home-made wine.

Claret is also frequently mentioned at an early period, but as the meaning of the word was probably *clear cup*, it must not be inferred that it was necessarily the wine of Bordeaux.

These beverages usually consist of the fermented juice of the grape, and should therefore have the elements which are found in grape juice, subject to the changes of fermentation and the effect of age and treatment. There are always alcohol, grape sugar, or glucose, bitartrate and malate of potash, tartrate of lime, chloride of sodium, and tannin, besides various essential oils, which give flavour, and œnanthic ethers, which give bouquet or aroma to the wine. According to Franck the salts consist of bitartrate of potash, tartrate of lime, tartrate of aluminium, tartrate of iron, chloride of sodium, chloride of potassium, sulphate of potash and phosphate of aluminium.

In the preparation of wine in France and Germany the grapes which are gathered during the day are usually pressed at night, and the juice is immediately set aside for fermentation. Such as are particularly fine and ripe are selected to make the choicest quality, or *Auslese* wine, and the remainder are used for the ordinary quality. When making sparkling wines from black grapes the grapes are at first pressed gently, so as not to squeeze out the colouring matter of the skin, but afterwards they are pressed more sharply for the inferior white wines, or added to the red grape in making red wines. The proportion of *Auslese* wine which can be made varies with the season, but it is small in comparison with the whole, and consequently obtains a much higher price.

The juice having been placed in a vat produces a froth upon the surface in the course of the night, which, after attaining a certain degree of thickness, is skimmed off and the process is renewed with a second or third layer of froth. At length all the remaining scum rises to the surface after the process of fermentation has fully set in, and is rapidly and completely skimmed away, when the clear liquor is transferred to barrels to complete the process and to ripen. The fermented juice is allowed to remain until about the middle of winter, viz., until February, when it is racked off from the lees and renewed fermentation with the return of the warmer weather is thus prevented or greatly lessened.

When making the ordinary wines, whether white or red, the grapes after having been macerated in water are trodden in layers of more than a foot in thickness, and left to ferment, while the vat is usually covered over. When the violence of the fermentation has subsided and the liquor is becoming clear and no longer sweet, it is racked off and run into tuns for perfect fermentation and ripening. The murc, or remains of the bruised grape and foot stalks, is then pressed, and the clear wine is distributed through the vats, whilst the thicker residue after complete pressing is made into inferior wine.

Wines thus produced vary in every constituent according to the locality, season, and age, but generally the produce of each vineyard retains its own leading characteristics. They are not usually drunk until they have undergone important chemical changes, some of which take place rapidly in cask, and others much more quickly in bottle.

The quantity of alcohol in wines varies with the nature of the grape, the season, and the vineyard. The French wines in the department of the Gironde vary from 7 to 15 per cent., but the majority have 9 per cent. and upwards, and the white wines are the stronger. The strongest wines are grown at Sauternes, Barsac, Podensac, Carbonnieux, Bommes, Preignac, Castlenau, and Queyries, all of which have over 11 per cent., whilst the well-known wines of Ch.-Lafite and Ch.-Margaux have less than 9, and Larose, St.-Éstèphe, and St.-Émilion have between 9 and 10 per cent.

The proportion of alcohol in the stronger wines has been determined by Brande, as follows:—

No. 137.

Wine	Sp. Gr.	Alcohol, per cent.	
		Sp. Gr. 0·825	Absolute
Port—mean	·97460	23·49	21·75
Madeira ,,	·97571	20·38	20·31
Sherry ,,	·97806	19·04	17·63
Bordeaux Claret—mean	·97251	14·61	13·53
Lisbon	·97846	18·94	17·45
Bucellas	·97890	18·49	17·22
Marsala—mean	·98095	16·26	15·14
Champagne ,,	·98529	12·05	11·65
Burgundy ,,	·98420	13·24	12·20
Hermitage—white	·97990	17·43	16·14
,, red	·98495	12·32	11·40
Hock	·98290	14·37	13·31
,,	·98873	8·88	8·
Vin de Grave	·98450	12·80	11·84
Côte-Rotie	·98495	12·27	11·36
Roussillon	·98005	17·24	15·96
Constantia	·97770	19·75	18·29
Tinto	·98399	13·30	12·32
Tokay	·98760	9·88	9·15
Lachrymæ Christi		19·70	18·24

When the quantity of sugar which is formed during the process of ripening wine is not very large, all of it may be transformed into alcohol, but in larger amount a portion produces about 20 per cent. of alcohol, which is sufficient to retard or prevent the transformation of the remainder, and the result is a sweet wine. This will be further referred to when describing the process of manufacture of the wines of the Peninsula; and when it is desirable to fortify the wine with a view to prevent further fermentation additional alcohol is added.

The acid or sub-acid flavour is due to the malates and tartrates already mentioned, and not necessarily to the presence of acetic acid; neither are the acid eructations which sometimes follow the use of imperfectly fermented or sweetened wine due to that acid in the wine, but to the transformation in the stomach of the sugar which had been added. When wine has been exposed to a limited quantity of air a substance is produced which having one atom more oxygen than alcohol is intermediate between alcohol and acetic acid, viz., aldehyde—one of the forms in which alcohol is subsequently emitted from the body. If it is desired to remove the natural sub-acid flavour, the practice is to add Paris plaster or sulphate of baryta, by which the natural vegetable acid salts are thrown down, and certain proportions of sulphates substituted for them. Such a course, however, robs the wine of one of its agreeable and useful elements, and substitutes that which is less agreeable. This practice seems to be of old date, for Falstaff discovered lime in the sack, and although he did not regard it as the worst of evils, it is clear that he did not approve it when he added that, 'Yet a coward is worse than a cup of sack with lime in it.' (1 Hen. IV. 11. iv.) Sherry which has been thus treated is apt to be bitter

When a large quantity of wine is distilled certain essential oils and œnanthic and other ethers come over at the last. The former are of the nature of fatty acids, and when pure are of the consistence of butter at 60°, whilst the latter are oily substances, and are said to form $\frac{1}{40000}$th of the wine.

The following quantities of those ethers are found in 500 grammes of the wines of the Gironde, the quantities being given in *grammes* and *centigrammes* :—

No. 138.

Ch.-Margaux	1·25
Ch.-Lafite	1·20
Cos-d'Estournel	1·15
Ch.-Latour } Ch.-Haut-Brion Léoville Barsac	1·10
Sauternes	1·05
Gruaud-Larose	·90
St.-Estèphe-Phelan	·85
St.-Émilion	{ ·50 ·70

France, Germany, Spain, and Portugal have hitherto been the chief sources whence our markets have been supplied with wine. In France the principal districts, or centres around which the different classes of wines are grouped, are Bordeaux, Burgundy, Rhône, Champagne, and the South of France; whilst in Germany the districts are the Rhine, Moselle, and Main.

The Bordeaux districts, including the wine of the Médoc and the department of the Gironde, supply the greater proportion of the light red wines which we receive. The wines are very complete in their fermentation, and show less acidity than most other French wines. Their alcoholic strength is moderate and quite natural, and they are pure and light and pleasant to the taste. They have also good body, colour, and bouquet, with a little roughness on the palate. The red wines of Burgundy have fuller body and aroma, with less roughness, and more delicacy of flavour, but they retain a larger proportion of saccharine elements, and are more liable to a second fermentation. The same class of wines on the Rhône have not so marked a character, but they are very agreeable having much of the body and colour of Burgundy, with the lightness of the Bordeaux wine. The white wines of both the Bordeaux and the Burgundy districts are generally stronger in alcohol than the red wines, and are full of flavour, as the Graves, Barsac, Sauternes, and Chablis, and few excel the *Château Yquem* in lusciousness and aroma, and we may add in price.

Champagne is chiefly produced in the district of that name on the two sides of the river Marne, where are the well-known vineyards of Ay, Bouzy, Verzenay, Épernay, and Avize; but it is made in many other departments, as those of the Jura and Loire, and is prepared from both black and white grapes, either mixed or separate as body and aroma or lightness may be required. It is classed as still, sparkling, and demi-sparkling, and is either sweet or dry according to the completeness of the fermentation, and has a greater alcoholic strength than the wines of the Bordeaux district. Other sparkling French wines are less known, but are of delicious flavour, as sparkling Hermitage, and red and white Burgundy and St.-Péray.

The wines from the South of France more resemble the Spanish and Portuguese wines than those of France, as they are fortified by the addition of alcohol, so that Roussillon and Masdeu are more like a Port than a Claret, and not so good as either.

The wines of Germany are generally of a higher class, drier and fuller of aroma, so that they are amongst the most enjoyable wines which are placed on the table. Their fermentation is perfect, and they are therefore not termed acid wines, yet they have a slightly acid flavour, which renders them fresh and cooling. The Hock and still Moselle may be regarded as perfect wines, but their quality necessarily varies with the season as well as the vineyard. Hochheim, Rudesheim, Marcobrunn, Rauenthal, Johannisberg, and Steinberg well represent the several classes. They are prepared with great care, and require to be kept in cask for some years before they are bottled, and even then they are rarely perfectly clear. The firm, B. J. Mayer, of Mayence, are growers and shippers of high repute.

Sparkling Hock is lighter than Champagne, and is preferred by many. It has a cleaner taste, with a delicate flavour; and it more rarely disagrees with the stomach than the sweeter kinds of sparkling Champagne.

The wines of Spain and Portugal are very numerous, and, as in other countries, embrace both white and red, but all that are sent to this country are fortified with alcohol. Their flavour is in part due to the vintage of the year, but it is much more commonly varied by the addition of the wine of other years, which may have an aroma and quality preferred in the market. Hence wines of any quality, flavour, and strength can be prepared to order. The chief wine exported from Spain is Sherry, which is the product of many districts; but sweet wines, as the Rota Tent and Malaga, are also made, and are luscious and full of body.

There are, however, wines of a light character produced in Spain, some of which are not largely exported, and of such Montilla should perhaps be placed first. A fine quality of Montilla has a fulness and delicacy of bouquet which is unsurpassed by any wine, but it must be drunk in the country of its production, and in vineyards where it is carefully prepared. The red wines, known as Catalonia, Tarragona, and Valencia are coming into the market, and will find their own position, but they cannot displace either the German or the Bordeaux wines, whilst they may largely aid in cheapening Port wine.

Port wine is a mixed, and not a natural production of the grape, and is in fact more of a cordial than a wine. The best vineyards are on the Upper and Lower Douro, in the neighbourhood of Lisbon, and their products are shipped from Oporto. After the juice has been pressed in the usual manner of all countries, whilst the fermentation is going on, a certain quantity of grape or other spirit is added so as to impede the process and to retain some of the saccharine matter as well as the flavour of the grape, and by that means a wine of a sweeter character and of fuller body than French wines is obtained. Hence a wine is prepared, which is sweet and strong and only partially fermented, whilst its alcoholic strength is raised from 35 to 42 per cent. of proof spirit, and it requires to be first kept in cask for some years and then in bottle to throw down a crust, and to become clear and fit for use. Wine thus prepared varies very much in quality and colour according to the year.

The qualities of Port wine are especially body and aroma, with moderate fruitiness and strength, but it should neither be sweet nor rough. Its colour should be full and rich, but if kept too long in cask it loses its colour, body, and alcoholic strength, and becomes a light, tawny, aromatic, and yet agreeable wine. The Port-wine drinker demands body as well as flavour and bouquet, but the wine which is old in wood, if of fine quality, suits other persons better. As, however, certain flavouring ethers are produced only in bottle or more rapidly in bottle than in wood, and are both agreeable and useful, it is necessary that Port should have been bottled and have thrown a crust before it is used, and within limits the longer it has been in bottle the more it is improved in fulness and delicacy of flavour.

The white wines of Portugal are most agreeable and delicate, and it is doubtful whether a *fine* Bucellas may not hold its own against the French wines—certainly a white wine fuller of body and aroma can scarcely be obtained when it is fine and old. Of other wines it will suffice to mention sweet Lisbon, which is not largely imported, and white Port, which was formerly consumed more abundantly than at present, and which was so highly esteemed, that fine old samples were sold at a guinea a bottle.

Whilst the consumption of Port wine is decreasing, and that of Claret increasing, competitors are arising which bid fair to force their way into the market and may ultimately gain a place amongst us. Of these, the wines of Hungary and Greece seem to occupy the first rank, and in point of quality, will hold their own against the wines of other parts of Europe.

The red wines of Hungary are not unlike the Rhône wines of France, since they are as full of body and flavour, and yet light and agreeable; but as now manufactured have somewhat more astringency, whilst the white wines are full of aroma, with a soft and mellow flavour. The whole country may be said to be a wine-producing country, and has been so for ages, and with proper cultivation and patronage might even rival France. The following are the best known districts:—

Red Wine: Buda, Egri, Visonta Vittarry, Karlovitz, Menes, and Magyarat.

White Wine: St. György, S'Oprony, Somlo, Tokay, Tetney, Bösing, Badasconyer, Pées, Kobánya, and Neszmély.

None, until lately, have been so well known in England as Tokay; but since the Great Exhibition in 1851, Hungarian wines have been kept before the public, and it may be predicted that many will have established their name and fame before the end of the present century.

The alcoholic strength of some of the best known Hungarian wines is as follows:—

No. 139.

	Proof Spirit
Somlauer	23·60
Erlauer	23·10
Menes	22·20
Bakator	21·03
Karlovitz	21·00
Ofner	20·30
Szegszard	20·20
Sweet.	
Tokay	18·04

The Greek wines successfully introduced by Mr. Denman of Piccadilly are less known than the merits of Greece deserve; but until a powerful government is established, and there is safety to person and property, the great resources of the country in this direction cannot be developed. No country in reference to soil, elevation, sun and climate, can excel it, and with capital and intelligence, the wines may equal, if not surpass, those of Central Europe.

These, like all other natural wines, have a character of their own, so that they can only in a very general manner be compared with wines in ordinary use, but all have this characteristic, viz., that when new they more or less resemble the white and red wines of France or Portugal drunk at dinner, and when ten years old in bottle have acquired qualities which make them resemble dessert wines, and in flavour are not inferior to many liqueurs. This contrast in quality of the same wine and rapidity of maturation is most remarkable.

They are produced from grapes which abound in saccharine matter, and in all the elements of grape juice, and being perfectly fermented are probably the strongest natural wines in the market, and when drunk at dinner, will allow a dilution of one third to lower them to the strength of the French and Rhine wines. Hence, whatever may be the value of wine, they possess it intrinsically in the highest degree, and in point of economy cannot be surpassed.

The following table shows the natural alcoholic strength of the Greek as compared with French wines:—

No. 140.

Greek.

	Proof Spirit		Proof Spirit
Thera	26·70	Cyprus	23·66
St. Elie	26·00	Red Mont Hymet	23·40
Santorins	25·92	Red Kephisia	23·03
White Patras	25·84		
White Kephisia	25·63	*Sweet.*	
White Mont Hymet.	25·14		
Como	24·54	Lachrymæ Christi	17·13
Red Patras	24·00	Vinsanto	15·61

French.

Hermitage	22·03	Sauternes	17·06
Pouilly	21·00	Graves	16·10
Chambertin } Clos-Vougeôt }	20·80	St.-Éstèphe } St.-Émilion }	16·06
St. George	18·30	Médoc } Ch.-Lafite }	15·70
Chablis	18·02		

The new red wines are somewhat astringent, and should be drunk with water, but the white of the same age have a milder flavour, and being without acidity, may be enjoyed equally with or without dilution.

Of all wines with which I am familiar none excel old bottled Thera in the delicacy, fulness and lusciousness of its aroma and flavour, and being a completely fermented wine made from the fresh grape, it is worthy to be regarded as a perfect wine, and the representative of the Nectar of ancient Greece. It is scarcely possible to make a selection of these wines, which when young, would be equally appreciated by all persons, yet perhaps the White Kephisia, St. Elie, Mont Hymet, Patras and Thera would be the most generally approved. The St. Elie developes an Amontillado character, whilst the Patras more nearly resembles Hock, and the Kephisia Chablis, but with a much greater fulness of body and flavour. Of the red wines, the Noussa, Patras, and Kephisia may be mentioned, all of which resemble the unfortified Rhône or Burgundy wines, and become less astringent when they have deposited a portion of their tartar and tannin by age. These full-bodied wines, whether in wood or bottle, develope various ethers, and this is particularly observable in the St. Elie. The White Kephisia is a very fine full-bodied dry dinner wine.

On the whole, I am of opinion that if these wines should continue to be prepared by perfect fermentation, without being fortified, and with the body and aroma which they now possess, they must occupy a very high place—perhaps the highest place among natural unfortified wines, and if the price should be wisely kept down, they must be admitted to general use to the yet further exclusion of fortified wines.

There is yet another class to which reference must be made, viz., the sweet wines prepared from the dried grape, of which the Lachrymæ Christi, a red, and Vinsanto, a white Santorin, from the island of that name, are delicious dessert wines.

M. About, in 'La Grèce contemporaine,' page 114, writes thus of the wines of the Isle of Santorin:—

'Le vin de Santorin se conserve longtemps, il résiste aux plus longues traversées. Il flatte les yeux par une belle couleur topaze, et satisfait le goût par une saveur franche. Il porte l'eau à merveille; je n'ai pas bu d'autre vin à mes repas pendant deux ans. Il se sent de son origine. Né sur un volcan mal éteint, il est le lacryma-Christi de la Grèce.

Les Russes sont très-friands du vin de Santorin, ils en achètent tous les ans pour cinquante mille drachmes, mais ils préféreraient l'avoir pour rien et boire sur place.'

In 'Le Roi des Montagnes,' page 13, the same author writes of 'un petit vin de Santorin':—

'Je crois pouvoir affirmer que ce vin-là serait apprécié à la table d'un roi; il est jaune comme l'or, transparent comme la topaze, éclatant comme le soleil, joyeux comme le sourire d'un enfant.'

The wines of Sicily are represented by Marsala, grown at Catania and fortified, which has established a reputation, and when fine and old is much esteemed. The characteristics of Sicily correspond with those of Greece as a wine-growing country, and on the slopes of Etna is a soil and climate particularly adapted to the vine. The sale of Marsala has never been so large as its quality and price might have warranted. It is not equal to a light Sherry in flavour, nor to a Rhine wine in bouquet and refreshing qualities; but other and choicer Italian wines will, no doubt, make a name in our market now that the country is so largely adding to its resources and wealth.

The wines of Madeira were known in this country before those of Spain. They have held their own in public estimation at the head of all white fortified wines, and in fulness of flavour, aroma, and bouquet, combined with a degree of sweetness and body, have never been surpassed. The flavour is the peculiar characteristic of the wine, and is obtained from the special quality of the grape growing in the island, so that with the loss of the grape a few years ago the supply of the wine ceased; but wine eight years old is again in the market, and before the stores of the old wine have entirely disappeared a new supply equal in quality to the old will be ready for use.

The saccharine quality of the wine shows that its preparation is very like that of the Portuguese wines, and that the fermentation is not complete. The aromatic quality increases with age, and may out-live the saccharine, so that very old Madeira is priceless for the mixer, and gives a bouquet and quality to newer wine, which enables the dealer to sell it as old. For the connoisseur in sweet wines, as well as for the invalid, there is none of the class to be preferred to old Madeira.

Of our own colonies, Australia is the first in the race in the production of wines, although second in time to the Cape of Good Hope, and has had a choice variety of grapes, and a good method of cultivation and of manufacture. There were 272 specimens of full-bodied and light, white, and red wines exhibited at the International Exhibition at Sydney in 1872, including thirty-five from New South Wales, sixty from South Australia, and thirty-seven from Victoria, derived from well-known grapes, and having much of the character and quality of their European congeners. Thus there were Hermitage and Tokay, as well as Roussette, Mataro, Gamai, Reisling, Chasselas, and Verdeilho. One grower in South Australia has, on a small farm of about 140 acres, a dozen kinds of vines, and produces excellent wines, but of a quality not to be exactly compared with the known vintage wines of

France, Germany, and Portugal. It is probable that ere long it will be a great wine and brandy producing country, and give wines which, with a character of their own, will obtain a fixed place in our markets.

The wines of the Cape of Good Hope have been known for some years by the Pontac, which is a fortified red wine, manufactured somewhat after the manner of Port, but differing from that wine in flavour. It differs also from Port, inasmuch as it is perfectly fermented before being fortified with alcohol. Constantia is also well known as a full-bodied and fine-flavoured white wine; and others have been added of late years from Paarl, Drakenstein, Stellenbosch, and other wine-growing districts. In general character they are generous and moderately strong; but the Sherry which has been imported into England for some years past has not been able to compete successfully in flavour and variety with that of Spain. Improvement both in the quality and cultivation of the grape, as well as in the manufacture of the wine, is required, and will doubtless be attained; but the commercial condition of the colony has for many years been unfavourable to the development of any branch of trade. With renewed attention it may be expected that the class of wines to be produced will rather resemble those of Spain and Portugal than of France or Germany.

The Americans have for some years been celebrated for the Catawba wine, which is produced on the banks of the Ohio and in Illinois, Indiana, Missouri and other Western States. The best known varieties of the grape growing there are the Iowa, Delaware, Diana, Catawba and Isabella. The sparkling wines which were exhibited at the International Exhibition were very good, and sustained the high reputation which they have in their own country.

It is, however, necessary to go further west in that great continent, and to reach the elevated and rocky districts of California in order to find a soil and climate perfectly adapted to the growth of the vine, and already there are large farms devoted to this product, one of which yielded 126,000 gallons and 40,000 bottles in one year. Over six millions of gallons of wine, and half a million of gallons of brandy, besides a great quantity of dried raisins, are produced annually. The production and profit per acre are enormous, and must attract capital and attention to that branch of industry.

It may, however, be a long time before the confidence of Europeans in the quality and purity of the American wine is obtained, but in the meantime there is an almost unlimited market in the United States itself.

Wines are produced from a great variety of substances besides the fresh grape juice.

One of the most *recherché* wines of the day, Tokay, is produced from raisins, as is also the Muscatel, and wines known as raisin wines. They are sweet and strong wines, and perhaps not adapted to daily use.

The home-made wines of this country, such as the elderberry, ginger, orange, cowslip, and coltsfoot, need only to be mentioned, but it may be added that they are, with the exception of the first perhaps, weak wines, and require the addition of brandy. Palm wine is obtained in Central Africa by tapping the palm tree, and fermenting the exuded juices. The illustration on page 405 represents an Indian palm so used.

The factitious wines of Hamburg and the South of France have lately attracted much attention, and the more so that by the addition of a portion of genuine wine or the admixture of an inferior with a superior wine, or the manufacture of various ethers and colouring matters, it becomes more and more difficult to detect their unreal character. The trade in them is very large whilst the variety which they manufacture is increasing yearly.

Our space will not permit us to cite general observations on the value of wine, but it may be interesting to extract a sentence from a curious *jeu d'esprit* published by John Groue 'at Furnival's Inn Gate in Holborne, 1629,' in the form of a dialogue, entitled ' Wine, Beere, and Ale, together by the Eares : '—

' I, Wine, comfort and preserue; let that be my character. I am cosen-german to the blood; not so like in my appearance as I am in nature. I repaire the debilities of age, and reuiue the refrigerated spirits, exhilarate the heart, and steele the brow with confidence.'

Again :—

' I am a companion for princes. I am sent for by the citizens, visited by the gallants, kist by the gentlewomen. I am their life, their genius, the poeticall Fury, the Helicon of the Muses.'

The action of wine upon any great vital function must depend upon the quantity consumed, and when only a glass is taken, which may contain from one quarter to half an ounce of alcohol the effect cannot be very marked. It was, therefore, difficult in my experiments to determine the dose which would accord with the habits of the community, but it was at length determined to select three fluid ounces or a glassful and a half as a moderate dose to be taken on an empty stomach. (*Phil. Trans.*, 1859, and No. 136.)

In a great number of experiments three ounces of tolerably good Sherry taken alone caused an average increase of 0·19 and 0·3 grain of carbonic acid expired per minute in myself and of 0·926 and 0·21 grain per minute in my friend, with maxima of 0·36, 0·44, 1·44, and 0·82 grain per minute. On some occasions, how-

No. 141.

The Palmyra Palm (*Borassus flabelliformis*), yielding Palm Wine.

ever, there was a decrease in both of us which amounted to an average of 0·32 and 0·32 grain and maxima of 0·52 and 0·9 grain. The quantity of air inspired at each inspiration, and the rate of respiration always fell, whilst the rate of pulsation was either increased or unchanged (No. 136).

Whilst, therefore, the results were neither great nor uniform, the general tendency of such a dose was to increase the vital actions.

The aromas of wines, which are attributed to volatile ethers of imperfectly known chemical nature, are greatly valued by wine drinkers, and developed by age. These, like the aroma of tea, may be a good guide when fixing the price in the market, but have little or no influence in wine drunk as a food, so that it became necessary to determine their effect apart from that of the other elements. This seemed possible only by inhalation, but that process caused an uneasy sensation in the lungs, and could not with propriety be continued beyond ten minutes at a time. Four ounces of the wine were placed in a Woolfe's bottle and frequently shook, so that the air which was drawn through the bottle might be charged with it.

The vapour of very fine old Port wine inspired during ten minutes at a time on five occasions within seventy minutes gave an average decrease in the quantity of carbonic acid evolved of 0·53 and 0·42 grain, and maxima of decrease of 0·87 and 0·58 grain per minute. The quantity of air inspired was decreased by 56 and 36 cubic inches per minute. The rate of respiration was decreased in myself. A moderately pungent sensation was perceived in the larynx, and particularly when the wine was shaken. The wine in the bottle lost 40 grains weight and the sp. gr. was increased. Good Sherry wine with a bouquet, which was not persistent, and which did not produce any irritation in the larynx, caused an average decrease in the quantity of carbonic acid evolved of 0·12 grain, and in the quantity of air of eight cubic inches per minute. The vapour exhaled was increased from 0·707 to 0·88 grain per 100 cubic inches of air (No. 136).

The general result of a long course of experiments showed that Port and Sherry wines of good qualities somewhat increased the vital actions when taken by the stomach, whilst their aromas, when inhaled, lessened those actions.

BEER AND PORTER.

These fluids have been known in all nations, and in all times, and that which is like our own was appropriately called wine of barley by Theophrastus. They are usually prepared from a decoction of barley, to which a proportion of hops is added, and by subsequent fermentation and fining; but it is believed that hops were not used prior to the time of Henry VIII. The trade in beer has increased so marvellously within the last thirty years that it is now one of the largest in this country, and occupies a prominent position in the manufactured products of other countries, and while other nations eat their meat the English are said to drink it. Neither the process nor the product is identically the same everywhere, but the following description may suffice to give a sufficiently clear idea.

The finest barley is grown in England, and particularly in Norfolk and Suffolk, where the weight of a bushel is 51 lbs. 100 volumes of such barley have been known to swell to 180 volumes after immersion in water; but this is greatly above the average, and the weight of 100 lbs. of dry barley becomes 147 lbs. after steeping.

Besides the component parts of barley already given, viz., starch, gluten, vegetable fibre, coagulated albumen, sugar, gum, phosphate of lime, and water, there is a peculiar volatile oil which is produced on distillation, and gives aroma to the beer. The female flowers of the hop have a yellow powder, which, on distillation, yields about 2 per cent. of another volatile oil, on which the flavour of the hop depends. It also yields the bitter principle termed *lupuline*, to the extent of from 8 to 12 per cent. Upon these two principles the chief value of the hop depends, and they are the most abundant in the hops of Kent. New hops, like new teas, have a larger proportion of volatile oil than old hops, and there is a strife amongst the growers to bring the earliest supply to the market.

From these two substances, almost exclusively, beers are produced; but, before they are both used, it is necessary that the barley should be malted. This process is effected in the following manner.

The barley is steeped from forty to sixty hours in a large cistern, in which the water is allowed to stand a few inches above the surface of the malt. The heavy and good barley sinks, whilst the light seeds remain at the surface and are skimmed away. During this period the grain imbibes about half its weight of water, and increases in volume by about one-fifth, whilst the skin becomes paler and the water yellower, and some carbonic acid gas escapes. When the grain will shed its flour on pressure between the thumb and finger, it has been steeped sufficiently long to cause the process of germination to commence. The water is then drawn off, and the grain, having been washed, is laid in heaps on the couch-floor for twenty-four hours. Gradually, the grain becomes dry, and then warmer by 10°, whilst it gives out an agreeable odour. The sweating stage, which induces germination, has now commenced, and the radicle sprouts, followed by the plumule. The greatest heat occurs in about ninety-six hours after removal from the bath, and the barley is now spread on the floor and turned over twice a day, to prevent too rapid germination. The grain loses about 5 per cent. in weight, whilst it absorbs oxygen and emits carbonic acid. The gluten and mucilage within the seed disappear, and the mass becomes friable and whiter, and the germinating process is completed within fourteen days, if the temperature be about 60°. The technical mode of determining this point is by the *acrospire*, or the growing-point, having reached the opposite end of the grain to that whence it sprang. The starch has become converted into sugar, precisely as when treated with diastase.

The grain is then removed to the kiln, after having perceptibly lost its moisture, and is rapidly dried, which arrests germination, and enables the barley to be kept for future use.

Hence, the whole process of malting consists in inducing germination in the seed, and in at length arresting it at the period when much of the starch has been converted into sugar. During the process the grain has lost 20 per cent. in weight, including the water in the natural seed, but the bulk has been increased by about 8 per cent.

The dried grain is then crushed or very roughly ground in a mill, and is ready for use.

The further process of mashing is partly physical and partly chemical. Thus it dissolves the sugar, and other soluble parts of the malt, whilst, by the aid of the diastase and gluten, it converts the remaining starch into sugar, gum, and dextrine, and by a temperature increasing from 157° to 160°, all the gum and dextrine are transformed into sugar. 13 quarters of malt with 2,400 gallons of water make 1,500 gallons of beer besides the mashed grains, which contain a large quantity of water, and are left after the liquor has been drawn off. Whilst the malt is thus macerated with the hot water, it should be constantly stirred. A portion of the water is first withdrawn, and then the malt, and after the wort has been drawn off in a clear state, a further quantity of hot water is added, and the operation of adding and withdrawing is performed thrice. The first quantity will be the best, and contains the largest amount of extract, whilst the second has usually half of the first, and the third half of the second.

The weight of the first will be about 84 lbs. in a barrel over that of water; but the wort, when drawn off, is about 1·112, 1·091, and 1·031 specific gravity at the three drawings. The hops are added when the wort which has been drawn is put into the copper, and is at or near the boiling-point. For ordinary beer $\frac{1}{4}$ lb. of hops to a bushel of malt is used, but for strong ale the quantity may be increased to $\frac{1}{2}$ lb. or even to 1 lb. per bushel. The effect of the hop is to coagulate the albuminous matter into the wort, and to convert the starch and hordeine into dextrine, whilst the tannin of the hop renders the gluten insoluble. By both means the beer is rendered fit for keeping, but the boiling must be continued for several hours, and the liquid will be concentrated by the loss of about one-seventh of its weight.

The boiling liquor is at length run into coolers through the hop-back, which strains out the hops, and the wort is cooled to 54° or 64° as rapidly as possible, whilst the liquid is thus further concentrated.

It is desirable that starch should not remain in the

wort, and if there be any it may be detected by adding a solution of iodine to it. The hotter it was when mashed, the less the hop, and the less time it was boiled, the more starch will remain.

The next process is that of fermentation by the addition of 1 gallon of yeast to 100 gallons of wort, during which a portion of the saccharine matter is converted into alcohol, but only a portion, for some must remain to prevent the acetous fermentation, and even the conversion of the alcohol into acid. The chemical change which occurs when sugar is entirely decomposed may be readily shown. Sugar is composed of $C_6 H_{14} O_7$, and those elements will produce 2 atoms of alcohol ($C_2 H_6 O$), 4 atoms of carbonic acid ($C O_2$), and 2 atoms of water ($H_2 O$). Some of the sugar is, however, not entirely decomposed, but is degraded by the removal of 4 atoms of oxygen and 4 of hydrogen to grape sugar and dextrine, which have the composition of starch. The temperature is kept at about 64° in winter and 55° in summer. The fermentation is active in six or eight hours, and carbonic acid gas is largely disengaged, and rises with the scum to the surface. The temperature at the highest point varies from 10° to 15°. During the process the beer is drawn off and cleansed. The quantity of yeast varies with the kind of ale, and is less as the temperature is high.

Table beer contains about five per cent. of malt extract, and has a specific gravity of 1·025. Medium ales have a density of about 1·040, and seven per cent. of extract, whilst strong ales have a specific gravity of 1·050 to 1·060.

The colour of the beer depends upon the colour or drying of the malt and the duration of the boiling.

The fining of the beer is a mechanical process, and is

best effected by means of isinglass, which being dissolved in water and added to the beer, combines with the tannin of the hops, and both together carry down the muddy particles.

The following table, by Allen, shows the quantity of saccharine matter remaining in beers according to the specific gravity:—

No. 142.

Sp. Gr. of the worts	Saccharine matter per barrel lbs.	Sp. Gr. of the ale	Saccharine matter per barrel lbs.
1·0700	65·00	1·0285	25·00
1·0780	73·75	1·0280	24·25
1·0829	78·125	1·0205	16·87
1·0862	80·625	1·0236	20·00
1·0918	85·62	1·0420	38·42
1·0950	88·75	1·0500	40·25
1·1002	93·75	1·0400	36·25
1·1025	95·93	1·0420	38·42
1·1030	96·40	1·0271	23·42
1·1092	102·187	1·0302	26·75
1·1130	105·82	1·0352	31·87

The quantity of alcohol which is present in beer differs extremely, but it bears a relation to the amount of saccharine matter which was fermented in the brewing. Brande in his day found 4·20 per cent. of alcohol (specific gravity 0·825), in porter; 8·88 per cent. in ale; and 6·80 per cent. in brown stout. At the present day there may be 7 to 10 per cent. in the strong East India pale ale, and 15 or 20 per cent. in many old home-brewed ales stored for private use; but usually the amount varies from 5 to 7 per cent. in good ales, and may be only 1 to 3 per cent. in small beer. Hence, one pint of strong home-brewed ale may contain as much alcohol as is found in two bottles of good claret wine; but, as a general expression, a pint of strong ale is equal in that respect to a bottle of fairly good claret.

It is well known that the quality of beer depends in some degree upon the water which is used in its preparation. That used by the Messrs. Allsopp contains 29 grains of lime and magnesian sulphate and 17 grains of earthy constituents, whilst that at Messrs. Bass & Co.'s has no less than $54\frac{1}{2}$ grains of sulphate of lime per gallon.

The adulteration of beer is a subject too technical, and perhaps too large, to be discussed at length here. It occurs almost exclusively at the retailer's, and has one of the following objects:—1. To increase the quantity; 2. To give intoxicating power; 3. To increase the colour and flavour; 4. To create pungency and thirst; and 5. To revive old beer. The first is effected simply by adding water or a weaker beer, and has the effect of lowering the proportion of all the constituents and of lessening the flavour. The second is effected by adding tobacco or the seeds of the *Cocculus indicus*; the third, by adding burnt sugar, liquorice or treacle, quassia instead of hops, coriander and carraway seeds; the fourth, by the addition of cayenne pepper or common salt; and the fifth, by shaking stale ale with green vitriol or alum and common salt.

The effect of beers will necessarily vary both with their composition and the quantity which may be taken; but for experiment it is needful to select good standard qualities and a moderate dose, and in our experiments it was as follows:—

Ten ounces, or half an imperial pint of good Dublin stout, gave an average increase of 0·85 and 0·81 grain, and maxima of increase of 1·56 and 1·02 grain of carbonic acid per minute in the expired air. The quantity of air inspired was increased by 41 and 46 cubic inches per minute, whilst the rate of respiration varied somewhat. The rate of pulsation was increased by 4 and 7 beats per minute, and the depth of inspiration was increased by 1·6 cubic inch (No. 136).

On other occasions the maximum increase in the quantity of carbonic acid evolved was 1·16 and 0·98 grain per minute.

The same quantity of fine old Hertfordshire ale, which had become a little acid, gave an average increase of 0·6 and 0·27 grain and maxima of 1·4 and 0·36 grain of carbonic acid per minute in the expired air. The volume of air inspired was increased by 60 cubic inches per minute. The rate of pulsation and respiration was scarcely changed (No. 136).

Hence, we have in beer substances which in their action are very like good foods, although the amount of action is not equal to that produced from bread or milk.

Various light beers had a less degree of influence, but the effect was in the same direction, and for a time they increased the vital actions.

In the *jeu d'esprit*, already referred to on p. 404, there are two stanzas of a song, in which are described certain qualities of ale and beer as compared with wine.

' *Wine.* I, iouiall Wine, exhilarate the heart.
Beere. Marche-Beere is drinke for a king.
Ale. But Ale, bonny Ale, with spice and a tost,
In the morning's a daintie thing.
Chorus. Then let vs be merry, wash sorrow away,
Wine, Beere, and Ale shall be drunke to-day.

Wine. I, generous Wine, am for the Court.
Beere. The Citie calles for Beere.
Ale. But Ale, bonny Ale, like a lord of the soyle,
In the Countrey shall domineere.
Chorus. Then let vs be merry, wash sorrow away,
Wine, Beere, and Ale shall be drunke to-day.'

Cider and Perry.

These refreshing fluids, the ordinary drink of our forefathers, are prepared from the fermented juice of the apple and pear, and in their composition and in many of their properties are not unlike wines from the grape. They contain alcohol in proportions varying from 5 to 10 per cent., saccharine matter, lactic acid, and other products.

The manufacture is very simple, and consists in crushing the apples with a roller, and straining the juice and pulp through sieves, which, after a short delay, is put into barrels for fermentation.

They are drank very largely in the counties where they are made, as in Worcestershire, Gloucestershire, and Devonshire; and although new cider may be drunk in quantities of several pints without intoxication, older and better qualities are as intoxicating as good ale. When in prime condition they are sparkling, highly agreeable, and refreshing, but, without the tonic property of ale, and constitute, perhaps, no inconsiderable portion of the so-called champagne. Indeed, fine sparkling perry is a delicious beverage.

It is said that natural cider or perry will not keep if it be removed in cask after it has been prepared, and in order to fortify it to bear a journey in cask, it is common to add sugar. This so far injures it that it may renew the acetous fermentation, but it temporarily masks the acid flavour, and makes the fluid more agreeable to the palate of those not accustomed to its use.

It is the practice to give the labourers in the cider districts the large quantity of half a gallon or a gallon of cider daily, as a part of their wages; but it cannot be recommended on the ground either of economy or morals. It is said that rheumatism prevails where these lactic acid beverages abound, and, on the other hand, that cases of calculi in the bladder are almost unknown.

In my experiments on good bottled cider there was an increase in the carbonic acid evolved in respiration, as well as in the quantity of air inspired, very similar to that from a moderately good beer, but the effect was neither so great nor so enduring. The sense of warmth and comfort which followed the use of cider was not so great as from ale; but the results of the enquiry proved that the effect of good cider and perry is clearly that of a food, and nearly equal to that of beer.

Ginger-Beer—Treacle-Beer.

These familiar refreshing beverages are agreeable, and so far useful; but they are food only in a very limited sense. They, however, contain about one per cent. of alcohol, a proportion sufficient to bring them under the rule of the Excise, if it were strictly enforced.

Other Fermented Liquors.

Numerous preparations are made in various parts of the world which correspond in character with our spirits and beer.

Quass is a beer made in Russia from rye instead of barley.

Chica is a beer used from time immemorial in South America, and is prepared chiefly from maize, but also from barley, rice, manioc, pine-apples, or grapes.

Weissbier is made in Germany from a mixture of wheat and rye.

Bouza is a fermented drink, which is prepared by the Arabians and Abyssinians from teff, the seeds of the *Poa Abyssinica*, and from the seeds of some of the *Sorghums* or millets.

Marwa is a fermented liquor in use at Darjeeling and on the southern slopes of the Himalayas in India, of which 1 or 1½ pint is drunk at a time.

Honey wine is in constant use in Abyssinia.

Fermented drinks are frequently made from milk, as *Koumiss* in Tartary, *Leban* in Arabia, *Yaoust* in Turkey, and *sour milk* in many parts of America, Ireland, Scotland, and the Northern Isles.

Ava is prepared in the South Sea Islands, from the rhizome of *Macropiper methysticum*.

A sweet beer is made from maize or millet in South Africa.

In the preparation of some of these, the ferment used is the saliva, which, however disgusting to us, is very efficient.

From the statement of the chemical effects of the various members of the class of alcohols, it has been shown that as foods, beers occupy the first place, then cider and perry, and then wines, and as they sustain and increase vital action, they must be allowed to be true foods. Of ardent spirits, rum alone exhibits the action of a food, while gin, brandy, and whiskey, act as medicines by lessening vital action. But the whole class disturb the vital actions, and prevent a uniform course of change, and have much more the character of a medicine than a food, as was stated in my communication to the Royal Society already referred to.

They however seem to exert actions other than chemical, and although not those of food alone, they are frequently associated with and go in aid of the influence of food.

Thus, the tendency of all the class, but particularly of those members which abound in alcohol, is to lessen the action of, or, in other words, to dry the skin—a tendency which is as marked as the effect of the spirit upon the sensorium and vital actions. Thus, the hands and feet, and the skin generally, become hot and dry, and an intoxicated man in a state of perspiration would be a *lusus naturæ*.

With such an action (which, however, is universal in reference to the elimination of water from the body), other results must follow. The cooling of the body is lessened by the diminution in the quantity of fluid emitted by the skin, which is converted into vapour, with an enormous absorption of latent heat. The blood is diverted from the circumference towards the centres, so that the pulse becomes fuller and harder, and the liver and other large circulation-centres receive more blood. The tendency of both these effects is to increase vital processes, and therefore may be of the greatest service in a state of body in which such is needed, or of injury when not needed.

The internal secretions are diminished, so that the larynx, mouth, and throat are dry, and the bowels constipated, and thus the tendency to congestion of the circulation-centres is increased, with beneficial or injurious tendencies according to the requirements of the system.

The relation of these actions to food is such, that when they are required they cause a necessity for increased food, but when not required they lessen the necessity for food. The tendency of all food, but particularly of animal food, is in the same direction, so that the skin is drier after than before dinner, other things being equal.

The action upon the sensorium and nervous centres clearly depends upon the quantity of alcohol which is taken in a given time, other things being equal. When there is a perceptible effect, or an approach to it, there is relaxation of the animal tissues, and particularly the muscles, so that contraction is less easily and fully effected. Where there are flat thin muscles, which act upon the skin, as of the forehead and face, their relaxation is shown by the falling of the features. The capability to continue exertion is also lessened, at least in the degree in which the readiness to make it is lessened. The direct tendency of alcohol is to diminish muscular power in a state of health, but indirectly it may have the contrary effect by improving the tone of the system through the appetite and digestion of food. In the state of body in which alcohol has reduced muscular contractility, all the vital actions temporarily languish; and so far the action of alcohol is opposed to foods, and it is not a food. The tendency to congestion of the blood centres is shown by the *post-mortem* state of such as are internal, and by the fulness of the vessels of the face and head, which gives the man's face an unusual redness, and the cock's comb a remarkable brightness, fulness, and redness.

Whilst the food-action of beer and wine may be accounted for by their known nutritive elements, other than alcohol, which they contain, much difference of opinion exists as to the true action of alcohol itself, and the problem to be solved is whether it acts physically or chemically. It is presumed that the actions just described are physical in their character, as are also those upon food immersed in alcohol, or alcohol and water, when it is hardened and the process of digestion retarded; and if it be shown that alcohol, whilst in the system, is not transformed, and does not enter into new combinations, but leaves the body as it entered it, its action cannot be that of a food. Hence, the proof is diligently sought as to the transformation or non-transformation of alcohol in the system. Up to a comparatively recent period it was assumed that the alcohol was transformed within the system, and was therefore a food, yet it had been shown by Dr. Percy and others, that after alcohol had been given to rabbits, it was found unchanged in the brain and other internal organs after twenty-four or thirty-six hours—a fact opposed to the idea of change.

MM. Lallemand, Perrin, and Duroy instituted special experiments, from which they showed that after alcohol had been administered it passed off by every outlet of the body for many hours, giving the reaction of alcohol or of aldehyde, and thence they thought that they had proved that it did not suffer any chemical change within the system.

I repeated and enlarged these experiments, and found alcohol in the transpiration from the skin and lungs, and in the urine and fæces for more than twenty-four hours after taking 2 oz. of brandy in water; all of which supported the conclusions of M. Lallemand.

But it had not been proved by actual collection of the whole of the alcohol that none had been transformed, and the argument went only to the length of showing that some alcohol was eliminated by every outlet, and so continued for thirty-six hours, whilst after that period unchanged alcohol had been found by Dr. Percy in the brain and other organs. To collect all the products of respiration and perspiration for so long a period as thirty-six or forty-eight hours was a Herculean if not impossible task, and if collected it would be most difficult so to isolate it as to measure and weigh it. Still it may not be impossible to effect this object by a sufficiently complete apparatus which should enclose the body and take account of the expired air.

These experiments have been repeated by several

chemists, and some of them, as Dr. Dupré, have denied the inference drawn by M. Lallemand, because only a portion of the alcohol could be recovered; whilst others. as M. Subbotin, whilst admitting the fact, denied that it proved that the remaining portion must be transformed and act as food. M. Subbotin experimented upon rabbits enclosed in a proper apparatus, and found that during the first five hours only about 2 per cent. was eliminated by the kidneys and 4 per cent. by the lungs and skin, whilst during twenty-four hours only 16 per cent. could be recovered. He, however, regarded it as essential that a food must aid in the transformation of living material, and he denied this power to alcohol.

The determination of the question may in some part, at least, depend upon the meaning of words, and particularly as to what is a food. I have given my definition of a food at the commencement of this work; but M. Voït, to whom physiological chemistry owes so much, has given another, and one with which I cannot concur. He writes as follows :—

'I do not agree entirely with Dr. Subbotin in his views on the importance of alcohol as a nutriment. I define a nutriment as a substance which is capable of furnishing to the body any of its necessary constituents or of preventing the removal of such constituents from the body. To the first class belong such substances as albumen (since it can be deposited as such in the body), or fat or water or the mineral constituents of the body; to the second class belong such substances as starch, which hinders the loss of fat from the body. If a nutriment is defined as a substance which by decomposition furnishes living force to the body, the definition would not be exhaustive, for it would exclude water and the mineral constituents of the body. Alcohol must, therefore, to a certain extent, be regarded as a nutriment, since under its influence, fewer substances are decomposed in the body. It plays in this respect a similar (though quantitatively very different) part to that of starch, which also protects fat from decomposition and, when taken in excess, causes deposition of fat in the organs or fatty degeneration. If a part of the alcohol is decomposed in the body into lower forms of chemical combination it *must* give rise to living force, which either benefits the body in the form of heat or may perhaps be used for the performance of mechanical work; the same is true of acetic acid, which is also not to be considered as an ultimate excretory product, and from which, therefore, in decomposition potential force passes into living force.

It is another question, however, when we ask what importance alcohol has for us as a nutriment, and whether we take it in order to save fat from decomposition and furnish us with living force, in other words, to introduce a nutriment into the body. Since alcohol, when taken in considerable amount, causes disturbances in the processes of the animal economy, we cannot introduce it in quantities sufficient for nourishment as we do other nutriments, and in the amount which we can take without injury its importance as a nutriment is too small to be considered. In this point, then, I agree entirely with Dr. Subbotin; we use alcohol not on account of its importance as a nutriment, but on account of its effects as a stimulant or relish.'

It is quite true that the definition given by M. Subbotin is too restricted, and that water is truly a food, but we cannot admit that the claim of starch to be a food rests solely, or indeed in any considerable degree, on its power to restrain the consumption of fat, for surely the changes through which starch passes are accompanied at every step by the production of heat whether it forms sugar, which is directly emitted as carbonic acid, or first produces fat. Moreover, a substance which prevents the removal of the necessary constituents acts as a medicine—for in so preventing it developes no force—and not as a food.

The state of the argument is still very much as I put it in my communication to the Society of Arts, in 1857, viz.: that whilst it is desirable to complete the proof by collecting all, or nearly all, the alcohol which was administered, the task is a Herculean if not an impossible one; for if it were practicable to collect all that was emitted within thirty-six hours—the largest period at which it has as yet been determined—and to determine its nature in the smallest quantities, there would remain an amount within the body which, according to Dr. Percy, can be proved to be unchanged. To ask for so much proof is scarcely reasonable; and may we not add, from the analogy of other foods, that any large portion passing off unchanged is a strong argument that all is unchanged, and particularly when, after so long a period as nearly two days, some remains in the body unchanged? If we enter on the consideration of the subject with the belief that it is a food, we may still think it possible that a portion may be transformed and act as a food when another portion was proved to be not transformed, but the probability is on the other side if we regard alcohol as a medicine. One of the deductions which Dr. Parkes has drawn from his experiments, is that the capillary circulation is increased by alcohol, because the vessels were fuller, and the tissues appeared more vascular; but, in my opinion, it was precisely the reverse, and whilst there was more blood in the capillaries there was less circulation. This is quite in accord with experiments which have shown that division of the sympathetic nerve caused congestion of the capillaries—that is to say, fulness of the vessels with little circulation; and it cannot be doubted that this action of alcohol is chiefly upon the sympathetic system, and lessens its influence. There is for a time increased fulness of the pulse, because it is due to increase of the *vis a tergo*, that is to say, to lessened capillary circulation, whilst, at the same time, the action of the heart itself is irritable and unsteady.

It has been popularly believed that the drinking of alcohols was peculiar to man, but there is every reason to believe that the brute may descend to the level of man and acquire a taste for them. Thus the horse enjoys ale with oatmeal and water, drinking it to the last drop; and experiments recently made show that the common fowl soon acquires a taste for it, which is followed by congestion of the comb, deterioration of the vital processes, and death within about two months.

Other examples of this kind are cited by Dr. B. W. Richardson, in a very agreeably written article in the 'Popular Science Review' for April, 1872; and as I concur in much that he has therein stated, I will transcribe his conclusions:—

'1. In the first place we gather from the physiological reading of the action of alcohol that the agent is a narcotic. I have compared it throughout to chloroform, and the comparison is good in all respects save one, viz., that alcohol is less fatal than chloroform as an immediate destroyer. It kills certainly in its own way to the extent, according to Dr. De Marmond, of fifty thousand persons a year in England, and ten thousand a year in Russia, but its method of killing is slow, indirect, and by painful disease.

2. The well-proven fact that alcohol, when it is taken into the body, reduces the animal temperature, is full of the most important suggestions. The fact shows that alcohol does not in any sense act as a supplier of vital heat, as is so commonly supposed, and that it does not prevent the loss of heat as those imagine " who take just a drop to keep out the cold." It shows, on the contrary, that cold and alcohol in their effects on the body ran closely together, an opinion more fully confirmed by the experience of those who live or travel in cold regions of the earth. The experiences of the Arctic voyagers,

of the leaders of the great Napoleonic campaign in Russia, of the good monks of St. Bernard, all testify that death from cold is accelerated by its ally alcohol. Experiments with alcohol in extreme cold tell the like story, while the chilliness of body which succeeds upon even a moderate excess of alcoholic indulgence leads direct to the same indication of truth.

3. The conclusive evidence now in our possession that alcohol taken into the animal body sets free the heart, so as to cause the excess of motion of which the record has been given above, is proof that the heart, under the frequent influence of alcohol, must undergo deleterious change of structure. It may, indeed, be admitted in proper fairness, that when the heart is passing through this rapid movement it is working under less pressure than when its movements are slow and natural; and this allowance must needs be made, or the inference would be that the organ ought to stop at once in function by the excess of strain put upon it. At the same time the excess of motion is unquestionably injurious to the heart and to the body at large: it subjects the body in all its parts to irregularity of supply of blood; it subjects the heart to the same injurious influence; it weakens and, as a necessary sequence, degrades both the body and the heart.

4. Speaking honestly, I cannot, by any argument yet presented to me, admit the alcohols by any sign that should distinguish them from other chemical substances of the exciting and depressing narcotic class. When it is physiologically understood that what is called stimulation or excitement is, in absolute fact, a relaxation, I had nearly said a paralysis, of one of the most important mechanisms in the animal body —the minute, resisting, compensating circulation—we grasp quickly the error in respect to the action of stimulants in which we have been educated, and obtain a clear solution of the well-known experience that all excitement, all passion, leaves, after its departure, lowness of heart, depression of mind, sadness of spirit. We learn, then, in respect to alcohol, that the temporary excitement it produces is at the expense of the animal force, and that the ideas of its being necessary to resort to it, that it may lift up the forces of the animal body into

true and firm and even activity, or that it may add something useful to the living tissues, are errors as solemn as they are widely disseminated. In the scientific education of the people no fact is more deserving of special comment than this fact, that excitement is wasted force, the running down of the animal mechanism before it has served out its time of motion.

5. It will be said that alcohol cheers the weary, and that to take a little wine for the stomach's sake is one of those lessons that comes from the deep recesses of human nature. I am not so obstinate as to deny this argument. There are times in the life of man when the heart is oppressed, when the resistance to its motion is excessive, and when blood flows languidly to the centres of life, nervous and muscular. In these moments alcohol cheers. It lets loose the heart from its oppression, it lets flow a brisker current of blood into the failing organs; it aids nutritive changes, and altogether is of temporary service to man. So far alcohol is good, and if its use could be limited to this one action, this one purpose, it would be amongst the most excellent gifts of nature to mankind. Unhappily, the border line between this use and the abuse of it, the temptation to extend beyond the use, the habit to apply the use when it is not wanted as readily as when it is wanted, overbalance, in the multitude of men, the temporary value that attaches truly to alcohol as a physiological agent. Hence alcohol becomes a dangerous instrument even in the hands of the strong and wise, a murderous instrument in the hands of the foolish and weak. Used too frequently, used too excessively, the agent that in moderation cheers the failing body, relaxes its parts too extremely; spoils vital organs; makes the course of the circulation slow, imperfect, irregular; suggests the call for more stimulation; tempts to renewal of the evil, and ruins the mechanism of the healthy animal before its hour for ruin, by natural decay, should be at all near.

6. It is assumed by most persons that alcohol gives strength, and we hear feeble persons saying daily that they are being kept up by stimulants. This means actually that they are

being kept down, but the sensation they derive from the immediate action of the stimulant deceives them and leads them to attribute lasting good to what, in the large majority of cases, is persistent evil. The evidence is all-perfect that alcohol gives no potential power to brain or muscle. During the first stage of its action it may enable a wearied or feeble organism to do brisk work for a short time; it may make the mind briefly brilliant; it may excite muscle to quick action, but it does nothing at its own cost, fills up nothing it has destroyed as it leads to destruction. A fire makes a brilliant sight, but it leaves a desolation; and thus with alcohol.

On the muscular force the very slightest excess of alcoholic influence is injurious. I find by measuring the power of muscle for contraction in the natural state and under alcohol, that so soon as there is a distinct indication of muscular disturbance, there is also indication of muscular failure, and if I wished, by scientific experiment, to spoil for work the most perfect specimen of a working animal, say a horse, without inflicting mechanical injury, I could choose no better agent for the purpose of the experiment than alcohol. But alas! the readiness with which strong well-built men slip into general paralysis under the continued influence of this false support, attests how unnecessary it were to put a lower animal to the proof of an experiment. The experiment is a custom, and man is the subject.

7. It may be urged that men take alcohol, nevertheless, take it freely and yet live; that the adult Swede drinks his average cup of twenty-five gallons of alcohol per year and yet remains on the face of the earth. I admit force even in this argument, for I know that under the persistent use of alcohol there is a secondary provision for the continuance of life. In the confirmed alcoholic the alcohol is in a certain sense so disposed of that it fits, as it were, the body for a long season, nay, becomes part of it; and yet it is silently doing its fatal work: all the organs of the body are slowly being brought into a state of adaptation to receive it and to dispose of it; but in that very preparation they are themselves undergoing physical changes tending to the destruction of their function

and to perversion of their structure. Thus, the origin of alcoholic phthisis, of cirrhosis of the liver, of degeneration of the kidney, of disease of the membranes of the brain, of disease of the substance of the brain and spinal cord, of degeneration of the heart, and of all those varied modifications of organic parts which the dissector of the human subject so soon learns to observe—almost without concern, and certainly without anything more than commonplace curiosity—as the devastations incident to alcoholic indulgence. Thus, the origin of such a Report as that of Mr. Everrett on the Census of America in 1860, related by Dr. De Marmon in the ' New York Medical Journal ' for December 1870:—

"For the last ten years the use of spirits has—1. Imposed on the nation a direct expense of 600,000,000 dollars. 2. Has caused an indirect expense of 600,000,000 dollars. 3. Has destroyed 300,000 lives. 4. Has sent 100,000 children to the poorhouses. 5. Has committed at least 150,000 people into prisons and workhouses. 6. Has made at least 1,000 insane. 7. Has determined at least 2,000 suicides. 8. Has caused the loss, by fire or violence, of at least 10,000,000 dollars' worth of property. 9. Has made 200,000 widows and 1,000,000 orphans." '

It seems desirable before concluding our observations on this subject, to refer for a moment to the contrast in the action of alcohol, and alcohols on the one hand and of tea and coffee on the other. The leading action of alcohol is as a narcotic upon the sympathetic nervous system, whilst that of tea and coffee is as an excitant upon the cerebro-spinal and particularly the respiratory system, whilst the action upon the cerebrum of the one is directly opposed to that of the other. As this question has remained unchanged since the period of my experiments in 1858, I will quote the conclusions at which I arrived, from the ' Philosophical Transactions ' for 1859 :—

'*Abstract of the Effects of Alcohols.*

'1. That the presence of alcohol, being one amongst many elements, and that one varying greatly in quantity, is an insufficient ground for classification, and does not give a common action to the members of this class.

2. The *direct* action of pure alcohol was much more to increase than to lessen the respiratory changes, and sometimes the former effect was well pronounced. Small doses repeated had a more uniform and persistent effect than would have followed the administration of the whole at once. The *indirect* action, as, for example, in lessening the appetite for food, and the mode of its action, I have not investigated.

3. Brandy, whiskey, and gin, and particularly the latter, almost always lessened the respiratory changes recorded, whilst rum as commonly increased them. Rum and milk had a very pronounced and persistent action, and there was no effect upon the sensorium. Ale and porter always increased them, whilst sherry wine lessened the quantity of air inspired, but slightly increased the carbonic acid evolved.

4. The volatile elements of alcohol, gin, rum, and sherry and port wine, when inhaled, lessened the quantity of carbonic acid exhaled, and usually lessened the quantity of air inhaled. The effect of fine old port wine was very decided and uniform; and it is known that wines and spirits improve in aroma and become weaker in alcohol by age. The excito-respiratory action of rum is probably not due to its volatile elements.

5. The quantity of vapour exhaled from the lungs was increased during the inhalation of the volatile elements of wines and spirits, without the quantity of air having increased. When gin was drunk, the quantity of vapour in the expired air was lessened, whilst it was increased under the influence of alcohol, in about the same degree as during the inhalation of that substance. Hence the exhalation of vapour and carbonic acid are not parallel acts.

6. The rate of respiration was in almost all instances lessened in both of us, whilst that of pulsation was as constantly increased in myself, but not in Mr. Moul.

7. The relation between the quantity of carbonic acid expired and the volume of air inspired was usually increased at the period of maximum influence.

8. The variation in the results was greater than the statement of the average and maximum effects indicates, as may be seen in the Tables and Plates.

The *general effects upon the system* of these substances may be thus epitomised:—

1. There is not an exact correspondence in time and intensity of the effects upon consciousness, sensibility, and respiration, and their principal influence is not upon the respiratory function. They *disturb* the vital actions.

2. There were two sets of effects in each of the enquiries on spirits.

 A. The early effects, consisting of—

  Lessened consciousness, with cloudiness, swimming or giddiness, beginning in less than 10 minutes, and increasing during about 30 minutes.

  Lessened sensibility to light, sound, and touch.

  Wavy or buzzing sensation passing through the whole body; and a semi-cataleptic state, in which there was indisposition to move any part of the body from the then existing position.

  These occurred at the same period as:

  Lessened voluntary muscular power and control, with sensation of stiffness and hanging of the upper lip, and stiffness of the face and forehead, beginning in 8 minutes, and continuing about 45 minutes. The dartos was relaxed, and the erector penis and the sphincter of the bladder were rendered less effective. The action of the heart and arteries was increased, as was that of the muscles of inspiration, with a sensation of sudden and forcible action, to a greater degree than the quantity of air inspired accounted for. There was certainly a difference in the effect upon the muscles subject to, and not subject to, volition.

  Lessened transpiration of vapour from the lungs

during ¾ to 1 hour, with dryness of the skin (as if it had been induced by an east wind), and particularly with rum. Increased arterial action near to the surface in 8 minutes, with heat, tingling and swelling of the skin, and a dry state of the whole mouth, with whiskey; and dryness, redness, and soreness of the tip of the tongue with rum.

Pleasant dreaminess and talkativeness, particularly with rum, in 13 to 15 minutes.

B. The later effects.

Taciturnity in from 18 to 80 minutes, followed by depression and a miserable feeling in from 60 to 90 minutes.

Sensation of cold often occurred suddenly and apart from the temperature of the air in about 50 minutes.

The principal influence over consciousness and sensibility was often lessened suddenly, and the effects of the alcohol nearly disappeared at the following periods: 71 to 73 minutes with alcohol; 43 to 120 minutes with rum; 66 to 84 minutes with whiskey; 46 to 80 minutes with brandy, and 68 minutes with gin.

1. That tea, coffee, chicory and cocoa are respiratory excitants, whilst coffee leaves depress the respiratory function.

2. The uniformity in the direction of the results is exceedingly striking, whilst the degree of influence is to a certain extent variable.

3. Tea is the most powerful, then coffee and cocoa, and lastly, chicory.

4. The rate of respiration was sometimes a little increased and at others a little decreased, but the depth of inspiration was always largely increased. The rate of pulsation was usually slightly increased.

5. With the addition of an acid the effect was somewhat lessened, and the rate of both functions was increased to a greater degree than with tea alone.

6. The addition of an alkali also lessened the effect of tea, and a fixed alkali totally destroyed its influence.

7. The action of acids and alkalies varies with the state of the system and in different persons.

8. The addition of sugar and milk in the ordinary way increased the effect.

9. Small doses of tea, frequently repeated, have much greater effect than the total quantity taken at once.

10. Cold tea, and tea infused and kept 24 hours, has as much effect as when hot and recently made.

11. Green tea has somewhat more influence than black tea, and particularly in lessening the rate and increasing the depth of respiration.

12. The proportion of the carbonic acid to the quantity of air inspired was always increased at the period of maximum influence.

13. Mr. Moul experienced much greater effect from tea than myself. He is exceedingly fond of tea, is not fond of coffee, and dislikes acids, and in the above experiments the results corresponded.

14. The influence of both tea and coffee is exerted almost immediately, viz., in 5 minutes, and the maximum is attained in from 25 to 60 minutes. The duration varies from 1 to 2 hours. In all these particulars there is a variation in different persons.

15. With tea we frequently found nausea in 10 minutes, and sometimes to a very unpleasant degree, but it left in 10 or 15 minutes. There was also a soothing or narcotic effect at first on several occasions, and when it had been taken with an alkali this effect was continued to the end; whilst on the other hand the influence was more stimulating with the acid. There was great freedom of inspiration, and sometimes of expiration also, in about 40 to 70 minutes, and with this there was a feeling of lightness and clearness. The pulse was always soft, and the skin moist or soft.

16. With coffee there was no nausea or soothing; the pulse was sometimes feeble, and the pulsation in the head and hands more perceptible. There was often an uncomfortable sensation in the small intestines and forcing at the rectum, and not unfrequently a sense of constriction about the diaphragm in from 16 to 40 minutes. There was more action upon the kidneys than with tea. The skin was often hot and dry.'

ALCOHOLS. 433

I cannot, perhaps, close this account of the action of alcohol, and the popular belief in the antidotal effect of tea, better than by the following amusing incident which occurred at Worship Street, one of the Metropolitan Police Courts :—

'WORSHIP STREET.

'*Henry Bass*, a middle-aged country gentleman, who bore most unequivocal marks of having been off his perpendicular, was charged before Mr. Knox with drunkenness, but not disorderly conduct.

The police having proved the offence,

Mr. Knox asked : Well, Mr. Bass, are you sober now ?

Defendant : Thank you, sir, very.

Constable : The gentleman was very bad ; but I think that he is pretty right now.

Defendant : What you say is very correct. I was bad, but I am right now—quite right ; in fact, all right. (Oscillating slightly.)

Mr. Knox : Indeed, I doubt it.

Defendant : Oh, I assure you I am perfectly compo. The fact is, I travelled 100 miles yesterday by train, and afterwards took some ale—pale ale ; it had an undue and corrupt influence on my system for a time certainly, but what you object to now is not proceeding from that. No, it's the roll of the carriage in me, not the ale.

Mr. Knox : I hope not.

Defendant : Thank you, sir.

Mr. Knox : But I fear that to part with you at present would be subjecting you to robbery ; therefore the constable will take——

Defendant (imploringly) : No, no, don't lock me up ; let me go home—pray do.

Mr. Knox : I don't purpose locking you up. I wish to save you from being robbed, and to restore you to your friends.

Defendant : Thank you ; yes, save me from my—no, send me to my friends.

Mr. Knox: The constable will see that you have some strong tea and——

Defendant (approvingly): That's it—that's it; tea is the thing—better than ale, ain't it, eh?

Mr. Knox: With a little rest afterwards, and then—let me see, I presume I may give a despotic order in this instance—then, constable, have a cab, and see him safe to the train.

Defendant (highly pleased): That's it—that's it—tea, rest, and train. That's it—just the thing.

Mr. Knox: In fact, constable, take care of the gentleman till he's safely off.

Constable: I will, sir.

Defendant: Yes, he shall. I'll make him.

Mr. Knox: I dare say he has plenty of money, and will defray expenses.

Defendant: Yes, yes; I'll defray anything, only don't confine me—because I want to get home—come along. Saying which Mr. Bass moved off, but quickly returning, addressed the magistrate in a grandiloquent manner: Allow me to return you my thanks, sir, for the sympathy you have shown in my most painful position, and to—to—wish you good day; after which this quaint individual left the office as steadily as the influence occasioned by the roll of the carriage would permit.'

PART III.

GASEOUS FOODS.

CHAPTER XXXVII.

ATMOSPHERIC AIR.

WE have now arrived at the concluding subject of our work, and although some of those which have preceded have not been devoid of difficulty or controversy, perhaps this exceeds them all in one or both of those particulars.

There is an analogy between air and water as to their relative composition, for both may be produced from the combination of only two gases, and yet such products are not found in nature. Both, in like manner, have other component parts, which some call impurities, but which are so far essential to their constitution that they cannot be obtained for the use of man without them; and in both the quantity of these additional matters, and not their presence or absence, is the present test of pure and impure air and water; or, as it would be much less unsatisfactory to say, of normal and abnormal air and water.

We are commonly in the habit of thinking and speaking of the air that we breathe, as if it were always the same, and could be described by a set phrase, whereas it varies with every hour of the day, every change of temperature and atmospheric pressure, every change of the wind and season, every degree of latitude and longitude, and with a hundred other circumstances. It is as unexact to imagine a gallon of air or a gallon of water to have a fixed composition, as to describe the characteristics of men and women, or of nations, by a phrase. We can obtain in that manner only a general and not a particular idea.

As a general expression, it may be stated that the atmosphere is mainly composed of two gases, with watery vapour, whilst there are other gases, and some solid particles; but for particular ideas we must know the state in which the two principal components are found at a given moment, and the quantity of the vapour, under exact conditions of elevation, temperature, and barometric pressure, as well as the quantity of additional gaseous and solid matters in a given locality, besides various electrical and magnetical conditions. As, however, this is not a work on Physics or Chemistry, it would be out of our province to attempt to write a complete description of the atmosphere, and we must limit our observations to such questions as bear upon the use of air as a food.

Atmospheric air, at a temperature of 60°, and with the barometer at 30 inches, contains in every 100 parts by volume about 21 of oxygen and nearly 89 of nitrogen. Those gases are not combined as oxygen and hydrogen combine to form water, but are simply mixed together; and although various influences tend to mix them so perfectly that the proportion in a given volume at different places is nearly the same, it is not absolutely so, and one of the gases may, with the greatest ease, be abstracted from the other. It will, therefore, be inferred that the proportions will be the least uniform where the causes which abstract one of them are the most operative, and where such modified air is the least mixed by the aid of the winds with air of a more normal type.

But where shall we look for such conditions? In towns there is the process of respiration in man and animals, by which oxygen is absorbed from the air, and carbonic acid, (or carbon and oxygen combined) given to the air, and the production of light and heat from every kind of fuel, by which oxygen is abstracted and various gases, containing or not containing oxygen, added to it, besides a thousand operations in manufactures, all of which have the same general action. Thus, in towns we look for a diminished proportion of oxygen.

In the country, vegetables have the property of absorbing nitrogen, and thus leave the oxygen relatively in excess, whilst, at the same time, they eat or drink the carbonic acid, and convert both it and nitrogen into their own substance. Moreover, the proportion of oxygen is further increased by the decomposition of a part of the carbonic acid, under the influence of sunlight. Hence, in reference to country localities, so far as there is a difference between them and towns in the abstraction of oxygen will there be an excess of oxygen; and, moreover, the country removes some of the useless or injurious gases which were supplied by towns.

This description is, however, apt to give a wider difference to the characters of the atmosphere in town and country than actually exists, for the country is not devoid of animals which respire, or of fuel which consumes oxygen; and the spaces between towns in our little island are generally so small that it is not easy to say where country begins, and town or village ends. But when we have in our view the great expanse of the ocean, covering two-thirds of the whole globe, great tracts of sandy desert, immense ranges of mountains, bearing practically neither animals nor vegetables, and certain great moorland tracts, where the proportion of animal to vegetable life is a mere vanishing-point, we observe that there are great masses of air, where the conditions which abstract oxygen and nitrogen are insignificant, and to utilise which it is only requisite to mix the smaller quantity of changed air with it to reduce the change to a minimum.

Hence, arises the desirability of man seeking localities where the change is small, and the absolute necessity of those agencies which carry the deteriorated air of towns to the country, and the whole to the sea, and of bringing back the unchanged air from the great stores to supply its place.

It is clear, that if a given composition of the atmosphere be essential to the health of man, the necessity for a great store of normal air, on the one hand, and the replacement of abnormal by normal air, on the other, are equally essential. Sea and mountain air must be brought to the homes of men, and the air of towns carried far out to sea.

The necessity for oxygen as a food is absolute and unintermittent. When the mixed gases of the atmosphere are received into the lungs, a portion is absorbed by the blood, and the oxygen combines with the carbon, nitrogen, and hydrogen of food, and in all the vital processes, to form compounds, which may be called generally carbonates, nitrates, and hydrates. Some of the compounds remain for a time in the body, and form a part of its substance, but a far greater proportion, after producing heat, leave the body as water and carbonic acid, or other compounds, and are called *excreta*. The body is a great oxidising apparatus, by which it sustains its bulk, produces heat, and modifies the composition of the atmosphere; and when it has cast off that which, having been used, is no longer useful to it, it not only deteriorates the atmosphere, but renders it impure. It is not too general an expression to say that every thought and act of man, as well as every action within his body, is accompanied by the consumption of oxygen and deterioration of the surrounding air.

Nitrogen is believed to play the part of a diluent only, and not to sustain or enter into any chemical combinations within the body, and is, in fact, the water in the glass of toddy. When it has conveyed the oxygen into the blood, and has removed the carbonic acid and other gases and vapours which are no longer needed, it is itself excluded, and must be discharged from further duty until it has detached itself from the new materials with which it has become associated.

But as the normal dilution of the oxygen is desirable, lest the action of the latter should be too violent, it is necessary that the abstraction of nitrogen, when the air is purified, should not be carried too far, and hence, in its negative character, it plays an important *rôle*, and must not be too much undervalued. At the same time there is a tolerating power in the body, by which variations, in the proportion of the diluent, do not cause immediate mischief. Nay, for a time, pure oxygen may be inspired, and although, by so doing, the amount of vital action is unduly increased, it is not increased at the five-fold rate in which it is supplied.

Experiments in the inhalation of pure oxygen by animals, and even by man, have been very numerous, and in me the effect was to increase the vital changes by about 10 per cent. When the air was mixed with a larger proportion of oxygen than is natural to it, it increased the carbonic acid expired in proportion to the quantity of oxygen inspired.

It is also possible to replace nitrogen by another gas, and thereby to retain the normal proportion of oxygen, but this could be tolerated for only a very short period. Thus, in my experiments, when hydrogen and oxygen were mixed together in the normal proportion of nitrogen and oxygen, the effect was very similar to that of common air.

It is needful to add, that such an experiment is not unattended with danger, for if the combined gases were ignited, an explosion, followed by fatal consequences, would result.

The important question then arises at what point is the oxygen of the air so reduced in quantity as to impede vital changes, and thereby diminish nutrition.

This has not been satisfactorily ascertained. There can be no doubt that when the proportion by weight is reduced to 20·7, air-privation has made much progress.

If we then take as our maximum standard of healthy air-feeding that the proportion of oxygen in 100 cubic inches of air shall be 21, we may proceed to consider the effect of those agencies which vary the bulk of the air, and thereby the quantity of each element viz., temperature and barometric pressure.

1. *As to Temperature.*

It is established that temperature increases the volume of a gas in the proportion of $\frac{1}{491}$ of its bulk at 32° for every degree of temperature.

The proportion of oxygen and nitrogen at 60° is that which is, perhaps, the most commonly respired within doors, as the temperature indicated is that which we endeavour to maintain there. In summer weather, however, the temperature is often 80°, when 100 volumes at 60° become 104·07 volumes; or, to put the subject in the more practical light, 100 volumes at 80° contain an amount of oxygen equal only to 20·16 per cent., as compared with 21·0 per cent. at 60°. In India, and other Eastern climes, the temperature is frequently 90°, and sometimes 95° in the shade, when 100 cubic inches of air at 60° would increase to 106·11 and 107·13 cubic inches, and the proportion of oxygen would be lessened, so that it would be 19·79 and 19·60 per cent., as compared with 21·0 at 60°.

Hence it follows, that if the standard quantity be necessary at 60°, the body with the higher temperature would require less, or vital actions which are required must languish, or the amount of air inspired must be greater. Experiments have shown that the two former propositions are correct, and that, under these circumstances, less heat is required, and less exertion is made, so that less oxygen is consumed, whilst there is, at the same time, a sense of languor, which implies that the body is not so vigorous, or the vital actions performed so fully as is desirable.

This was well shown by an experiment, which has been already referred to at page 11, by which it was proved that with sudden increase of heat, as well as with the slower increase with the seasons of the year, the quantity of carbonic acid, or the product, and therefore a measure of vital changes was greatly reduced.

On the other hand, with the temperature reduced to the freezing point (32° F.), 100 volumes of air at 60° become only 94·3, and the oxygen at 32° is increased to an amount equal to 22·27 per cent. instead of 21 per cent. at 60°. But in North America, and the Arctic and

Antarctic regions, a temperature of 50° below zero is not uncommon in the winter season, and then the 100 cubic inches at 60° become only 77·6 cubic inches, and the oxygen is increased to an amount equal to 27·06 per cent., as compared with 21· per cent. at 60°.

With such a temperature, the contrary expectations will be realised: for there is greater necessity for heat, greater activity of vital functions and muscular organs, and a greater consumption of oxygen. In this extreme, also, we find the body tending to defective vital action, as is shown by the torpidity which is experienced when the temperature is reduced to so low a degree.

2. *As to Pressure.*

The effect of the pressure of the atmosphere is such, that the air expands in a geometrical ratio as the height increases in arithmetical ratio, so that 100 volumes of air at the sea level become by expansion 200 at an elevation of 3 to 4 miles. The height of the column of mercury in the barometer may be said to be 30 inches at the sea level in this latitude, and the variations rarely exceed 1 inch each way, viz., 2 inches from 29 to 31 inches. 100 cubic inches of air under a pressure of 30 inches of mercury became 96·8 cubic inches at 31 inches, and 103·45 cubic inches at 29 inches; so that, as a general expression, it may be stated, that there is an increase or decrease of 0·345 cubic inch for every 10th of a degree.

With the barometer at 31 inches the amount of oxygen in 100 cubic inches will be equal per cent. to 21·7, as compared with 21 per cent. at 30 inches, whilst it will be 20·3 at 29 inches, with the same basis of comparison.

This action may intensify or oppose the corresponding action from temperature, so that with a falling barometer and increasing temperature the quantity of oxygen diminishes more than when there is increasing temperature and increasing pressure, and when this occurs at a very hot season, and particularly in hot climates, when sudden and violent storms are accompanied or immediately preceded by a great fall of the barometer with high temperature, and the limits of endurance by the body are nearly attained, the conjunction of both influences may induce absolute air-starvation.

There is also a physical action from atmospheric pressure by which the fluids of the body are unduly retained or rapidly emitted, which has an indirect bearing upon the food required.

Such are the ever-acting causes which influence the quantity of oxygen in a given volume of air, and which vary its food-power. The quantity is increased as the necessity for the consumption of food with which it combines increases, as in cold weather and cold climates, and *vice versâ*. This proceeds, as already intimated, on the assumption that the total amount is not materially varied by the quantity of air which is inspired, whether by the increased fulness of each inspiration, or by increased frequency of respiration; but this is not strictly correct, for there can be no doubt that on the sea level there is a parallel action as to quantity of air, so that in conditions where the air is expanded and the proportion of oxygen lessened, the quantity of air inspired is lessened also, and *vice versâ*. Hence by both means the food-power of oxygen is increased or decreased.

How far this applies to the state of persons living at high elevations is not accurately known. It is said that the natives of those altitudes have more capacious chests than those on the sea level; but it has not been proved.

In my experiments at different elevations in Switzerland up to the summit of the Brevent, at a height of about 9,000 feet, the quantity of air inspired at an inspiration and by the minute differed but little from that on the sea level. As, however, the expansion of the air increases with elevation, so that at 3·4 miles it is doubled, and thereby the quantity of oxygen in 100 cubic inches is reduced by one half, it will be readily understood that at great elevations the respiration must be seriously impeded, and that the languor, stupor, and blueness of the skin, which occurred in Mr. Glaisher's ascents, to about 23,000 feet, were natural results. Whether, however, this obtains simply because the usual depth and frequency of inspiration are insufficient to supply air, or that the fatigue of deeper and more frequent inspirations is so great as to induce exhaustion and thereby insufficient respiration, does not appear.

We cannot forget, that whilst the lungs usually receive about 30 cubic inches of air at each inspiration, they are capable of inhaling at one inspiration more than 200 cubic inches, but with greater effort and at longer intervals. An increase of double the usual amount per minute is far less than that which accompanies gentle exertion, such as walking one mile per hour, and might be borne without great fatigue. The probable explanation is, that the person rising suddenly to so great an altitude does not with sufficient readiness increase the depth of respiration, and also that the smaller per-centage of oxygen in the air admitted into the blood is insufficient to carry on the vital changes with proper rapidity, although the rapidity of the circulation might be materially accelerated. It is possible that at the highest inhabited parts of the Himalayas there may be a greater volume of air inspired per minute and at each respiration, than at the sea level, and that the normal rate of pulsation may be also increased; but it must not be forgotten that at such altitudes the effect of temperature moderates that of elevation, for the air is much colder during at least the greater part of the day and night, and by so much is the proportionate quantity of oxygen increased.

There is, however, we venture to think, a very general misapprehension as to the degree in which this occurs, for it is very probable that the diminution in the temperature at high altitudes causes the air to be richer in oxygen than at the average temperature on the sea level. On this supposition alone, can we account for the robust health and great vigour of those who occupy high mountain ranges, and who are capable of making and sustaining exertion beyond those living at the sea level. At the same time, it is clear that with great rarefaction of the air by both elevation and heat, an increased quantity of air must be inspired per minute to maintain the vital processes in their integrity.

We will now proceed to show what is the actual quantity of oxygen which exists in the atmosphere as stated by different observers and under different conditions, citing first those which were made on air obtained from out-door and good, or tolerably good, sources, and for some of which, as for other extracts, we are indebted to Dr. Angus Smith's excellent work:—

GASEOUS FOODS.

No. 143.

Oxygen.

	Oxygen per cent.
Gay-Lussac and Humboldt, from many experiments, found that it varied from 20·9 to 21·2 per cent. . . . mean	21·0
Gay-Lussac, in the air from mountains and from fens . ,,	21·49
De Saussure, in the air from Chambeisy, found variations from 20·98 to 21·15 . . . . ,,	21·05
Berthollet	21·05
Thom. Thomson	21·0
Davy	21·0
Vogel, on the Baltic	21·59
Hermbstädt, on the Baltic.	21·59
Dalton, at Manchester, variation from 20·7 to 21·15 . mean	20·87
Regnault, in the air of Paris . . . . . . ,,	20·96
,, ,, ,, ...	20·913 to 20·999
,, ,, Lyons and around . . .	20·918 ,, 20·966
,, ,, Berlin	20·908 ,, 20·998
,, ,, Madrid	20·916 ,, 20·982
,, ,, Geneva and Switzerland . .	20·909 ,, 20·993
,, ,, Toulon and Mediterranean .	20·912 ,, 20·982
,, ,, Atlantic Ocean	20·918 ,, 20·965
,, ,, Ecuador	20·960
,, ,, Higher than Mont Blanc . .	20·949 ,, 20·981
Bunsen, at Heidelberg . . . . average 20·924 ,,	20·840
Graham	20·9
Liebig	20·9
Dr. Angus Smith, sea-shore, Scotland	20·9990
,, tops of hills, ,, ...	20·9800
,, suburbs of Manchester . . .	20·9470 and 20·9800
,, St. John's, Antigua	20·9500
,, London, open places, summer . . .	20·9500
Frankland, air from Chamounix	20·804
,, ,, top of Mont Blanc . . .	20·963
,, ,, Grands Mulets	20·802
Brunner—Faulhorn	20·91
Berger—Jura and other mountains	20·3 to 21·63
Miller, from balloon ascent, 1,800 feet high . . .	20·88
,, ,, near the earth . . .	20·92
Dr. Angus Smith—London, N., N.E. and N.W. districts average	20·857
,, ,, S. and S.W. ,, ,,	20·883
,, ,, E. and E.C. ,, ,,	20·86
,, ,, W.C. and W. ,, ,,	20·925
,, ,, N.W., S., S.W. and W., Park, &c. .	20·95
,, Mountains of Scotland, top . . . mean	20·98

		Oxygen per cent.
Dr. Angus Smith—Mountains of Scotland, bottom	. . mean	20·94
,, Many parts of Scotland	. . . ,,	20·96
,, Worst parts of Perth .	. . . ,,	20·935
,, Workhouse wards, London .	. 20·88 to	20·93
,, ,, four best wards .	day average	20·92
,, ,, ,, ,, .	midnight ,,	20·881
,, ,, ,, ,, .	morning ,,	20·880

The following analyses show conditions of the atmosphere varying from starvation point to a state simply unfavourable to health :—

No. 144.

	Oxygen per cent. volume
Dr. Angus Smith—Air very difficult to remain in many minutes .	17·2000
,, In mines: the worst specimen yet examined	18·2700
,, ,, when candles go out . . .	18·5000
,, ,, in sumps or pits . . .	20·1400
,, ,, under-shaft of metalliferous mines . . . average	20·4240
,, Court of Queen's Bench, Feb. 2, 1866 .	20·6500
,, Mines, large cavities in . . average	20·7700
,, About backs of houses and closets in London	20·7000
,, Pit of theatre, 11.30 p.m.	20·7400
,, Gallery ,, 10.30 p.m.	20·8600
,, Sitting-room which felt close, but not excessively so	20·8900
,, Tunnel on Metropolitan Railway . . .	20·60
Frankland—Air in laboratory of Owens College, Nov. and Dec. mean	20·873
Regnault—Air at Toulon harbour	20·85
,, ,, Algiers	20·42
,, ,, Bengal Bay	20·46
Leblanc—Close stable, École Militaire	20·39
,, Salle d'Asyle, with 116 children	20·53
,, Salle d'École Primaire	20·65
,, Sleeping-room at the Salpêtrière	20·36
,, Another	20·44
,, Chemical theatre at the Sorbonne . . .	20·28
,, Bed-room in the new wing	20·74
De Luna—Air of Madrid, outside the walls . . average	20·75
,, ,, inside the walls . . . ,,	20·74
,, In hospitals	20·55

Ozone.

An interesting question has attracted attention of late years as to the nature and properties of Ozone, to which we must refer, for if the commonly accepted views be correct that substance has greater food-properties than oxygen itself.

It is not necessary to name those who have been so active in our day in their investigations into the subject further than to quote the well-known name of Schönbein, to whom all subsequent enquirers owe obligation.

We are all familiar with a peculiar odour in the air surrounding an electrical machine which has been in motion, and when the sparks have passed through the air even for a minute. The same is noticed, on a wider scale, under certain electrical conditions of the atmosphere, accompanied by a sense of closeness. The like also occurs after the electrolysis of acidulated water or after electrical sparks have been passed through oxygen, or a stick of phosphorus has been left for half an hour in a jar of oxygen. In all these instances the odour is the same, and although it cannot be described, its identity under different conditions is not doubted by those who have perceived it. This is *Ozone*, which is said to *ozonise* the air, and as it decomposes iodide of potassium, and sets free the iodine to act upon starch, many have set themselves to determine its existence in time and place, and its relative amount, with test-papers.

The properties of this substance are marked and important. Thus it oxidises many moistened metals, including mercury, copper and iron, and is decomposed to an unlimited extent, by dry silver leaf or silver filings. It corrodes organic matters, bleaches vegetable colours, oxidises black sulphide of lead into white, makes red ferro-cyanide of potassium out of yellow, renders moist sulphide of manganese brown, converts moist silver into the peroxide, and decomposes the peroxides of hydrogen and barium with evolution of oxygen. It is insoluble in water and acid solutions, whilst it is freely absorbed by pyrogallic acid and iodide of potassium, is destroyed at a temperature at or above 250°, and is naturally associated with moisture.

Its presence in the atmosphere is believed to be very general, but its quantity very variable. It is said to be produced by the action of the sun on the leaves of plants, and on the juices of plants and their products, and to exist in all forms of fermentation, putrefaction, or decay.

What, then, is the nature of this new and powerful agent? Williamson says it is a tri-oxide of hydrogen, but nearly all other investigators affirm it to be oxygen in an allotropic state—an oxide of oxygen—which may be removed from oxygen without alteration of bulk. Schönbein affirmed it to be a negative oxygen, because it is evolved at the negative pole of a battery, and he termed another form which appears at the positive pole *Antozone*, but this has since been proved to be peroxide of hydrogen, H_2O_2.

In reference to its action upon man, it is assumed that air is the better food when ozonised, and ozonised oils have been introduced in the belief that they are more effective agents, but for this there is not, as yet, any sufficient proof. It is not the fresh and invigorating air which contains ozone in abundance, but the oppressive and electrical atmosphere, in which vital changes are rather diminished than increased. It is, however, said that ozone is more readily absorbed and given out by the blood corpuscles than ordinary oxygen, and so far facilitates vital action.

So far we have referred to abnormal states of the atmosphere leading to air-starvation from one manifest cause, viz., the absence of a sufficient supply of oxygen, and we will now proceed to consider the carbonic acid and other components of atmospheric air.

Carbonic Acid.

Carbonic acid is found in all known specimens of air, and is therefore a normal constituent of the atmosphere; but it may nevertheless be in a quantity which shall indirectly tend to starvation or directly induce poisoning. The normal quantity is very small, and scarcely exceeds three parts in a thousand, but it is believed to be in greater proportion at high elevations.

Dr. Angus Smith found as follows:—

No. 145.

	Carbonic Acid Volume per cent.
On hills in Scotland, from 1,000 to 4,000 feet high	·0332
At the bottom of those hills	·0341
On hills of various elevations, from 1,000 to 3,000 feet	·0332 to ·0337
Frankland found on the Grands Mulets	·111
,, ,, Summit of Mont Blanc	·061
,, ,, Chamounix	·063
De Saussure found on Lake of Geneva	mean ·0439
,, ,, at Chambeisy	,, ·0460
Da Luna found at Madrid	,, ·0501
Pettenkofer found in the air in and around Munich	·05

The quantity of this gas is sometimes very greatly increased. Thus, according to Angus Smith:—

No. 146.

In Manchester streets	·0403
,, ,, during fog	·0679
About middens	·0774
In workshops, down to	·3000
In theatres, worst part of	·3200
In mines, average of 339 analyses	·7850
,, extreme amount found in worst	2·5000

The presence of carbonic acid is very readily determined by shaking the air with lime water or baryta water, when the solution becomes cloudy, and the quantity may also be readily ascertained by weighing; but a yet more simple plan has been arranged by Dr. Angus Smith, after Pettenkofer, on the volumetric system.

Thus he finds that by taking a fixed quantity of lime water and varying the capacity of the bottle in which it is placed, and therefore the quantity of air used, he can very closely determine the proportion of carbonic acid in the contained air. The bottle used by him is of sufficient size at the top to admit the hand for the purpose of cleansing, and he draws air through the bottle with a flexible bellows pump until the contained air represents that which is to be analysed. Taking half an ounce of lime water as the standard quantity, he has constructed the following table:—

No. 147.

Carbonic Acid per cent.			Size of bottle	
·03	without precipitate in		20·63 oz. avoirdupois	
·04	,,	,,	15·60	,,
·05	,,	,,	12·58	,,
·06	,,	,,	10·57	,,
·07	,,	,,	9·13	,,
·08	,,	,,	8·05	,,
·09	,,	,,	7·21	,,
·10	,,	,,	6·54	,,
·15	,,	,,	4·53	,,
·20	,,	,,	3·52	,,
·25	,,	,,	2·92	,,
·30	,,	,,	2·51	,,
·50	,,	,,	1·71	,,
1·00	,,	,,	1·10	,,

The air must be well shaken with the water, and no precipitate will occur unless the quantity of carbonic acid be greater than that which is placed opposite to the capacity of the bottle.

The question now occurs: Does the presence of carbonic acid gas add to or take from the food-qualities of air? It is quite clear that its presence in *large* quantity cannot be beneficial, for it does not support animal vital action, and if we take the least important view of its action it must be indirectly hurtful, for in proportion as it is present, so will free oxygen be absent, and we have seen that any gas which diminishes the quantity of oxygen by 0·1 per cent. interferes with the vital processes, and is injurious to health. But there is little doubt that carbonic acid cannot be used as a mere diluent and be substituted for nitrogen in the air, or that respiration can be maintained with 79 per cent. of carbonic acid and 21 per cent. of oxygen, and therefore that its action is not negative simply, but according to the amount of it poisonous. There is no known vital process in animals by which carbonic acid becomes fixed in the body after the manner of vegetables, or that if introduced it undergoes any transformation during which it could produce heat or otherwise act as a food. It is the product of vital changes, and is cast out of the body at least as useless if not positively injurious.

The late Sir James Simpson of Edinburgh administered it as a medicine by allowing a patient to inhale it from a flask in which it was produced, but great care was taken that an almost unlimited admixture of atmospheric air should take place during inspiration.

Solid Particles.

It is most usual to find solid particles or dust in the air when a considerable quantity is washed in clean distilled water, and this is made strikingly evident by looking at a sunbeam passing through the air of a darkened room. This necessarily varies much with the locality, and will be greater where there is smoke from mineral fuel, manufactories, dusty roads, or large tracts of sand moved by strong winds, and is sometimes so great as to excite coughing. In the same way the pollen of plants and other minute particles of matter travel along the atmosphere and cause hay fever. Minute living organisms, and even animalcules, are also found in the air, and it is very probable that this kind of impurity is a far more frequent cause of *Malaise*, and even of disease, than is at present known, but there is no evidence to show that such additions to the atmosphere act as foods.

An interesting discussion occurred some time ago on Professor Tyndall's assertion that not only particles of dust, but organic germs may be arrested by passing the air through cotton-wool, so that an apparatus may readily be prepared to be worn over the mouth by those who are specially exposed to either, or in conducting various trades, or in presence of infection.

Nitrogenous Compounds.

There is another class of substances to which reference must be made, although in their nature they are rather poisons than foods, viz., the organic matters which are emitted by the lungs and skin of animals, and from animal and vegetable decomposition. The sense of foulness in the atmosphere which is due to this cause is well known, and the degree increases with the number of persons and the duration of their stay in an enclosed space, but the exact estimate of its quantity and importance has not been made. It is, however, quite clear that the deterioration of the air is not even in great part due to this emanation, but to the consumption of oxygen when the quantity of air is limited. Also that the foulness of the smell is not sufficient evidence of the degree of deterioration of the air, for there are certain emanations from the body in the bed-room resulting from the absence of due ablution which are very penetrating and offensive, whilst the atmosphere may not be otherwise unduly deteriorated.

There are, however, the same grounds for believing that a portion of this matter in the air is as injurious to health as the organic nitrogenous matters in water, and that it acts as a ferment by which diseases of the nature of fever are engendered.

In reference to the presence of organisms, Dr. Angus Smith writes:—'I mentioned some time ago that I had got a quantity of organic matter from the windows of a crowded room, and I have since frequently repeated the experiment. This matter condenses on the glass and walls in cold weather, and may be taken up by means of a pipette. If allowed to stand some time it forms a thick, apparently glutinous mass; but when this is examined by a microscope it is seen to be a clearly marked confervoid growth.'

The same observer has done more than any other chemist to enable us to ascertain the presence of organic matter quantitatively by the use of a graduated solution of permanganate of potash which is decolourised by nitrogenous matter, and although it does not profess to indicate absolute quantities, its comparative indications are very valuable. He has recently improved the method, and has given the following results of experiments in his work (p. 417):—

No. 148.

			Oxygen required instantly	Oxygen required in a few minutes with acid
			Grains per million cu. ft. of air	Grains per million cu. ft. of air
York place		Jan. 21 .	. 70·23	735·62
Front of laboratory		,, 24 .	. 277·51	479·31
,,	,,	,, 27 .	. 227·05	484·37
Yard behind {		,, 20 .	. 224·36	527·58
		,, 27 .	. 655·92	908·20

Ammonia is present in the air almost universally, and particularly where there are decomposing animal and vegetable matters, as manure. It is recognised in two forms, viz., as ammonia, and as organic or albuminoid ammonia, in precisely the same manner as in water. The process of Nessler for free ammonia and that of Wanklyn, as already described in the chapter on water for albuminoid ammonia, are required for the examination of air.

The following results of experiments are extracted from the tables published by Dr. Angus Smith:—

No. 149.

Manchester.

		Ammonia free or with acids	Organic or Albuminoid Ammonia
		Grains per million cu. ft. of air	Grains per million cu. ft. of air
Laboratory and yard, and street adjoining	1869 and 1870 . average	53·582	116·544
Bed-room	„	44·305	104·118
Midden	„	146·911	181·524

London.

Chelsea	Nov. 4 . . . „	19·936	48·180
Hyde Park	„ 5 . . . „	12·655	37·965
Woburn-square and off Regent-street	„ 11 . . . „	16·614	58·148
Islington, Hoxton, Dalston and Hackney	Nov. 8 . „	26·701	65·286
Bethnal Green and Stepney	„ 9 . „	41·534	83·086
London Bridge	„ 10 . „	27·419	61·166
Embankment of Parliament houses Nov. 13	„	21·119	71·805
Back street near Lambeth workhouse Nov. 10	„	88·700	105·595
Near Vauxhall bridge Nov. 6 . . .	„	16·614	66·455
Average of many places in and about London	.	26·780	65·947
Underground railway, Nov. . . . average	31·561	163·167
Glasgow, Feb. and Mar. . . . . „	34·169	
Shore—Innellan, Frith of Clyde, Mar. . „	22·845	

There are also various mineral acids which are obtained by washing air in pure distilled water, the amount of which is commonly small, but it varies extremely, as shown by Angus Smith's Reports. The following are a few examples in absolute quantities:—

No. 150.
Manchester.

	Hydrochloric Acid Grains in 1 million cu. ft. of air	Sulphuric Acid Anhydrous Grains in 1 million cu. ft. of air
Grosvenor-square, March	49·43	682·80
,, ,, Oct. 7	155·50	75·
Greenhay's-fields ,, 12	79·5	978·2
St. Helen's, open space, Oct. 15	170·3	1236·3
Hill near Poole's Cavern, south of Buxton } Oct. 29	31·78	971·0
The cliffs near Blackpool, Dec. 1	37·36	259·45

The following comparative tables by the same observer are full of interest :—

No. 151.
Blackpool, taken as 100.

	Total acid	Hydrochloric Acid	Sulphuric Acid Anhydrous
Blackpool	100	100	100
Didsbury	282	277	282
Buxton	327	415	315
London	348	320	352
St. Helen's	488	516	484
Manchester	498	396	513
Metropolitan Underground Railway	1483	974	1554

No. 152.
Innellan, on the Clyde, taken as 100.

	Total Ammonia	Ammonia	Albumenous Ammonia
Innellan	100	100	100
London	112	117	109
A bed-room	179	194	173
Glasgow	202	150	221
Inside and outside of office	205	235	193
Metropolitan Underground Railway	234	138	271
A midden	395	643	301

WATERY VAPOUR.

Besides the foregoing, there is one other compound of the atmosphere which is never absent in nature, but the quantity varies with all the conditions which influence the amount of oxygen, viz., watery vapour. It adds to the volume of the air, and it may usually be added to, and may always be removed from, the air by artificial means.

There is a capacity in air to receive and retain moisture which is inherent to its nature, but it has limits, so that air which under given circumstances cannot receive a further increment of moisture is said to be saturated, and the point of saturation is indicated by the deposition of the vapour as water. This power is directly associated with the temperature of the air, in such a manner that the higher the temperature the greater is the capacity of the air to receive and retain vapour. Air at a given temperature which is not saturated becomes so on lowering the temperature to a given point. This is seen familiarly in a glass holding water at the ordinary temperature on a warm day by its remaining clear on the outside, but on the introduction of ice into the water, by which the temperature of the water and surrounding air is lessened, there comes a deposition of moisture upon the outside, because the surrounding air at the lower temperature is saturated with vapour.

Hence it follows that while warm air may not show any signs of the presence of vapour, it may have a much greater amount than air at a lower temperature which is saturated. The point of saturation at different temperatures may therefore be an indication of the quantity of water which the air then contains, and thus become a ready mode of admeasurement.

Taking this as a guide, tables have been constructed which show the weight of water in a cubic feet of air at the dew-point, a point of saturation of which the following is one.

GASEOUS FOODS.

No. 153.

Weight in grains of a cubic foot of watery vapour under a uniform pressure of 30 inches of mercury, in every degree of temperature from the freezing point to 100° F. The temperature is the dew-point, and the weight of vapour is the weight which can be sustained at that temperature without being visible—that is to say, without being deposited.

Temperature, Fahrenheit.	Weight in grains of a cubic foot of vapour.	Temperature, Fahrenheit.	Weight in grains of a cubic foot of vapour.
°	grs.	°	grs.
32	2·13	67	7·27
33	2·21	68	7·51
34	2·30	69	7·76
35	2·48	70	8·01
36	2·48	71	8·27
37	2·57	72	8·54
38	2·66	73	8·82
39	2·76	74	9·10
40	2·86	75	9·39
41	2·97	76	9·69
42	3·08	77	9·99
43	3·20	78	10·31
44	3·32	79	10·64
45	3·44	80	10·98
46	3·56	81	11·32
47	3·69	82	11·67
48	3·82	83	12·03
49	3·96	84	12·40
50	4·10	85	12·78
51	4·24	86	13·17
52	4·39	87	13·57
53	4·55	88	13·98
54	4·71	89	14·41
55	4·87	90	14·85
56	5·04	91	15·29
57	5·21	92	15·74
58	5·39	93	16·21
59	5·58	94	16·69
60	5·77	95	17·18
61	5·97	96	17·68
62	6·17	97	18·20
63	6·38	98	18·73
64	6·59	99	19·28
65	6·81	100	19·84
66	7·04		

The importance of this in its bearing upon our subject, is the effect which it has on the removal of water from the body, whether that water be produced by chemical action within the body, or has been introduced as a food.

It is evident that if the temperature of the inspired air be that of the body, and when inspired the air is saturated with moisture, it cannot receive any water from the body; but if it be saturated at a lower temperature, it will be capable of absorbing more when inspired and raised to the temperature of the body, and may then remove some superfluous water. The degree of dryness of the air is manifestly of the utmost importance, for if the air be very dry it will remove an undue amount of water, and locally or generally induce water-starvation, whilst if it be too moist it cannot perform one of its most necessary functions, viz., the removal of water. The same actions take place upon the skin as in the lungs, for in proportion as the air is dry perspiration is promoted, and the perspired fluid removed, whilst when it is saturated at the temperature of the body it almost arrests that action. In both alike it influences the necessity for and use of food, whilst the vapour which is introduced into the lungs, may, in a certain sense, be regarded as food.

CHAPTER XXXVIII.

VENTILATION.

It has become evident in the course of the preceding chapter that the atmosphere surrounding a living man must be so deteriorated and vitiated, that it is rendered less fitted for food, and more like a poison; but the degree depends not only on the man, but upon the removal of any or all of the expired air from his immediate vicinity. Hence, it will be greater in an enclosed than in an open space, and without than with much wind. It ordinarily is at a minimum in an open space, either because the man moves from the expired air, or the expired air is removed from him; but when the air is stagnant or without perceptible motion, the removal of the expired air is so slow as to be almost prevented in a limited time; a sense of oppression occurs, and air-privation is at hand. That which is the rare exception in an open space is the rule in an enclosed one, unless means are taken to bring the conditions of the latter nearer to those of the former. Air-privation in an open space may seem a paradox, yet it is nevertheless possible, and occurs with crowds of people closely packed together, or with a few, under certain atmospheric conditions; but in an enclosed space it is an occurrence of every night, if not of every day.

The conditions of the atmosphere to which we now refer are as follows:—

1. Deterioration of the air by the consumption of a portion of its oxygen.

2. Vitiation of the air by the addition of ammoniacal and other nitrogenous products.

3. A nearer approach to saturation of the air by aqueous vapour.

4. Increased heat of the air.

It is true that all this will occur with one single expiration in temperate climates; but in hot climates the heat of the atmosphere at the moment may be greater than that of the body, and cannot therefore be increased by respiration. The degree of vitiation then produced would be inappreciable, and the moment when the degree is worthy of notice must depend upon the conditions as to enclosure and space and number of persons. There will, however, be a point at which measures should be taken to remedy the evil, as well as a period antecedent in which the evil is unimportant; and hence some agreement should be arrived at as to the state of the atmosphere which marks the proper period of sanitary action. Let us now endeavour to show the conditions of the atmosphere at that moment.

1. *As to Deterioration.*—The normal amount of oxygen has already been stated to be 21 per cent. by volume, and as that applies to the best sources, it may be admitted as the maximum requirement; but there are a thousand places where the proportion actually consumed by multitudes of men living in open spaces is much less. Thus at Manchester, Dalton found the mean to be 20·87, and the ordinary minimum 20·7 per cent. In Paris, according to Regnault, the mean is 20·96, and the ordinary minimum 20·913 per cent. In Berlin the ordinary minimum is 20·908, and even at Heidelberg it is 20·924 per cent. Dr. Angus Smith found in the ordinary atmosphere around and within London that the average was so low as 20·857 to 20·95 per cent. in different districts.

There can, therefore, be no question that a proportion of 21 per cent. is high enough. But Regnault found the oxygen at Toulon harbour to be only 20·85, at Bengal Bay 20·46, and at Algiers 20·42 per cent. under the ordinary conditions of the atmosphere. Even in the largest and purest sources, viz., the sea and mountain air, the proportion is below the standard of 21 per cent.; so that, according to Frankland, it was only 20·802 at the Grands Mulets, and 20·804 per cent. at Chamounix, which is situated at an elevation of upwards of 4,000 feet. Nay, Berger states that on the Jura and other mountains it was sometimes so low as 20·3 per cent. Miller found in air at more than three miles above the level of the sea, a proportion of only 20·88; and on the Atlantic Ocean Regnault found only 20·918 per cent. With such variations, where shall we draw the line, and say that this degree is healthful, and that is not?

Then let us cite another fact. Those living in the same locality are subject to the varying pressure of the atmosphere, so that the weight of oxygen in any given volume of air will increase as the barometer rises and decrease as it falls.

In the application of these facts it is to be first allowed that we cannot draw the line above the proportion existing in any locality in question—at any rate, so far as refers to present supply, for no better could be obtained; secondly, that the usual allowances which may be due to atmospheric causes in the same place must be admitted; and thirdly, that when we attempt to indicate a point below, we must bear in mind the lower proportion in which such great masses of people in whole towns or counties live. It is clear, therefore, that in London we cannot take 21 per cent. as the standard, and certainly not more than 20·95, whilst it is probable that 20·9 would be nearer the mark. From this deduct a proportion for varying pressure and temperature, and bear in mind that the average in the N., N.E., and N.W. districts is 20·857, and we infer that the standard may be reduced to 20·8 per cent. in the free open air. Let us then fix upon 20·8 per cent. as the minimum of pure London air, and then arises the question, Is every lower proportion injurious to health? If the answer rested upon the quantity of oxygen alone, we should answer, No; for the power of adaptation which exists in the body could, by a very slight increase in the respiration and circulation, prevent evil results; but whatever may be deficient in oxygen is supplied, not by nitrogen, which is powerless, but by other gases which may be injurious. Still there is manifestly a space below the line of health in which air-privation does not occur injuriously, and it is mere theoretical refinement to insist that every degree below the normal standard must inflict an injury upon health.

With the facts now known, we think it should be the aim of all persons to breathe air which contains the highest proportion of oxygen in their several localities, and not to allow it to fall below 20·8 in the enclosed spaces in which they may live, and it is desirable that it should not fall below 20·9. Yet Angus Smith found it below 20·8 in the Court of Queen's Bench and in many other parts of London, whilst in and about Madrid it was 20·74; and Leblanc found it in certain schools and bed-rooms in hospitals in Paris 20·74, 20·65, 20·53, 20·44, 20·39, and even 20·36 per cent. As in the last-mentioned facts the proportion was far below the average of the external air in Paris, it showed an unnecessary privation of air; but it also shows that the dividing line between sufficiency and insufficiency of air is a tolerably broad one.

As to Vitiated Air.—The normal amount of carbonic acid in the open air varies from 0·02 to ·07 per cent., and it is by common consent represented at from ·03 to ·04 per cent. Yet can we fix upon that as a hard and fast dividing line, when in Manchester it has been proved to be ·0679 in a fog, and at Munich and Chamounix ·05 and ·063 per cent. ordinarily? Angus Smith found it to be nearly ten times the normal standard, viz., ·3 and ·32 per cent., in various workshops and in the worst part of theatres occupied by men and women, so that we may fairly extend our range to some point between ·06 and ·3 per cent. This is a very long range; and if we adopt the mean, it would still be very much above the standard of ·04 per cent.

There is no reason to fear evil results from air which, otherwise proper, supplies 20·9 per cent. of oxygen, and 0·1 per cent. of carbonic acid, provided this be not the result of respiration, and facts show that it would be within the actual limits of health if the proportion of the latter were ·15, or even more, in places (such as soda-water factories) where the persons respiring it were not entirely restricted to it at any time, or for more than the usual intervals between meals, and where organic respiratory impurity is absent.

When, however, carbonic acid is produced by the respiration of animals it is attended by the nitrogenous principles already mentioned, and they are usually proportionate to each other. Hence, a material increase in the carbonic acid means also increase in the most injurious elements of vitiated air, as well as decrease in the vital component of air. Such air has a close, if not foul odour, which is peculiar to over-crowded places; but which nevertheless is variable as other elements are present or absent, such as the smell from unwashed skin and feet, and the strong odours which many persons emit. The smell therefore should not necessarily be taken as the absolute measure of the active substances which we are now considering; but it has nevertheless much importance. It has, for instance, furnished a basis for calculating the amount of air required for good ventilation, by observing the composition of air which does not differ sensibly from the outside air.* It is, however, to be noted that the sense of smell is very quickly dulled, and that by remaining even a short time in a foul atmosphere we become more or less insensible to its character, so far as smell is concerned. To judge of it in this way we must come into it from a purer

---

* See 'The Theory of Ventilation,' by F. de Chaumont, Proc. R. S., 1875.

atmosphere, the outer air if possible. Diminution in the quantity of oxygen, and even the addition of carbonic acid, are unattended with any odour, and cannot be detected by the sense of smell. It is, therefore, possible that they may occur without man being aware of it, as in the snow huts of the Polar regions; but when produced by combustion there is usually some smell, as that of sulphur, and when by respiration the odour referred to, which shows, at least, that the vitiated air has not been sufficiently renewed, and that the two former conditions probably exist. This is, therefore, the tell-tale or indicator of air-privation, and referring only to the state of the atmosphere which is due to respiration, we may be assured, on the other hand, that if this smell be absent there is not a degree of deterioration which need cause alarm. The sense of the presence of these substances is much more oppressive in hot than in cold weather, both because they are less potential in the latter, and the necessity for maintaining warmth overrides other considerations.

There is no fixed limit as to the quantity of these substances which may be permitted to remain, since their nature and effects are so various; but the use of the permanganate of potash test is a general guide, if the sense of smell be not already sufficient for that purpose.

3. *As to the Vapour.*—The air which is emitted by the lungs is saturated with vapour, and being at a high temperature gives out a portion to the neighbouring and cooler air. Hence, the tendency of the air in an enclosed space is towards saturation, and thereby to diminish greatly the value of the air for the use of man.

It may be affirmed that the air in an enclosed space should not be nearer saturation than the free surrounding air outside, and since the air within is for the most part of higher temperature than that without it is relatively drier. It may therefore be drier than the outside air, and yet too nearly approach saturation.

This, however, varies much with the period of the year, and the temperature of the house. When the temperature of air is high, say 80°, the higher temperature of the body allows only a very slight increase in the absorbing power of the inspired air, therefore very little moisture can escape with it; but when it is at 40°, if it be then saturated, the increased temperature acquired in the lungs gives it a great absorbing power.

4. *As to the Heat.*—This is a very varying quantity according to the number of persons respiring, the space occupied, the closeness of the enclosure, and the external temperature. Experience has shown that a temperature of from 56° to 65°, is the most agreeable and healthful, and we therefore strive to maintain it in our houses; but the difficulty of doing so varies with the season, and the duration of our exposure to it. Moreover, the sensation with the same degree of temperature is very different at different seasons and by persons of different ages. The young and old need warmth, as do also the sick and those who are unable to take much exercise, and the poor and ill fed, and to them warmth is life—almost more necessary than pure air. As to season, the sensation of the air at 56° in winter is that of warmth, and equal to that of 65° to 70° in summer. The greatest sensation of cold which we ever experienced was in the morning, at 5 o'clock, with the thermometer at 56°, in Texas, where we were accustomed to ride under a sun-heat of 150° during the day.

It is quite clear, that whilst the well-known regulation is the most correct and general expression, it is of very indefinite character in its effects.

Such is an outline of the effects of respired air on the atmosphere, and in seeking remedies we must not attempt too much refinement, or be guided by any fancy of the fitness of things; but take a sober and practical view, and recognise the fact that millions of men live under conditions which are theoretically wrong, and yet continue to live, and to live as well and as long as others more favourably placed. On the other hand, it is no evidence of wisdom to be content with things as they are, if there be means of amendment; but it is wise, in seeking as high a standard as can be practically attained, to proceed cautiously and slowly, so that the prejudices of men may not be needlessly assailed. It may be doubted whether this subject has not more to fear from its friends than its foes, by placing assumptions in place of proof, regarding only one aspect of the question, ignoring the actual experience of men, and exciting hostility by dogmatism based chiefly on theory.

The remedies cannot be solely artificial, for it would be impossible for man to purify or restore the air which he has contaminated. As in the case of water, the subject is too huge for any action short of that of nature, and we must turn to natural agencies to effect our purpose. The essential action is the removal of the used air, and as it leaves, its place will necessarily be supplied by the air immediately surrounding. Hence the problem of ventilation is to remove the used air and supply new air with the least inconvenience to those who require it.

It is very remarkable that scarcely any house which was built ten years ago (and indeed we may apply the observation to our own day) has any special means for ventilation, and persons sit in rooms for hours and pass whole nights in bed-rooms with a minimum change of air. There are doors for entrance and exit, windows for light, and chimney-flues for smoke, all of which may be ventilators; but the doors and windows may be shut, as in winter, and the chimney-flue closed, as in summer, and then the inhabitant of the room is in a closed box. To show yet more strongly the absurdity of the present state of things, the bed-room door and window will probably be open during the day, when no one occupies the room, but at night they will be carefully shut and fastened, the register of the stove will be lowered, and the curtains of the bed will be extended as if it were intended to suffocate the incomer rather than to afford him healthful and refreshing food during eight hours of sleep.

Even in public buildings, as in some of our great hospitals, the same defect exists, and it was seriously charged by a renowned surgeon against workhouses, that they must be ill ventilated because the nurses were not sufficiently intelligent to have the opening and closing of the windows confided to them. It is scarcely credible that the ventilation of a sick ward should in his opinion be dependent upon the opening and closing of windows.

It must, however, be allowed that in the erection of good houses and large public buildings at this moment, special means of ventilation are adopted, and during the next generation it may be nearly as healthful to sit in a handsome room as in the open air; but whether it will generally descend so low in the social scale as the workman's room remains to be seen. Up to this period the better the house has been built the more nearly it has represented a series of closed boxes, and the heavier the drapery of the bed-room the less refreshed was the sleeper, and the only improvement which has in any way modified this result has been the greater capacity of the rooms.

It will not be expected that we shall here do more than sketch an outline of the more ready modes of obtaining fresh air, as this is not a work on Ventilation, and although it is a subject with which we have long been connected, we must be content to limit ourselves to a few directions.

The golden rule in reference to this, as to many other evils, is to prevent rather than to remedy. Let there be means of ventilation in every inhabited room, and such as shall prevent the contamination of the air beyond the limit already agreed upon. This implies the incessant removal of the used air from the place immediately occupied by the consumer, and from the surrounding space, until we arrive at the outer air. With stagnant air and an immovable person, this is impossible, for the only mode by which it can then be removed is that of diffusion of gases—a process of immense importance, but slow in operation. Hence movement of the air is essential, whether by mechanical means, as fans or punkahs, or by the influence of heat, or by the movement of the external air.

Mechanical means of the kind named are too cumbrous for general use, and are limited in the space within which they act, and moreover they so disturb the mass of air that they are apt to mix both the good and the bad air together, and are not perfect ventilators. When a system is devised by the aid of proper channels and pumps, it is possible to direct the current of the air at pleasure, whether from or toward the consumer, and such a system well devised may be very efficacious, provided all other inlets and outlets of air be closed, as, for example, in prison cells and in bed-rooms, but as the doors of rooms occupied by day are frequently opened, the system is not of general application. The best example with which we are acquainted is the Lunatic Asylum at Michelover, near Derby, which was erected by Mr. Dewsbury, and is a model of efficiency.

The influence of temperature is unceasing, for whenever there is a difference between two bodies of air the hotter will ascend and the cooler descend or fill up the void laterally. The only question is the degree of influence, for it may be so small as to be practically inoperative, or so great as to excite a hurricane. As to natural heat, we may observe that it is a rare occurrence for the air within a room to be of the same temperature as that without, and whichever may be the hotter will cause movement in the other. Hence there is a constant tendency to an exchange of air within and without a room in all directions where there are communications between them. So also within a room, if there be an artificial source of heat it will cause the ascent of the air in its neighbourhood, and if there be no exit the warmer air will accumulate above, but if there be an exit a current will be established which will draw into it portions of cooler air adjoining; but such a current may not have much lateral extension.

When a system of channels has been devised, such as that referred to in the action of pumps, a fire may be the most convenient mode of setting a current in motion and of maintaining it. Such is the system adopted in the Houses of Parliament, and there the fire is placed in an air chamber, towards which all the existing channels converge, and from which one common exit is made.

But with this admitted influence of heat, there are many who overlook the fact that one or two occupants of a room cannot generate heat sufficient to cause the removal of the vitiated air by that means alone, and in fact that such an action is at the most only subsidiary to others. One person may vitiate the air of a room, whilst the heat generated by ten would not suffice to cause its removal.

The movement of the external air is the chief agent in a system of natural ventilation whether in open spaces or closed rooms, and without it all other means would be ineffectual. It is said that man cannot live on the wind; he certainly cannot live without the winds.

The admission within any space of the air as moved by the winds, is therefore the essence of air-purification or air-renewal, and as the winds blow from every quarter in turn, and not from one at all times, there must be means of admission in more than one direction.

Such, then, are the objects and agencies, and the whole art of ventilation consists in so applying the latter that their action shall be unintermittent, gentle, effectual. The mechanical system aims to act alone, and the tendency of its defects is to create currents which may be felt, and to rarefy the air. The natural system embraces the two latter agencies, and makes use of mechanical means for controlling and directing them. The greatest difficulty to be overcome is not the removal of vitiated air or the admission of bad air, but to do so without so lowering the temperature as to induce a sensation of cold, or if artificial heat alone be used, to prevent great variations of temperature in the same room and an undue dryness of the air.

We will now add a few general directions :—

1. Inhabited rooms should, if possible, have external walls on two sides, so that air may be admitted through both.

2. The openings should be small and many, rather than few and large, defended by perforated zinc, and

placed as distant as possible from those who inhabit the room. Hence the cornice above and the skirting below are convenient positions, but rooms of less than ten feet in height are not easily ventilated without draughts. The connection between the inside and the outside of the room should not be direct, but at an angle, so that a direct current may not be produced.

3. Such ventilators as direct the current to the ceiling are useful in a degree, but the cold air thus admitted will descend before the air is warmed.

4. Channels which are divided by a perpendicular diaphragm on the theory that there will be an ascending current in one and a descending current in the other, are for the most part based on a fallacy, and when the heat of the air is very great there will be an upward and a downward current in both.

5. An air-flue by the side of the chimney-flue, into which the exit tubes lead, will act in some degree so long as there is fire in the chimney to rarefy the air, but if an extracting cowl be placed at the top of such a flue it will induce an upward current when there is no fire.

6. The use of the chimney-flue as an exit for the air is liable to allow the smoke to enter the room through it when there is a down draught, notwithstanding the excellent contrivances which have been devised to prevent the return current.

7. Whenever it is proposed to remove air, means for supplying a larger quantity of air should be provided, or the attempt at ventilation will be ineffectual.

8. It is often desirable to warm as much of the air which is admitted as possible, and for that purpose stoves have been designed with an exit from a special channel into the room. When the same object is

effected by allowing cold air to enter by an opening through the wall, and then to direct it over heated pipes, the distribution is less satisfactory; but such pipes should be heated with water, and not with steam of an uncertain temperature.

9. Stoves which are placed in the middle of a room should be fed with wood and enclosed in an outer skin of metal or pottery. With coal and a furious fire, the air is not only dried to an injurious extent but burnt.

10. It is more than doubtful whether in our climate, or in any moist climate, the ordinary open fire-grate can be supplanted with advantage. It is otherwise in countries where the air is very cold and dry.

It is now pretty generally admitted that the amount of respiratory impurity, measured by the excess of carbonic acid in a room over that of the outer air, should not be more than 0·02 per cent. As each individual gives off on an average about 0·6 of a cubic foot of carbonic acid per hour, it follows that 3,000 cubic feet of fresh air per hour ought to be supplied if possible. This is what ought to be aimed at, although it may not be often attained.

But whatever may be the quantity of air required per hour, it is most difficult to determine the quantity actually allowed, for men do not live in boxes with one inlet and one outlet, or with any number of inlets and outlets under perfect control, but partly in the open air, where the quantity is unlimited, and partly in rooms which have doors opened without regulation, windows opened occasionally, and an ascent of air in the chimney-flue varying with the fire which is in the grate, and all these are influenced by the direction and force of the wind and variation of the temperature. It is not possible to estimate with complete accuracy the quantity of air which is allotted to a man under such circumstances, and useless to pretend that it can be at all times the same. If, however, we take closed rooms, such as prison cells, and bed-rooms in actual use, and quite close every door, window, and chimney, whilst the whole ventilation is arranged by mechanical means, it will be possible to effect this object.

It is important to remember that the same amount of fresh air must be supplied whatever be the size of the room; but there are of course limits to size in both directions. Too large a space is difficult to warm, whilst too small a space is difficult to ventilate.

Dr. Angus Smith, in his examination of the sick wards of London workhouses, where 500 cubic feet of air-space were allowed to each inmate in well-ventilated wards, found as follows:—

No. 155.

	Oxygen per cent.	Carbonic Acid per cent.
Day	20·893	0·0568
Midnight	20·875	0·0780
5 A.M.	20·869	0·0802

On selecting the four best-ventilated wards, the proportions were more favourable, viz.—

No. 156.

	Oxygen per cent.	Carbonic Acid per cent.
Day	20·92	0·0463
Midnight	20·886	0·0677
Morning	20·884	0·0694

These results show that it is sometimes possible with small spaces to have fair ventilation, and it is always to be remembered that it is less the size of the space than the renewal of air that is required. The shape of the room is also important, because the floor space must bear a certain proportion to the total cubic space—not less than *one-twelfth*. Thus a space of 1,000 cubic feet ought not to have less than 84 square feet of floor space, for ordinary dwellings.

The object of this is to prevent crowding laterally, as mere height can never take the space of lateral or floor space.

INDEX

ABO

ABOUT, M., on Greek wine, 400
Acetic acid, 232
— ethers, 232
Adulterations :
— beer, 413
— milk, 319
— sugar, 256
— tea, 340
— wheat flour, 176
Albumen, 91
— composition of animal and vegetable, 91
Alcohols, 371
— in wines, 390
Allsopp's ale, 413
Almen, Prof., on Liebig's extract, 89
Almond, 227
— milk, 163
America, butter-making, 129
American wine, 402
Ammonia in water, 284
— in air, 455
Amidon, 146
Anderson's analysis of cabbage, 207
Angus Smith, Dr., on air and water, 273, 474
Animals storing up flesh, 59
Antelope, 72
Antiseptic gases, 27
Ants, 86
Apple, 218
— Alligator, 219
— Custard, 218
— Elephant, 223
— Mammee, 219
Apricot, 222
— Dingaan's, 223
Arabs: Frumenty, 177

BED

Ardent spirits, 373
Arlidge, Dr., on tea, 35
Aroma of wines, 406
Arrack, 387
Arrowroot, 239
Artichoke, 202
Artificial milk, 326
Ass as food, 72
Assam tea, 338, 344
Atmospheric air, 435
— — in water, 275
— — acids in, 455
Auslese wine, 389
Australian wines, 401
Ava, 417

BACON, 64
— composition of, 67
— curing, 65
— use to poor, 65
Ballyshannon eels, 110
Bananas, 219
Baobab, 236, 239
Bareilly: Barley and wheat flour, 195
Bargout, 169
Barley, 193, 407
— malting, 408
— sugar, 263
Bass, Mr., before Mr. Knox, 433
Bass's ale, 413
Bates on turtle, 118
Bats, 86
— prohibited, 83
Beans, 153
Beaumont's experiments on digestion. See each food
Bedouins, 206
— eat locusts, 86

Beef, 46
— composition of, 47
— cooked, composition of, 49
— offal, 47
— proportion of lean and fat, 47
Beer, 407
Beer Street, 372
Bermuda arrowroot, 242
Beetroot, 204
— sugar, 257
Betel pepper, 237
Bhât, 165
Bilberries, 217
Birds, graminivorous and carnivorous, 101
— kinds used as food, 104
Birds' nests, 94
Biscuits, 188
Bison, 71
Bisulphite of lime, 28
Blackberries, 217
Black puddings, 83
Black Sea wheat, 172
Blood, 18
— as food, 83
— composition of, 84
— diseased, 85
— of fowl, 103
— prohibition of, 84
Boar's head, carol on, 68
— — recipe, 15th century, 68
Boaz, 194
Bone, 38
— composition of, 39
— in carcass, 48
— nutritive value, 40
Bordeaux wines, 392
Boston milk, 313
Bouza, 162, 416
Bran, 174
Brande on alcohol, 390
Brandy, 381
Braxy mutton, 54
Brazilian nuts, 228
— tea, 338
Bread, 181
— weight and price of, in 1775, 192
Breadfruit, 205
Brine, composition of, 35, 36
Broccoli, 207
Broth, composition of, 49, 53, 61
Brown bread, 176
— flour, 175

Brwchan, 169
Buckland, F., on fish, 108
Budram, 169
Burckhardt, 206
Burgoul, 178
Burgundy wine, 393
Burton, Capt.: Cassava, 245
Butter, 127
— factories, 129
— in workhouses, 134
Buttermilk, 328
Butter-nut, 220
Buzzard, 101

CABBAGE, 207
Californian wine, 403
Calvert, Prof., analysis of wheat, 181
Calves killed, 51
— — in Boston, 51
Camel's flesh, 58
— milk, 318
Canterbury gurnets, 109
Cape of Good Hope wine, 401
Capelin, 108
Capitone, 111
Carbonic acid in air, 450
Cardamoms, 235
Carissa, 223
Carolina rice, 163
Carrion birds, 101
Carrots, 203
Cartilage, 40
Casein, 94, 121
Cassareep, 239
Cassava, 244
Cassia, 236
Castes of India eating dead animals, 55
Caviare, 113
Cayenne pepper, 235
Cellulose, 151
Ceylon: Cassia, 236
Champagne wine, 393
Chapman's process, 287
Charlock, 208
Charqui, 35
Chaucer and Wiclif, 233
Chavica, 237
Cheese, 120
— known to ancients, 120
Cherries, 217

Chestnut, 228
Chica, 416
Chick pea, 153
Chicory, 368
Chili potato, 198, 203
Chinese sugar grass, 256
Chlorine in water, 297
Chocolate, 369
Christison, Sir R., on salmon, 108
Chych, 153
Cider, 415
Cinnamon, 236
Claret, 388
Clark's process, 280
Classification, 1-3, 15
Climate on vital functions, 11
Cloves, 236
Cocoa, 369
Cocoa nut, 227
Coffee, 359, 362
— leaves, 358
— planting, 360
— pots, 364
Cognac, 381
Cold to preserve meat, 25
Composition of structures of the body, 7
Condiments, 229
Congou tea, 345, &c.
Connection of animals and vegetables, 3, 4, 8
Cooked meat, composition of beef, 49, 54
Cooking egg, 96
— fish, 111
— flesh, 19
— starch, 148
— loss of weight, beef, 49
— — — — mutton, 53
— — — — pork, 61
Cotgrave on amydon, 146
Cotton-seed oil, 248
Couscousou, 246
Crab. *See* Shell fish
Craig's prepared fat, 29
— pea-meal food, 156
Crane, 102
Crane's, Dr., factory, 321
Cream, 327
— cheese, 126
Cubebs, 234
Cucumbers, 209
— oil of, 248

Cumberland bacon, 66
Curlew, 102
Currants, 216, 222
Curry powder, 238
Custard apple, 218
Cyprus wine, 163

DALY, Dr., on preserved milk, 323
Darby's fluid meat, 89
Darjeeling tea, 344
Dates, 217
Dauglish, Dr., bread making, 183
Day and night on vital actions, 11
Dead animals eaten, 54, 55
De Caisne, milk of suckling women, 316
Denman, Mr., on Greek wines, 397
Derbyshire oatmeal, 168
Dhal, 153
Dhye, 136
Dried meat, 23
Dripping, 141
Dumplings, Somersetshire, 189
Dupré on wine, 421
Dynamic force, 6. *See* also each food

EARTH-NUT, 228
Economy in joints, 44
Edward IV. regulating sale of bread, 192
Eels, 110
Effect of foods on respiration : Beef, 49
— — — — — casein, 94
— — — — — eggs, 92
— — — — — isinglass, 93
Eggs, 92, 95
— composition of, 99
— cooked, 96
— importation of, 98
Eland, 72
Elgin factory, 322
English lakes, 284, &c.
Exertion on vital functions, 12
Extract of meat, 86
— — — composition of, 87
— — — Liebig on, 88
— — — small nutritive value, 89

FALSTAFF on wine, 391
— Fat cells, 138
Fat, composition of, 45
— in animals, 137
— liked by consumptives, 129
— refined, 29
— to preserve meat, 28
— vegetable, 246
Fattening animals, 59
Feet, 80, 81
Fennel, 208
Fermentation of malt, 411
Factitious wine, 403
Figs, 217
Filberts, 228
Filters, 308
Fish, 105
— composition of, 107
— white and red blooded, 107
Fleetings, 329
Flesh, anatomical composition of, 16, 17
— juice, 18
— tender, 18
— effect of cooking, 19
Fowl, blood of, 103
— composition of, 103
Frankland's experiments. *See* each food
— — on water, 288, 294
French wines, 392
Fresenius: Analyses. *See* each food
Frijoles, 153
Frogs, 86
Fruits, succulent, 213
— albuminous, 225
Frumenty, 177

GAME, 100
— kinds of, 100
Garlic, 238
Gas used in making bread, 184
Gaseous foods, 435
Gelatin, 92, 197
German wines, 393
Ghee, 136
Gin, 383
Gin Lane, 372
Ginger, 224, 237
— beer, 416
Gironde wines, 390
Gluten, 179, 197

Goatsbeard, 208
Goat's flesh, 57
Golden syrup, 261
Gooroo nut, 229
Gooseberries, 222
Grains of Paradise, 235
Gram, 249
Graminivorous birds, 101
Granadilla, 220
Grape juice, 216, 388
Greek wines, 397
Green vegetables, 207
Groats, 169
Gruyères cheese, 122
Guava, 219
Gulliver, Prof., on Canterbury gurnets, 109
Gunpowder tea, 345

HAGGIS, 78
— Hamburg beef, 24
— wine, 403
Hard water, 279
Hare, 104
Hart's brown meal, 180
Hassall's dried meat, 23
Hazel nuts, 228
Heat to preserve meat, 30
— — — — defects of method, 33
Henley's method of preserving meat, 38
Herbs, 207
Herepath: Potatoes, 200
Heron, 102
Herring, 110
Hertfordshire ale, 414
Hog and hominy, 59
Holland, 383
Home-made wine, 403
Hominy, 159
Honey, 263
— wine, 417
Hops, 408
Horse-flesh, 72
— — dinner of, at Langham Hotel, 74
Hungarian wines, 396
Hurreah, 387

ICE-FIELDS, 26
— Ice-making, 26

I

Iceland moss, 213
Iguana, 85
Impurities in water, 272
Indian corn, 156
— tea, 338
Infants' milk, 316
— preserved milk, 323
Irish buttermilk, 329
Isinglass, 93
— effect of, 93

J

JAPAN TEA, 339
Jatropha, 244
Joints of meat, 42, 48
Jonathan, 161, 171
Jones's method of preserving meat, 28, 31
Jowaree, 162
Juice of flesh, 18
— — — extracted by cooking, 21
Julpaun, 165

K

KAFFIRS: Umbila, 160
Kangaroo, 72
Kean on selection of meat, 52
Kemmaye, 206
Ketchup, 239
Kola-nut, 229
Kopf's pea-soup, 156
Koumiss, 319, 417

L

LALLEMAND, M., on wine, 421
Lamb, 57
Lamb, C., on sucking pig, 64
Lamprey, 110
Langham Hotel, dinner at, of horse-flesh, 74
Lard, 139
Laurus, 236
Laver, 213
Lawes and Gilbert. *See* various foods
— — — on carcass, 47
— — — on milk, 314
Lean, composition of, 47
— and fat meat, proportion of, 43
— — — — properties of, 45
Leban, 417
Legs and shins of beef, 48
Leipsic soup-kitchen, 49, 53, 61
Lentils, 153

Leveret, 104
Levitical prohibition of food, 55, 58, 84, 85, 86, 103, 104, 111
Lichens, 212
Liebig's extract of meat, 86
Lignine, 151
Lime in wine, 391
Linseed, 248
Liver, 78
— composition of, 79
Liverpool workhouse milk, 313
Lizards, 85
— prohibited, 86
Llymru, 167
Lobster, 117
Locusts, 86
Loquat, 220
Loss of weight in cooking beef, 48
Lungs, 79
Lupuline, 408

M

MACCARONI, 190
McDougall's Phosphatic flour, 180, 185
Mace, 234
Madeira wine, 400
Maize, 156
Malai, 327
Malting, 408
Mammee apple, 219
Mango, 223
Mangosteen, 219
Mangrove swamps, 249
Manihot, 245
Manioc, 244
Manna, 266
Maple sugar, 256
Marwa, 387, 417
Maslin, 196
Maté tea, 358
Mead, 387
Measly pork, 62
Meat, extract, 86
— fluid, 89
— lean and fat, 41, 43
— preserved. 22
Medlock and Bailey's method of preserving meat, 28
Médoc wine, 392
Melon, 219
Metheglin, 387
Mexico: Beans and frijoles, 153

MEX

Mexico : tortilla, 159
Mignot's machine, 26
Milk, 312
— sugar, 259
Miller, Prof.: Water, 275
Millet, 152, 161, 246
Mineral waters, 310
Minerals in water, 276
Mint, 236
Mocha coffee, 362
Molasses, 261, 384
Monkey, 86
Montilla wine, 394
Morgan's method of preserving meat, 35
— — — — fluid, 36
Mulberries, 217
Muscle of fly, 17
Mushrooms, 210
Mussels. *See* Shell fish
Mustard, 235
Mutton, 52
— composition of, 54

NAPLES maccaroni, 190
Nessler's reagent, 290
Nettle, 208
Neufchatel cheese, 127
Nicaragua tasajo, 153
Nitrates and nitrites in water, 287, 297
Nitrogenous compounds in air, 453
— foods, animal, 15
— — vegetable, 143
Nutmegs, 223, 234
Nuts, 226

OATS, 166
Offal, proportion of, 47
— various kinds, 75
— composition of, 76
Oils, animal, 142
— olive, 247
— from seeds, 248
Omentum, 79
Oolong tea, 344
Organic matter in water, 282
Organic matter in air, 454
Oxen storing up flesh, 59
Ox-head, 27
Oxygen in atmospheric air, 446

POR

Oyster beds, 115, 116
Oysters, 115
— in India and Ceylon, 115
Ozone, 448

PADDY rice, 164
Palaver sauce, 239
Pale ale, 412.
Palm wine, 403
Pancreas, 79
Paraguay tea, 358
Parched rice, 165
— corn, 194
Parmesan cheese, 122
Parsnips, 203
Passover cake, 170
Payen's analyses. *See* different foods
Peach palm, 220
Peacock, 102
Pea meal, 156, 173
Pearl barley, 195
Pekoe tea, 339, 344
Pemmican, 29
Pepper, 233
Perry, 415
'Philosophical Transactions,' papers in, 49
Phosphates added to flour, 180
Pie at Haddon Hall, 102
'Pieces.' sugar, 254
Pig storing up fat, 59
— wild, 67
Pigeon pears, 220
Pig's head, 77
Pilchards, 113
Plantain, 206
Plaster of Paris, 173
Plums, 222
Poached eggs, 97
Pohl : Analysis of potato, 199
Polenta, 159, 246
Pollard, 179
Pomegranate, 219
Pork, 58
— composition of, 60
— collared, 81
— diseased, 62
— pickled, 63
— wild, 67
Porpoise, 177
Porter, 407

POR

Portuguese wines, 394
Potato, 198
— diseased, 201
Potted meat, 29
Poultry, 100
Preserved meat, 22
— — by coating, 28
— — by cold, 25
— — by drying, 23
— — by gases, 27
— — by heat, 30
— — by pressure, 38
— — by salt, 34
— milk, 320
Pressure of the atmosphere, 442
— to preserve meat, 38
Prickly pear, 219
Prince of Wales at Haddon Hall, 102
Privy Council Reports, 175
Prussian pea sausages, 156
Puddings and pastry, 189
Pulsation and respiration, daily, 9
— — — season, 11
Pumpkin, 209

QUASS, 416
Quince, 223

RABBIT, 104
Racoon, 105
Raisins, 216
Rapeseed oil, 248
Raspberries, 221
Recipes, 14th century:
— — — apple-tart, 218
— — — beans, 154
— — — blanc mange, 163
— — — collared pork, 81
— — — fish soup, 112
— — — frumenty, 177
— — — jelly of fish, 113
— — — macaroni, 191
— — — rice, 162
— — — salad, 207
— — — venison, 71
— — — wild pig, 68
Redwood, Prof.: Preserving meat, 28
Respiration, hourly, 11

SCH

Respiration, with exertion, 12
Respiratory foods, 259
Rhubarb, 210
Rice, 162
Richardson, Dr. B. W., on alcohols, 424
Rivers Pollution Commission, 284, 301
Roe of fish, 113
Royal Society papers. *See* various foods and 428
Rum, 384
Ruth, 194
Rye, 196

SAGO, 240
— palm, 241
Salmon, composition of, 107, 108
— dried, 112
— roe, 113
Salmonidæ, 108
Salsify, 208
Salt, 229
— in water, 298
Salted meat, 34
— — effect of salt, 36
Salts of blood, 85
Santorin wines, 399
Sausages, 80, 82
— Majorca, 83
— Prussian, 83
Scented teas, 334
Schlossberger: Analysis of mushrooms, 211
Schönbein, 449
Seagull, 101
Seakale, 209
Season on vital actions, 11
Seaweed, 212
Secale, 196
Seed oil, 248
Seeds, 145
Semolina, 245
Shakspeare on cheese, 123
Shea butter, 247
Sheep, breed of, on mutton, 53
— storing up flesh, 59
Sheep's head, 77
Shell fish, 114
Shepis talon, 81
Shor, 327
Schulze's process, 287

SIC

Sicilian wines, 400
Siliceous cuticle, 150
Simpson, Dr.: Fish in Dacca, 105
Skin, 76
Slugs in China, 85
Smelts, 110
Smith, Dr. Angus, on air and water, 447, &c.
Snails, 85
— prohibited, 85
Snakes, 86
Soap test, 280
Society of Arts, Papers at, 175, 345, 423
Solid particles in air, 452
Soup, ox-head, 78
Sources of food, 3
Sowans, 167
Spanish wines, 394
Sparkling hock, 394
Specific gravity of potatoes, 199
— — — beers, 412
Spices, 232
Squirrels, 105
Starch, 239
— cells, 147
Stevens's bread-making-machine, 185
Stewed flesh, 20
Stirabout, 159
Strawberries, 214, 221
Structures of body, 7
Subbotin, M., on wine, 421
Succan, 167
Sucking pig, 63
Suet, 139
Sugar, 249
— cane, 252
— candy, 262
— grass, 256
— making, 252
Sulphurous acid to preserve meat, 27
Swan, 101
Sweet potato, 202
Sweetbread of calf, 51, 79

TAPIOCA, 240
Tartary bouza, 162
Tasajo, 153
Tea, 330
— plant, 332
— classification, 339

WAT

Tea picking, 336
— preparation, 337
— substitutes for, 341
Teff, 416
Tellier, M., method of preserving meat, 24, 26
Temperature of atmospheric air, 440
Thames water, 275
Thermometric force, 5. *See also* each food
Tinned meat, 32
Tobacco, 237
Tokay wine, 397
Tomatos, 209
Tongue, 76
Tortilla, 159
Treacle, 261
— beer, 416
Trichina spiralis, 62
Tripe, 80
— composition of, 80
Tucker's analysis of cocoa, 370
Turkey, wild, 103
Turmeric, 238
Turnip, prairie, 203
— Swede, 203
Turtle, 118
Tyndall, Prof., on air, 453

UMBILA, 160
Ure's analyses, 262
Uses, 4

VAPOUR in air, 458
Veal, 50
— composition of, 51
Vegetable foods, 143
— — succulent, 197
— jelly, 94
— marrows, 209
Venison, 70
Ventilation, 459
Vetches, 153
Vinegar, 230
Voelcker, Prof., on cheese, 124
— — — whey, 330
Voit, M., on wine, 421

WANKLYN, 287, 291, 303
Water, 269

Water in butter, 134
Watts' Dictionary, 309
Weiss-bier, 195, 416
Wheat, 179
Wheaten flour, 171
— — grinding of, 178
Whey. 329
Whiskey, 378
White's butter casks, 131
Whitebait, 108
Wild animals as food, 72
— pig, 67
Willard, Prof., on butter factories, 129
— — — milk, 321

Wine, 387
— manufacture, 389
Wynkin de Worde, carol on boar's head, 68

YAM, 202
Yaoust, 417
Yelk of egg used in oil colours, 96
Yucca 202

ZANZIBAR lizards, 85
— wild pig, 67
Zebra, 72

For Product Safety Concerns and Information please contact our EU
representative GPSR@taylorandfrancis.com
Taylor & Francis Verlag GmbH, Kaufingerstraße 24, 80331 München, Germany

www.ingramcontent.com/pod-product-compliance
Lightning Source LLC
Chambersburg PA
CBHW071618230426
43669CB00012B/1978